ELECTROANALYTICAL CHEMISTRY

VOLUME 5

ELECTROANALYTICAL CHEMISTRY

A SERIES OF ADVANCES

Edited by

ALLEN J. BARD

DEPARTMENT OF CHEMISTRY
UNIVERSITY OF TEXAS
AUSTIN, TEXAS

VOLUME 5

1971

MARCEL DEKKER, INC., New York

MARCEL DEKKER, INC.

95 Madison Avenue, New York, New York 10016

LIBRARY OF CONGRESS CATALOG CARD NUMBER 66-11287
ISBN NO. 0-8247-1041-X

PRINTED IN THE UNITED STATES OF AMERICA

PREFACE TO VOLUME 5

Several aspects of the chapters in Volume 5 call for a brief introduction—a departure from previous practice for volumes in this series.

The concept of the involvement of the hydrated electron in electrode processes is rather new and still controversial. Chapter 1, by G. A. Kenney and D. C. Walker, presents an introduction to the history and properties of the hydrated electron and gives what evidence is available for its role in electrochemistry. It is hoped that this brief review will stimulate thought and experimentation in this developing field.

Although electrodeposition is one of the oldest fields in electrochemistry and forms the basis of the oldest electroanalytical techniques, the nature of the mechanism of electrocrystallization is still actively being investigated. The Newcastle group has made many contributions to the field and Chapter 2, by J. A. Harrison and H. R. Thirsk, reviews the methods and models employed in current research.

The involvement of chemical reactions in polarography has been the subject of numerous theoretical and experimental studies, but a general review of the nature of the processes or the mathematical methods for dealing with them has not been given. R. Guidelli has accomplished this in Chapter 3, using models taken from the thermodynamics of irreversible processes and the concepts of diffusion and reaction overpotential. Although this approach to problems of this type is not conventional in most of the electroanalytical chemistry literature, it forms a satisfying unification of the field and hopefully will be useful in the attack of new problems in this area.

A. J. B.

INTRODUCTION TO THE SERIES

This series is designed to provide authoritative reviews in the field of modern electroanalytical chemistry defined in its broadest sense. Coverage will be comprehensive and critical. Enough space will be devoted to each chapter of each volume so that derivations of fundamental equations, detailed descriptions of apparatus and techniques, and complete discussions of important articles can be provided, so that the chapters may be useful without repeated reference to the periodical literature. Chapters will vary in length and subject area. Some will be reviews of recent developments and applications of well-established techniques, whereas others will contain discussion of the background and problems in areas still being investigated extensively and in which many statements may still be tentative. Finally, chapters on techniques generally outside the scope of electroanalytical chemistry, but which can be applied fruitfully to electrochemical problems, will be included.

Electroanalytical chemists and others are concerned not only with the application of new and classical techniques to analytical problems, but also with the fundamental theoretical principles upon which these techniques are based. Electroanalytical techniques are proving useful in such diverse fields as electro-organic synthesis, fuel cell studies, and radical ion formation, as well as with such problems as the kinetics and mechanism of electrode reactions, and the effects of electrode surface phenomena, adsorption, and the electrical double layer on electrode reactions.

It is hoped that the series will prove useful to the specialist and non-specialist alike—that it will provide a background and a starting point for graduate students undertaking research in the areas mentioned, and that it will also prove valuable to practicing analytical chemists interested in learning about and applying electroanalytical techniques. Furthermore, electrochemists and industrial chemists with problems of electrosynthesis, electroplating, corrosion, and fuel cells, as well as other chemists wishing to apply electrochemical techniques to chemical problems, may find useful material in these volumes.

A. J. B.

CONTRIBUTORS TO VOLUME 5

ROLANDO GUIDELLI, Institute of Analytical Chemistry, University of Florence, Florence, Italy

J. A. HARRISON, Electrochemistry Research Laboratories, Department of Physical Chemistry, University of Newcastle upon Tyne, Newcastle upon Tyne, England

GERALDINE A. KENNEY, Department of Chemistry, University of British Columbia, Vancouver, British Columbia, Canada

H. R. THIRSK, Electrochemistry Research Laboratories, Department of Physical Chemistry, University of Newcastle upon Tyne, Newcastle upon Tyne, England

DAVID C. WALKER, Department of Chemistry, University of British Columbia, Vancouver, British Columbia, Canada

CONTENTS OF VOLUME 5

CONTENTS OF OTHER VOLUMES

VOLUME 1

AC Polarography and Related Techniques: Theory and Practice, DONALD E. SMITH, *Department of Chemistry, Northwestern University, Evanston, Illinois*

Applications of Chronopotentiometry to Problems in Analytical Chemistry, DONALD G. DAVIS, *Department of Chemistry, Louisiana State University in New Orleans, New Orleans, Louisiana*

Photoelectrochemistry and Electroluminescence, THEODORE KUWANA, *Department of Chemistry, Case Western Reverse University, Cleveland, Ohio*

The Electrical Double Layer, Part I: Elements of Double-Layer Theory, DAVID M. MOHILNER, *Department of Chemistry, the University of Texas, Austin, Texas*

VOLUME 2

Electrochemistry of Aromatic Hydrocarbons and Related Substances, MICHAEL E. PEOVER, *Ministry of Technology, Division of Molecular Science, National Physical Laboratory, Teddington, England*

Stripping Voltammetry, EMBRECHT BARENDRECHT, *Central Laboratory, Staatsmijnen in Limburg, Geleen, The Netherlands*

The Anodic Film on Platinum Electrodes, S. GILMAN, *General Electric Research and Development Center, Schenectady, New York*

Oscillographic Polarography at Controlled Alternating Current, MICHAEL HEYROVSKÝ AND KAREL MICKA, *J. Heyrovský Institute of Polarography, Czechoslovak Academy of Sciences, Prague, Czechoslovakia*

VOLUME 3

Application of Controlled-Current Coulometry to Reaction Kinetics, JIRÍ JANATA, *Department of Analytical Chemistry, The Charles University, Prague, Czechoslovakia;* AND HARRY B. MARK, JR., *Department of Chemistry, The University of Michigan, Ann Arbor, Michigan*

Nonaqueous Solvents for Electrochemical Use, CHARLES K. MANN, *Department of Chemistry, Florida State University, Tallahassee, Florida*

Use of the Radioactive-Tracer Method for the Investigation of the Electric Double-Layer Structure, N. A. BALASHOVA AND V. E. KAZARINOV, *Institute of Electrochemistry, Academy of Science of the U.S.S.R., Moscow, U.S.S.R.*

Digital Simulation: A General Method for Solving Electrochemical Diffusion-Kinetic Problems, STEPHEN W. FELDBERG, *Brookhaven National Laboratory, Upton, New York*

VOLUME 4

Sine Wave Methods in the Study of Electrode Processes, MARGARETHA SLUYTERS-REHBACH AND JAN H. SLUYTERS, *Laboratory of Analytical Chemistry, State University, Utrecht, The Netherlands*

The Theory and Practice of Electrochemistry with Thin Layer Cells, A. T. HUBBARD, *Department of Chemistry, University of Hawaii, Honolulu, Hawaii*; AND F. C. ANSON, *Gates and Crellin Laboratories of Chemistry, California Institute of Technology, Pasadena, California*

Application of Controlled Potential Coulometry to the Study of Electrode Reactions, ALLEN J. BARD, *Department of Chemistry, The University of Texas at Austin, Austin, Texas*; and K. S. V. SANTHANAM, *Tata Institute of Fundamental Research, Colaba, Bombay, India*

ELECTROANALYTICAL CHEMISTRY

VOLUME 5

Hydrated Electrons and Electrochemistry

Geraldine A. Kenney

and

David C. Walker

DEPARTMENT OF CHEMISTRY
UNIVERSITY OF BRITISH COLUMBIA
VANCOUVER 8, BRITISH COLUMBIA, CANADA

AUTHORS' PREFACE

"Curiouser and curiouser," said Alice in amazement at the White Rabbit's words and then hastily poured him another cup of tea so that he should not stop this exciting tale. Wiping his whiskers and paws very carefully the rabbit repeated grandly,

"Yes, *quite* an exciting discovery, in my opinion one whose impact still has to be explored. And the machines they used, you cannot imagine how complicated they were!"

"Yes, yes," stamped Alice a little impatiently, "but *what* did they discover?"

"Why the hydrated electron, of course."

"The what?"

Breathing heavily the White Rabbit pushed back his chair.

"The hydrated electron."

"What is that? It sounds a very curious thing to be excited about."

"Ah—how can I describe it—A species older than homosapiens, a transient blueness when the lightning plays upon the rain ...."

"That's all nonsense," interrupted Alice sharply, "you are making fun of me." With a supercilious shrug the White Rabbit continued in his prosaic vein, "... the hydrated electron is an elusive phantom charge that's never still in its watery trap, so quick to react that ...."

"Gobbledygook!" retorted Alice, "I won't stand for these ridiculous ideas."

The White Rabbit stood up, gave her a quelling look and tucking his paws into his white coat returned to the laboratory. Over his shoulder came the mutterings .... "What did she want me to say? That it is an excess electron bound in a self-induced stable quantum state of the polarized dielectric?"

"Well that *is* a little clearer and a lot more logical," admitted Alice to herself as she began to clear away the cups and saucers.

I. AN INTRODUCTION

The demand for a serious study of the possible involvement of hydrated electrons (e_{aq}^-) in the cathode discharge process in dilute aqueous solution can be justified for several reasons. First, some of the experimental data on the nature of the species involved finds a simple and logical explanation in terms of hydrated electrons. Second, these hydrated electrons have usurped the position of the hydrogen atoms as the primary reactive species produced during the reduction of water by most other physical and chemical means. A third reason comes from the analogy with liquid ammonia where solvated electrons are produced and observed visually at a noble cathode when the cations are nonreducible (*1*). Finally, since none of the presently advocated mechanisms for the evolution of hydrogen appears to be universally acceptable (particularly in accounting for data at pH > 9), there is clearly room for a fourth candidate to add to the more familiar slow discharge, catalytic and electrochemical explanations (*2*).

It should be noted at the outset that no attempt is made here to discuss the considerable wealth of information on the hydrogen evolution reaction at intermediate and high pH which resulted in the formulation of these more familiar mechanisms. This chapter presents only information which can be seen to be pertinent to the question of the existence of hydrated electrons, and thus the reader is left to assess its merit in the perspective of seventy years of electrochemical studies relating to this fundamental process.

Sixty-two years ago Cameron and Ramsey (*3*) suggested an analogy between the action of ionizing radiations and that of electrolysis, but decided that this could not be formulated as a grand generalization when they failed to deposit copper from $CuSO_4$ solution using radium α particles. Actually the Cu^{2+} would have been reduced to copper by radiolysis but not necessarily "deposited." Cameron and Ramsey's analogy would achieve considerable thrust if it were shown that the principal primary oxidizing (OH) and reducing (e_{aq}^-) species formed in the radiolysis of water were the very same species formed initially at the anode and cathode, respectively, during the electrolytic decomposition of water.

Hydrated electrons were first discovered in studies on the radiolysis of water (*4–8*). For several decades prior to this it was believed that the net effect of high-energy radiations on water was simply to create H atoms and OH radicals, but a transition in ideas spanning 1957 to 1962 led to the eventual unanimous acceptance that the principal reducing species was a new agent, the hydrated electron and not the hydrogen atom. Naturally, a reassessment followed not only of basic physiochemical interactions and diffusions in radiation processes, but also of the previous notions of reactivity and the mechanistic details of reductions in general. This discovery of e_{aq}^- was a very important step with consequences that were felt beyond radiation chemistry in the studies of fundamental reduction processes that occur in aqueous media.

Having acquired status and respectability as a self-sufficient chemical entity in at least one discipline, the hydrated electron has been subjected to numerous reviews in recent years (*9-20*) many of them both comprehensive and searching. Consequently, this chapter leaves many things unsaid about hydrated electrons, some intentionally, and the references listed are selective rather than exhaustive. Particular aspects are emphasized in different reviews: Anbar (*14*), for instance, deals very thoroughly with the reactions of e_{aq}^-; Schindewolf (*19*) highlights the analogies between electrons solvated in aqueous systems and those in ammonia; Dainton (*11*) makes useful comparisons between e_{aq}^- and electrons trapped in various media while discussing the thermodynamics of the different systems; in contrast, both Hart (*9*) and Thomas (*15*) concentrate on radiation chemical studies. Anbar and Neta (*21*) have compiled tables of rate constants for reactions of e_{aq}^- (also H and OH radicals) from which all quoted rate constants in this chapter are taken.

A. Hydrated Electrons

Hydrated electrons are a novel species of the genus solvated electron which appears in many forms; these range from excess electrons caught momentarily by short-range repulsive forces in nonpolar liquids such as liquid helium to those electrons permanently trapped as F centers at anion vacancies in ionic crystals. Despite the extreme differences in their immediate environments these electrons share several common characteristics. Their optical absorption bands are all intense, broad, structureless, and asymmetric (see Fig. 1) while the ESR absorption bands are very sharp and narrow. Their equivalent conductancies are well in excess of other ions in the same environment and they exhibit strong chemical reactivity as reducing agents. For our purpose electrons solvated in liquid ammonia—the first discovered and probably the best understood of these species (*24*)—provide the most useful analogies for hydrated electrons, even with such inherent striking differences reflected by their relative lifetimes, cavity size, and so on.

A wealth of information has been accumulated concerning the chemical behavior of e_{aq}^- in its role as a very powerful and reactive reducing agent, but the structure of e_{aq}^- and the nature of the electron binding remain a subject for conjecture. Perhaps a useful, if vague, description of e_{aq}^- is that of an excess electron associated with, and smeared over, a region of polarized and oriented solvent molecules such that it possesses diffusion and reaction rate parameters comparable to normal charged chemical species. Or, more succinctly, the hydrated electron is an electron caught in a stable quantum state of the polarized dielectric.

The recently reported ESR spectrum of e_{aq}^- (*25*) indicates that it cannot be described as a solvated H_2O^- ion because there is no preferred association

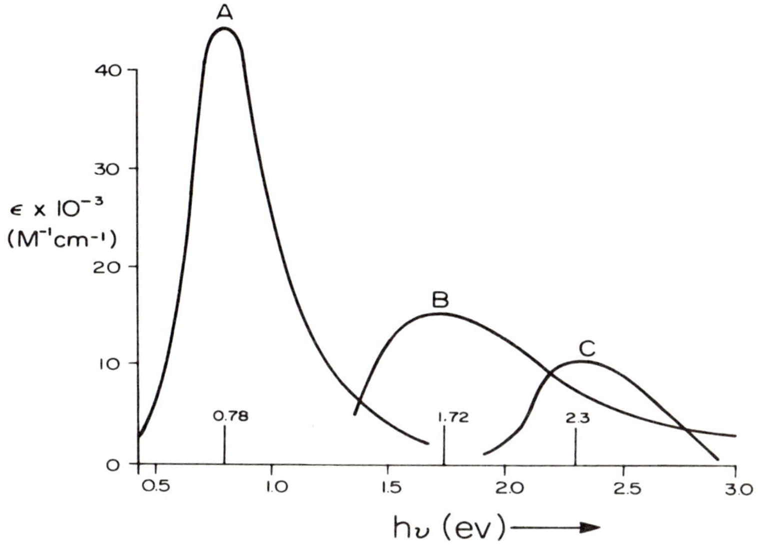

Fig. 1. Optical absorption spectra of solvated electrons in various media. A = liquid NH_3 (alkali metal solution (*19*)), B = liquid water (transient in pulse radiolysis (*22*)) and C = *F* center (KCl crystals (*23*)).

with any two protons and the splitting shows only very weak interactions. Nor is it likely, for reasons presented previously (*20*), that the e^-_{aq} is a true polaron (in the sense of inducing its own solvation sheath wherever it moves) but rather, it is suggested, e^-_{aq} exists in and jumps between pre-existing, suitably oriented sites, though this idea is far from proven. The mobility of e^-_{aq} is very high (1.8×10^{-3} cm^2 sec^{-1} V^{-1}) (*26*), substantially higher than all other ions in aqueous solution (*27*) except OH^-_{aq}(2.0×10^{-3} cm^2 sec^{-1} V^{-1}) and H^+_{aq} (3.5×10^{-3} cm^2 sec^{-1} V^{-1}). In comparison, the mobility of F^-, equivalent to that of K^+, is 7.5×10^{-4} cm^2 sec^{-1} V^{-1}. Since OH^- and H^+ derive their phenomenal mobilities from the presence of hydrogen bonded chains of molecules of liquid water (in which the transfer of a proton merely corresponds to the shift of an electron in the opposite direction), it is not unreasonable to suppose that the mobility of e^-_{aq} also may be derived from some mechanisms not available to other ions—such as the jumping or tunneling between the sites described above. Nevertheless e^-_{aq} quickly acquires an ionic atmosphere with which, for example, equilibrium is achieved in less than 3 nsec after its formation in a solution whose ionic strength is 0.23 *M* (*28*).

Despite this very high mobility, the diameter of the sphere of influence of

e_{aq}^-—its effective collision diameter which can be derived from kinetic data—is relatively large, being of the order 6 Å. A similar number has also been deduced from theoretical treatments of e_{aq}^- based on a polaron model (*29*) and in further consideration of the Heisenberg uncertainty principle the position of the electron could not be located with greater precision than a few angstroms because of its small momentum when in thermal equilibrium with its solvent surroundings at about 300°K. Thus e_{aq}^- has a size, corresponding in fact to a dispersed charge distributed over the solvent sheath, that is similar to an iodide ion. The I_{aq}^- ion has a comparatively small free energy of hydration consistent with its bulkiness and these features, too, are exhibited by e_{aq}^- whose free energy of hydration is only about 40 kcal mole^{-1}. This latter value and the size of the charge distribution for e_{aq}^- fit excellently the Noyes semi-empirical relationship (*30*) developed for univalent anions in aqueous solution.

$$\Delta G_h^\circ = 1.58r^2 - 0.568 - 163.89/r + 19.79/r^2$$

ΔG_h° also corresponds very closely to the photon energy of maximum absorption in the optical absorption band of e_{aq}^- (shown in Fig. 1, $\lambda_{max} =$ 7200 Å = 40 kcal mole^{-1}), which would be expected if the excitation process occurred as the direct photo release (i.e., photoionization) of the electron from its solvation shell. However, these values would be within 75% of each other on another model of electron binding formulated by Jortner in which the excitation of e_{aq}^- is considered to be a $(2p \leftarrow 1s)$ transition between levels described by hydrogenic wave functions.

In summary hydrated electrons are ordinary chemical entities to which reaction rate parameters (perhaps even energies of activation), diffusion constant, collision diameter, hydration energy, and third-law entropy can be assigned. If this were not so, then of course there would be no difficulty in distinguishing chemically between e_{aq}^- and hydrogen atoms, and the questions we ask in this chapter either would not have arisen or would have been answered long ago.

B. Hydrogen Atoms and Hydrated Electrons

As it is a problem *does* exist because hydrogen atoms and e_{aq}^- are barely chemically distinguishable, both being extremely reactive and very powerful reducing agents that frequently react with additives at comparable (often diffusion-controlled) rates to yield the same product. A distinction was possible, however, based on marked differences between their physical characteristics of charge, ESR, and optical absorption spectra.

These two species constitute a conjugate acid-base pair given by the equilibrium (1)

$$e_{aq}^{-} + H_{aq}^{+} \rightleftarrows H_{aq} \tag{1}$$

The hydrogen atom exhibits a pK_a of about 9.5 (*17*), which implies that although the hydrogen atom is rather a weak acid e_{aq}^{-} is also a weak base, being comparable to the cyanide or phenolate ions. Equilibrium (1) is seldom, if ever, established under experimental conditions because in most systems in which the chemistry of H or e_{aq}^{-} can be studied, either or both have lifetimes that will not exceed tens of microseconds. Consequently, they may be interconverted through pH adjustments but it is still improbable that process (1) as an equilibrium is ever attained. Their rate of interconversion can be calculated readily for any pH using the known rate constants for reaction (1), $k_1 = 2.3 \times 10^{10}\ M^{-1}\ \text{sec}^{-1}$, and reaction (2), $k_2 = 2 \times 10^{7}\ M^{-1}\ \text{sec}^{-1}$,

$$H + OH_{aq}^{-} \rightarrow e_{aq}^{-} \tag{2}$$

The spontaneous decomposition of e_{aq}^{-}, its natural lifetime being governed by the rate of reaction with its solvent molecules as given by the rate of reaction (3),

$$e_{aq}^{-} + H_2O \rightarrow H + OH_{aq}^{-} \tag{3}$$

is very slow—that is, $k_3 \leq 16\ M^{-1}\ \text{sec}^{-1}$ (*31*)—but probably many orders of magnitude faster than the analogous reaction of the solvated electron in liquid ammonia. The very small value of k_3 shows that e_{aq}^{-} is not simply a precursor of hydrogen atoms (except at pH <3 in a typical system) and cannot be dismissed as a transient of little chemical importance. Reaction (3) is one of the most crucial and fascinating reactions of e_{aq}^{-} since its rate constant with that of the reverse reaction are the cornerstones for calculations of the thermodynamic properties of e_{aq}^{-} including its standard potential. It is important to realize, however, that reaction (3) will not contribute significantly ($<10\%$) to the events in any chemical system when either (a) the local mean concentration of e_{aq}^{-} exceeds $\sim 2 \times 10^{-6}\ M$, because of reaction (4) (see below), or (b) when the pH is less than 6.3 at which stage reaction (1) dominates.

C. Reactions of the Hydrated Electron in Pure Water

Even in perfectly pure water a high concentration of e_{aq}^- cannot easily be reached because of reaction (4)

$$e_{aq}^- + e_{aq}^- \xrightarrow{k_4} (e_2^{2-})_{aq} \rightarrow H_2 + 2OH_{aq}^- \qquad k_4 = 0.5 \times 10^{10} M^{-1}\,\text{sec}^{-1} \qquad (4)$$

which has a rate constant only slightly less than that of the analogous bimolecular combination reaction of hydrogen atoms ($H + H \rightarrow H_2$, $k = 1.2 \times 10^{10}$ M^{-1} sec^{-1}). If e_{aq}^- were produced in water at a rate R (M sec^{-1}), then the steady-state concentration of hydrated electrons $(e_{aq}^-)_s$ would be given by the relationship,

$$(e_{aq}^-)_s \leqslant (2R \times 10^{-10})^{1/2} M$$

In order to produce a concentration of 10^{-4} M under steady-state conditions in a radiation chemical study with continuous radiations, we would require the incredibly high dose rate of $>2 \times 10^{10}$ rad sec^{-1}. For electrolysis, if one gram-mole of hydrated electrons could be generated at an electrode by the passage of a Faraday of charge, then in order to produce 10^{-4} M of hydrated electrons within a layer only 10^3 Å thick at the electrode surface, we would need a current density of 50 mA cm^{-2}. Since the mean lifetime of e_{aq}^- at 10^{-4} M would be about 1 μ sec, it is impossible to obtain a layer thicker than $\sim 10^3$ Å because that is the maximum distance e_{aq}^- could diffuse in 1 μsec (the diffusion constant D is 5×10^{-5} cm^2 sec^{-1} (*26*) and the electric field strength beyond the double layer will be insufficient to induce a mobility significantly greater than this).

Under pulsed conditions, however, a concentration of 10^{-4} M can be achieved with a radiation dose of 4×10^4 rad delivered in a period significantly shorter than the lifetime of e_{aq}^- (~ 1 μsec). This radiation intensity is readily available, yet in most pulse radiolysis studies it is unnecessary and often undesirable (*32*) to ultilize concentrations greater than 10^{-5} to 10^{-6} M. Likewise, a concentration of $\sim 10^{-4}$ M of e_{aq}^- could be achieved by pulse electrolysis using $\sim 10^{-7}$ C cm^{-2}/μsec pulse, but here the nonhomogeneity of the concentration of e_{aq}^- away from the electrode surface will cause difficulty in lifetime studies because this "concentration" will be changing by virtue of diffusion.

These rough data have been presented to emphasize the problem anticipated in any experimental investigation of e_{aq}^- that have been generated electrolytically. The higher the concentration of e_{aq}^- required for detection and study the shorter the e_{aq}^- lifetime and the thinner the layer of species available. This problem will be discussed more fully later in the chapter.

Reaction (4) has been written with the species $(e_2^{2-})_{aq}$ as an unstable intermediate. It is formulated in this way to imply, by analogy with F' centers, the pairing of two electrons in one solvation shell. Perhaps the liquid phase dielectron differs from those found in solids but the e_2 species in liquid ammonia is also thought to be diamagnetic and only slightly less mobile than the momomeric species (*11*). It has been shown unequivocally by isotopic substitution studies that the H_2 formed by reaction (4) did not have a hydrogen atom as a precursor and hence that H is not an intermediate (*33*). It is reasonable to suppose that the second step of this reaction produces a transient hydride ion, which reacts with water very efficiently to produce H_2.

There are tentative reasons for believing $(e_2^{2-})_{aq}$ is comparatively long lived (*34*)—perhaps having a lifetime greater than a hundredth of a second at pH 11—in which case it may be expected to play a very important role in electrochemical processes (particularly under ac conditions) for those systems in which e_{aq}^- is produced.

Hydrated electrons react at almost diffusion controlled rates with a vast array of reagents. So far we have outlined their reactivity towards each other, their reactions with H_3O^+ and H_2O. Some other reactions which may be of interest in electrochemistry are tabulated at the end of this chapter, and measured rate constants for reactions of e_{aq}^- with more than 220 inorganic and more than 280 organic chemicals may be located in Ref. (*21*).

II. HYDRATED ELECTRONS IN ELECTRODE REACTIONS

A. The Postulate

The occurrence of hydrated electrons in the radiolysis of liquid water is not a peculiarity of any of the succession of events characteristic of the interaction of high-energy radiations, except that of ionization which ultimately leads to the existence of thermalized electrons that rapidly become hydrated. There are good reasons for believing that whenever "free" electrons are generated in, or injected into, liquid water, e_{aq}^- are produced. For instance, the photoionization within a charge transfer band of ions such as I_{aq}^-, OH^-, $Fe(CN)_6^{4-}$, the reaction of a metal such as sodium with water according to reaction (5),

$$Na \xrightarrow{H_2O} Na_{aq}^+ + e_{aq}^- \tag{5}$$

the photoelectron emission from a metal surface such as mercury, the chemical oxidation of U^{3+} ions by water, and so on, all appear to produce e^-_{aq} (*20*).

Therefore it is reasonable to postulate that hydrated electrons will be formed in water adjacent to a cathode if the latter can be made to "eject" electrons by suitably adjusting its electric potential with respect to the water. This would be represented simply by the net overall reaction (6).

$$e^- \text{ (cathode)} \longrightarrow e^-_{aq} \tag{6}$$

The detailed mechanism may include electron tunneling or, alternatively, may be the net result of discharging the so-called "nonreducible" cations such as Na^+, which subsequently gives sodium atoms (either free or alloyed) that may be very rapidly dissociated by the water according to reaction (5) to produce e^-_{aq}. In the electrolysis of aqueous NaOH for instance, in which H^+_{aq} cannot participate, the cathode seems to have only two alternatives: either it transfers an electron directly to the water according to reaction (6) or the current-carrying cation is discharged and reduced to Na atoms, which lead to e^-_{aq} via (5) (rather than to hydrogen atoms via process (7) as previously advocated).

$$Na + H_2O \longrightarrow Na^+_{aq} + OH^-_{aq} + H \tag{7}$$

In acidic solution the net reaction (6) will be in competition with the primary discharge reaction involving H^+ to give adsorbed hydrogen atoms. Even if reaction (6) occurred as a primary step under conditions of low pH, the hydrated electrons would be converted rapidly to hydrogen atoms by reaction (1). For instance, the mean half-life of e^-_{aq} at pH 1 is less than 1 nsec.

The principal objection to what has been said so far is that the electrode will not normally be at a sufficiently negative potential to (a) eject an electron, (b) reduce Na^+, or (c) produce e^-_{aq}, the standard potential of which is about -2.7 V (*35*). Objection (a) may be countered by suggesting that the Fermi levels of the metallic electrons are significantly affected by the presence of the water. This is corroborated by the energy of the photoemission thresholds noted in Barker's experiments (*36*), which were substantially smaller than the "vacuum" photoelectric work function. The other two objections may be untenable because true thermodynamic equilibrium will not pertain in an electrolysis experiment, particularly involving sodium or e^-_{aq}, and therefore the electrodes cannot be considered to be reversible. On this basis an attempt will be made later to answer objections (b) and (c).

B. Evidence for e_{aq}^-

Although not plentiful the evidence is certainly varied and if at times the isolated case appears ambiguous collectively the data warrants serious consideration. Experimental findings are discussed in five separate but casually related approaches, and finally some conceptual problems, inherent in this field, are considered.

1. Competition Studies

In examining the identity of the precursor to molecular hydrogen in the reduction of water by various means (*37*) use was made of competitive scavenging between nitrous oxide and methanol as given by Eqs. (8) to (11).

$$e_{aq}^- + N_2O \rightarrow N_2 + O^-, \quad k_8 = 5.6 \times 10^9\ M^{-1}\ \mathrm{sec}^{-1} \tag{8}$$

$$H + N_2O \rightarrow N_2 + OH, \quad k_9 \sim 10^5\ M^{-1}\ \mathrm{sec}^{-1} \tag{9}$$

$$e_{aq}^- + CH_3OH \rightarrow ?, \quad k_{10} < 10^4\ M^{-1}\ \mathrm{sec}^{-1} \tag{10}$$

$$H + CH_3OH \rightarrow H_2 + ?, \quad k_{11} = 1.6 \times 10^6\ M^{-1}\ \mathrm{dec}^{-1} \tag{11}$$

It is evident from these rate constants that if methanol and N_2O are present in the reaction mixture in the ratio $[CH_3OH]/[N_2O] = 10$, then reactions (8) to (11) will lead exclusively (>99%) to N_2 if e_{aq}^- are produced and to H_2 if hydrogen atoms are produced. Confirmation of the validity of such a diagnosis could be obtained by changing the pH of the solution. Through reactions (1) and (2) any N_2 may be replaced by H_2, or H_2 by N_2 by working at low and high pH, respectively.

In this way it was possible to show that reaction (5) occurred when sodium amalgams were reacted with water (*37*)—a conclusion also arrived at independently by Hughes and Roach (*38*). It was necessary to use amalgams in order to slow down the reaction at the metal surface so that the scavengers could compete successfully with the bimolecular reaction (4). When current was passed between smooth platinum electrodes through a solution of Na_2SO_4 made from very pure water with added CH_3OH and N_2O, N_2 (plus a small amount of H_2 which increased with increasing current density) was produced at the cathode at pH 7 and 13, whereas only H_2 was produced at pH 1. The N_2O never exceeded 10^{-2} M and polarographic studies on N_2O solutions failed to show a half-wave potential that would correspond to the direct reduction of N_2O at the cathode.

Two conclusions were drawn from these results: (a) solvated hydrogen atoms were not precursors of H_2 in the cathodic decomposition of water and (b) the data were entirely consistent with the formation of e_{aq}^-. However, since the relative chemical reactivity of absorbed H atoms, or other possible species, towards CH_3OH and N_2O were not known, further deductions could not be made.

2. Spectrophotometric Study

If hydrated electrons are produced in the vicinity of a cathode in electrolyzed Na_2SO_4 solution, then it should be possible to detect and identify them spectrophotometrically since, in sharp contrast to hydrogen atoms, e_{aq}^- has a very intense optical absorption band spanning the entire visible spectrum (Fig. 1). However, experimental difficulties are encountered for the reasons outlined previously. Even assuming a perfectly pure system in which e_{aq}^- disappears only via reaction (4), we would be restricted by the following considerations: the lifetime of e_{aq}^- is limited by its steady-state concentration according to $\tau = 1/k_4(e_{aq}^-)_s$ and the lifetime limits the distance d that the species can diffuse from the surface; hence the thickness of the layer of absorbing species is given by $d = (D\tau)^{1/2}$ where D, the diffusion constant = 5×10^{-5} cm^2 sec^{-1}(*26*). Thus we have to select from combinations such as the following, remembering that these are the maximum values possible because first only reaction (4) is considered and second nonhomogeneous distributions will diminish the values further: $(e_{aq}^-)_s = 10^{-4}$ M, $d = 10^{-5}$ cm; $(e_{aq}^-)_s = 10^{-6}$ M, $d = 10^{-4}$ cm, and so on. The product $(e_{aq}^-) \times d$ will always be less than 10^{-9} M cm for any system in which values of $d > 10^{-5}$ cm are considered. (It would be meaningless to consider layers much thinner than this because e_{aq}^- may be formed > 100 Å from the surface and the high field strength of the electrical double layer will alter the diffusion process. Furthermore, the layer thickness would then be significantly smaller than the wavelength of light used.) At the wavelength of maximum absorbance (720 nm) of e_{aq}^- the molar extinction coefficient is $\varepsilon = 1.5 \times 10^4$ M^{-1} cm^{-1} (*39*); so for light passing perpendicularly through this layer once, the optical density (OD) would be OD < 1.5×10^{-5}, which is well below the sensitivity limit for steady-state spectrophotometry.

These calculations show that regardless of how high a current is passed through a solution, a very large number of repeated reflections through the layer would be required to obtain a useful optical de nsity. Very little could be gained by going to pulsed electrolysis, or pulsed discharge, for the reasons given earlier—the shorter the time, the thinner the layer and hence the shorter the absorption path length. These figures also draw attention to the difficulties encountered in trying to measure the absorbance by passing a narrow beam of light down the surface of a long, very straight cathode. Here the optical density will ultimately be limited to d/r multiplied by the optical density within d of surface, where r is the thickness of light beam. Even if we were to construct a very flat cathode one meter long and scan it with a perfectly parallel light beam having $r = 1$ mm, the optical density (OD) obtained would still be ≤ 0.015.

It was for these reasons that the equipment sketched in Fig. 2 was designed

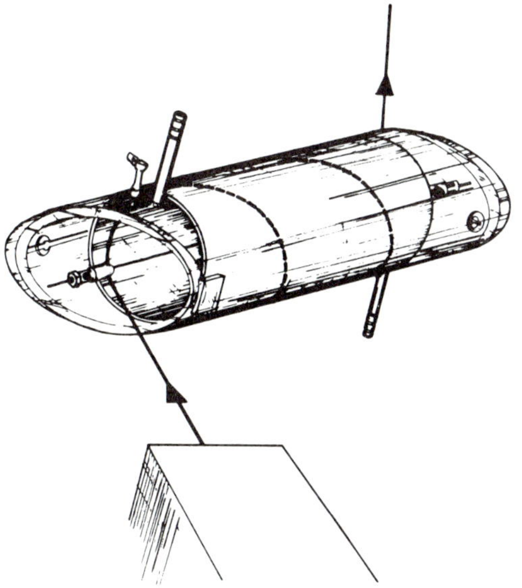

Fig. 2. Electrolysis cell used in continuous specular reflection technique, showing the laser beam directed tangentially onto the silver surface and spiralling down the tube.

(*40*) (and in accordance with trends in technical nomenclature dubbed the "continuous specular reflection" technique). It simply alleviates the difficulty discussed above by directing a narrow parallel beam of light tangentially onto an inner, cylindrical, highly polished silver surface so that some of the light stays within depth d the whole time and the rest penetrates region d innumerable times as the beam spirals down the tube. This optical arrangement is illustrated in Fig. 3. A 632.8 nm CW He/Ne laser was used as the light source in this work since this has the obvious advantages of providing a

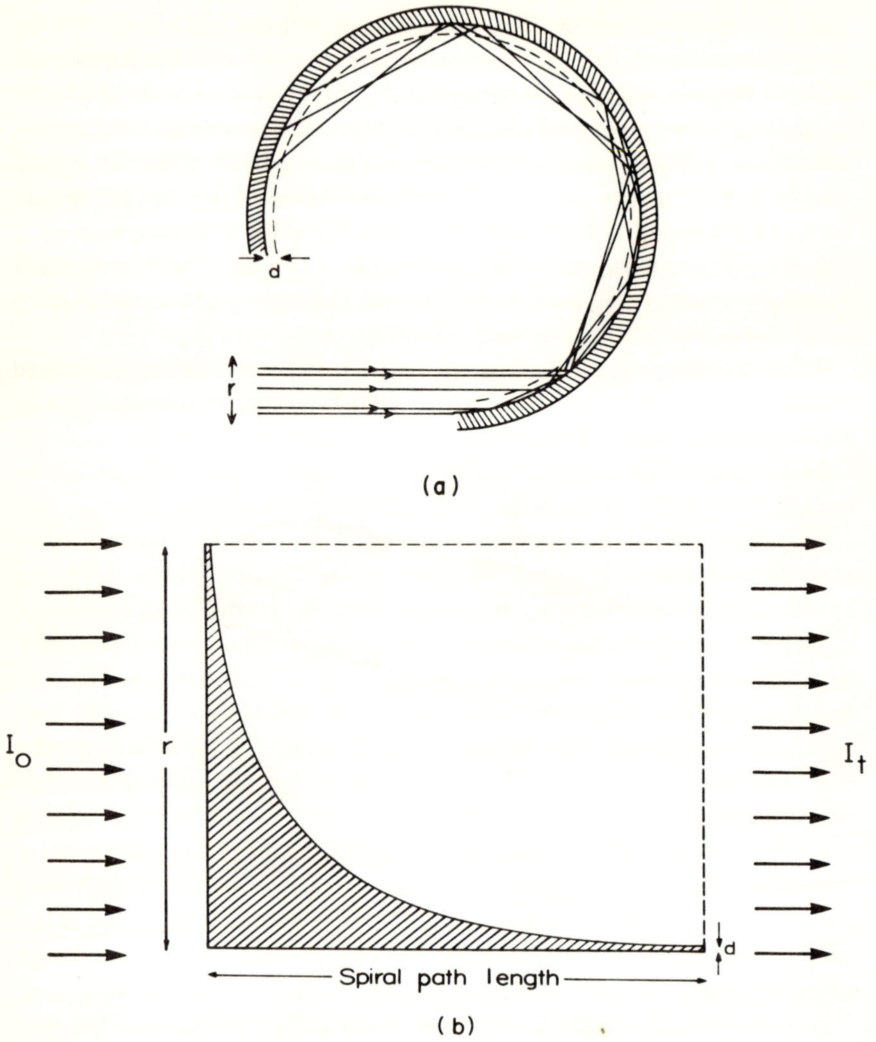

Fig. 3. Schematic representation of continuous specular reflection. (a) grossly exaggerated light beam of thickness r reflecting through region of absorbing species of thickness d. (b) Equivalent optical cell for this arrangement. The following relationship has been shown (*41*) to hold, $(I_0 - I_t)/I_0 \propto \varepsilon cd/r$ (where c is concentration, ε, d, and r are described in text).

very intense, parallel, narrow beam at a wavelength very close to the absorption maximum of e_{aq}^- ($\varepsilon_{633} = 1.25 \times 10^4\ M^{-1}\ cm^{-1}$ (*39*)). It has the disadvantage of being restricted to 633 nm, so these experiments could yield data for only one wavelength.

In this cell the cathode was made of highly polished silver deposited by vacuum evaporation. The anode was a thin platinum wire stretched down the center of the tube to give a radially uniform field. In order to avoid the build up of H_2 gas on the electrode surface, low current densities and alternating voltages were used. This ac arrangement also permitted greater detection sensitivity by the use of a lock-in amplifier that allowed the detectors' amplifier to be tuned to the frequency with which the current was alternating in direction. Experiments were also conducted under dc conditions when qualitatively equivalent results were obtained. Deoxygenated 0.25 *M* Na_2SO_4 solution was used and occasionally adjusted to pH 1 with H_2SO_4.

When the reflecting surface became the cathode, a decrease in light transmission was observed. No change in light intensity occurred when this surface was made the anode. The optical density decreased when the acidified solution was used for the same current and electrode voltage. Voltages across the cell of 0.8 to 3 V were used.

These data were interpreted as being entirely consistent with the formation of hydrated electrons in the close vicinity of the cathode. This was offered as the simplest satisfactory explanation. The incomplete suppression of e_{aq}^- at pH 1 was attributed to the fact that the $[H^+]$ within depth d would be depleted after a few seconds electrolysis and that H^+ ions diffusing into that region caused the observed decrease in optical density by reaction (1). An analysis of the equivalent optical cell (*41*) showed that if e_{aq}^- was formed, then it existed within $d = 1.7 \times 10^{-4}$ cm of the surface at the mean (but rather nonhomogeneous) concentration of $\sim 8 \times 10^{-7}$M.

If the change in light transmission was really caused by the presence of a short-lived transient generated at the cathode that absorbs red light very strongly, then it seems very likely that this should be e_{aq}^-. But what other effects could produce a change in transmitted light intensity? Let us briefly consider some possibilities.

(a) There was no permanent change in the surface reflectivity; but it is possible that a reversible transient oxidized or reduced state of the silver surface altered the net specular reflection efficiency. There are two factors against this: first, under dc conditions there was no gradual or continuous change in transmission; second, during one cycle of the alternating current the total charge passed was only about $1 \mu C\ cm^{-2}$, whereas something in excess of $300\ \mu C\ cm^{-2}$ is required to cause an oxidized layer on silver or indeed to reduce that layer (*42*). The presence of trace impurities in the solution such as cations or O_2 which could cause reversible surface oxidation seems to be excluded by the dc experiment.

(b) Changes of refractive index due to changes in the concentration of the electrolyte at the surface could alter the optics in such a manner as to cause a diminution in transmission. However, because the transport numbers of Na^+ and SO_4^{2-} are rather similar, a comparable effect should have been found during the half cycle when the silver was the anode and also in the dc experiments when the silver was the anode. In fact, however, no such effect was observed.

(c) Since the laser beam was not only plane polarized but polarized by specular reflection as well, it is conceivable that a Kerr electro-optic effect was involved. Again, however, this possibility seems to be excluded by the absence of an effect in the reversed polarity experiments.

(d) Electroreflectance effects will cause some change in the reflectance of the silver surface during the ac cycle. However, it appears to be a minor factor because it should work in the opposite sense to that observed; i.e. it should cause a decrease during the anodic phase. It is possible that the grazing angle of incidence used in these experiments makes this effect unimportant. The involvement of electroreflectance phenomena in this type of experiment is receiving attention (J. D. E. McIntyre, private communication).

(e) It has been suggested that hydrated electrons may be produced in this system and be peculiar to this type of experiment because of electrically assisted photoemission (*43*). There was sufficient power in the laser beam to create by photoemission from the silver enough e_{aq}^- in the light path to cause the observed absorption, without significantly affecting the total current flowing through the cell, even if only about 20% of the laser intensity was absorbed by the metal. However, the energy of the photons (1.95 V) seems to be much too small. Using Barker's data (*36*) on the threshold potential necessary for photocurrent production at various wavelengths and extrapolating it out to 633 nm (it was linear from 254 to 425 nm) would indicate that we need to polarize the cathode to the extent of -2.12 V versus NCE. The effects reported were obtained at much smaller voltages.

It is still possible, however, that the light was not without effect and that at least photoassisted electrolysis was occurring.

3. Pressure Coefficient of Hydrogen Electrode Reaction

Whereas Na_2SO_4 solutions at pH 7 were used in the two experiments just described, this work concerns the hydrogen evolution reaction at pH 1. Hills and Kinnibrugh (*44*) studied the pressure dependence of this reaction on Hg at 1500 atm pressure and from the results deduced that there was a negative volume of activation of -3.4 ml $mole^{-1}$ for the rate-controlling step. These authors argued that this result was not in accord with a reaction mechanism involving either adsorbed hydrogen atoms or H_2^+ ions; that is, it was contrary to the three popular explanations for this reaction—slow discharge, catalytic, or electrochemical mechanisms. They proposed that the negative volume of activation would be consistent with the emission and hydration of metallic electrons as represented by reaction (6). Since the publication of that paper the partial molar volume of e_{aq}^- (this will be almost equal to the volume of activation of reaction (6)) has been estimated in a radiation chemical system to be between -5.5 and -1.1 ml mole^{-1} (*45*), a range centering exactly on Hills and Kinnibrugh's value. (This negative value of $\bar{V}(e_{aq}^-)$ is in sharp contrast to the large positive value of ~60 ml mole^{-1} (*46*) for the solvated electron in liquid ammonia, where the density change can be seen to be associated with large cavities containing the electron. Jortner's theoretical study on the optical absorption spectra of solvated electrons does, however, predict that large cavities should exist in ammonia but extremely small ones in water (*29*). From purely electrostrictive factors we would certainly anticipate that the process of solvation of an electron would lead to a volume decrease, providing large cavities were not formed.)

It should be recalled that if reaction (6) does represent the primary discharge step in the hydrogen evolution reaction at pH 1, this will be followed very rapidly by reaction (1) leading to a hydrogen atom (or H_3O, or H_9O_4 radical). Combinations of these will then yield H_2, but possibly not before some have been caught by H^+ to give H_2^+ which then could migrate to the electrode surface where the second neutralization may proceed.

4. Neutralization of Metallic Ions at Macrodistances from a Cathode

A paper by Yurkov (*47*) of this title has been quoted (*44*, *48*) as examining the electrochemical consequences of electron emission. Yurkov showed by visual means, a projection of which is given in Fig. 4, that an opaque layer of colloidal Cu metal resulting from charge neutralization could be formed at distances from 0.1 to 1 mm from a cathode surface when a solution of $NaNO_3$ was electrolyzed at very high current density (~2 A cm^{-2}) using a copper anode and almost any metal as cathode. The distance from the cathode at which the layer formed was about 0.1 mm at 0.8 V but increased with increasing potential to about 1 mm with 2 V applied between the electrodes; the latter were ~1.5 mm apart and ~0.2 mm^2 cross-sectional area. The layer was detectable only after 30 to 50 sec duration of electrolysis, by which time the solution between the layer and anode acquired the color of $CuNO_3$ solution. No gas evolution occurred at the cathode or between the cathode

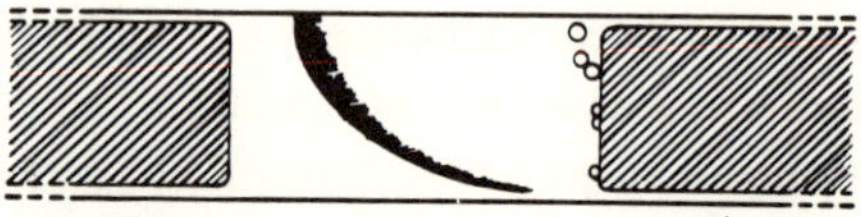

Fig. 4. (*a*) Layer of finely dispersed copper metal formed between copper electrodes during the electrolysis of $NaNO_3$ solution. Bubbles can be seen around the anode but not the cathode. (b) Similar experiment but with a magnetic field across the cell.

and the layer, nor did the cathode show signs of dissolution or deposition. A magnetic field applied across the cell caused a distortion of the layer formation as shown in Fig. 4(b).

Yurkov concluded that "electrons were the current carriers" in the region between the cathode and the layer. No suggestion was made as to whether these electrons were to be regarded as free or solvated, but we are now in a better position to comment on that.

Free electrons would not give the observed voltage dependence of the layer position since this could arise only from the migration of two species of comparable mobility, though the negative one must be somewhat higher. The mobility of free electrons is several orders of magnitude higher than that of solvated ions, and their lifetime is probably limited to 10^{-11} sec, the dielectric relaxation time of water. We can also rule out hydrated electrons as being responsible for carrying the current simply on the basis of lifetimes. Since the mobility of e_{aq}^- is 1.8×10^{-3} cm^2 sec^{-1} V^{-1}, it would take 4.2 sec for e_{aq}^- to travel the 1 mm to the layer under a field strength of 2/0.15 V cm^{-1}. Since 4 mA of current was passed, this means 4×10^{-8} mole sec^{-1} of e_{aq}^- would have been necessary, each taking 4.2 sec to traverse a solution of total volume 3×10^{-7} liter. This would have resulted in a solution of mean concentration 0.56 *M* in e_{aq}^-, which is quite impossible because of reaction (4) and because H_2 was not formed. For analogous reasons, and because they would migrate to the layer by random diffusion only, we can discount the possible involvement of H atoms.

It is just possible, however, that Yurkov's experiments do demonstrate the formation of e_{aq}^- in the cathodic discharge reaction and that these combine together according to the first step of reaction (4) very close to the electrode, to produce the dielectron species $(e_2^{2-})_{aq}$. The latter species in this quite basic environment might have a sufficient lifetime to migrate to the layer. Here it will meet Cu^{2+} migrating towards the cathode and initiate neutralization leading eventually to colloidal copper particles.

The fact that it took 30 to 50 sec before a layer was observed corresponds roughly to the time it took either to discharge or at least to transport completely all the Na^+ and NO_3^- of the original electrolyte from the appropriate electrodes. That this had occurred is evidenced by two factors; so much charge was passed that every ion in the very small cell would have been involved in charge carrying and discharge after about 20 sec electrolysis, even if the solution were 5 *M* in the first instance; (b) the fact that a layer of copper forms and the color of Cu^{2+} is observed indicates that copper was dissolving at the anode and was the principal charge carrier from the anode. Thus after 30 to 50 sec the solution close to the cathode would contain no NO_3^- (these ions would react rapidly with e_{aq}^- to form NO_3^{2-} ions (*49*) which are a possible alternative charge carrier resulting in layer formation at a distance from the electrode).

After a period of electrolysis, the solution would contain a high OH^- concentration (near the cathode) in which $(e_2^{2-})_{aq}$ might be expected to be long lived. In any event the absence of H_2 and formation of a layer of Cu particles away from the cathode seems to be comprehensible in terms of hydrated electron but not hydrogen atom formation.

We interpret in the same way some results of Uhlig and Krutenat (*50*) in which a temporary reducing species (possessing a diffusion constant greater than that of H_2) having an average half-life of $\sim$5 min was shown actually to be present *in the solution* around a platinum, nickel, or tin cathode in the electrolysis of NaCl solution at pH 10. Again, the data might indicate the formation and longevity of a solvated dielectron species formed by the combination of two e_{aq}^- in the first step of reaction (4). Certainly adsorbed hydrogen atoms or H_2^+ ions could not account for those results.

5. Photocurrent Production at Metal-Water Interfaces

Some 38 years ago Bowden (*51*) reported experiments in which he demonstrated an increased current density at a mercury cathode in contact with N/5 aqueous H_2SO_4 under the illumination of ultraviolet light. Bowden suggested that the light provided an activation energy to the oriented dipoles at the cathode surface, thus altering the interfacial potential, reducing the overvoltage, and increasing the current. A rather different conclusion was reached recently by Barker, Gardner, and Sammon (*36*) for the results of a very comprehensive and elegant study of this phenomenon.

These authors measured the photocurrents produced by light of various wavelengths directed onto a dropping mercury cathode as a function of the electrode potential and the concentration of various scavengers added to the electrolyte. They used advanced polarographic equipment capable of measuring minute photocurrents and illuminated the cathode with light delivered at constant intensity, intermittantly, or as single intense flashes. With N_2O, H_3O^+, or NO_3^- present as scavenger in the solution the photocurrent increased for a given light energy and cathode voltage and, in addition, the effect of neutral salts indicated that the photocurrent was caused by diffusing negatively charged species in equilibrium with an ion atmosphere. Some of these data are shown in Fig. 5. Their interpretation follows: Electron emission from mercury was induced by photolysis; the photoejected electrons traveled beyond the electrical double-layer before becoming thermalized and solvated; the electrode potential and the pressure of scavengers that react with e_{aq}^- determine the magnitude of the photocurrent by controlling the lifetime of the charged species in solution.

Similar types of experiments have been performed by Delahay and Srinivasan (*52*) using polychromatic light from conventional flash-photolysis equipment directed at mercury and thallium-mercury electrodes polarized by a potential below the faradic process level. They found that the charge produced depended on the concentration of H_{aq}^+ in solution and the polarizing potential. They, too, interpreted the data in terms of e_{aq}^- formation by photo-induced electron emission into water. Heyrovsky (*53*) also performed experiments along similar lines but attributed his results to the reduction of H_2O to give H and OH^-. Other studies of this phenomenon were reviewed recently in this series (*54*).

Regardless of whether these are examples of electrically assisted photoelectric emission or photo-assisted electrolysis, they demonstrate that the reduction of water near to the cathode appears to proceed by electron emission, thermalization, and solvation followed by reaction of e_{aq}^-, rather than by dissociation into $H + OH^-$ of a water molecule polarized at the interface. It is also interesting to note that the sum of the energy of the photon plus the cathodic voltage in these experiments was often less than the "vacuum" photoelectric work function of the metal by about 1.5 V—just about the solvation energy of the hydrated electron.

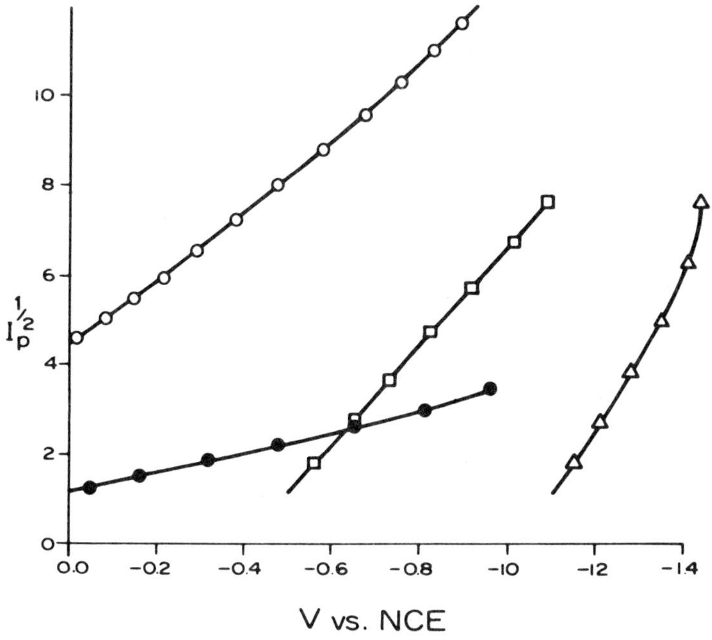

Fig. 5. Variation of photocurrent ($Ip^{1/2}$) as function of electrode potential for light of various wavelengths and different scavenger concentrations, for a 0.2 *M* KCl solution.

○ 253.7 nm, □ 360.0 nm, △ 425.0 nm: sat. N_2O ($\sim 2 \times 10^{-2}$ M)

● 253.7 nm, 3×10^{-3}M N_2O

C. Conceptual Considerations

A rigorous theoretical treatment and discussion of the mechanism of hydrogen evolution at high pH stemming from these experiments is clearly called for. Pyle and Roberts (*55*) have made the first attempt to test this mechanism on a limited theoretical basis and have found the outcome in general accordance with experimental data. However, the question still remains, can reaction (6),

$$e^-_{\text{cathode}} \rightarrow e^-_{aq}$$

followed by reaction (4),

$$2e^-_{aq} \rightarrow (e_2^{2-})_{aq} \rightarrow H_2 + 2OH^-_{aq}$$

or (1),

$$e^-_{aq} + H^+_{aq} \rightarrow H_{aq}(\rightarrow H_2)$$

adequately account for the isotope separation factor, slope of the Tafel plot, hydrogen overvoltage on various metals, and so on? Can this mechanism do as well as current theories? With regard to the separation factor, data on the thermodynamic and kinetic properties of the electron solvated in deuterium oxide are available and in general the reactivity is lower than that of e^-_{aq} in H_2O (*56*, *57*). Concerning the $\frac{1}{2}$ which appears in the slope of the Tafel plot, this may arise as a consequence of the bimolecular reaction (4) at pH > 7 or in acidic solutions from the combinations of two hydrogen atoms formed in reaction (1). Perhaps even the different overvoltages and the dependence of exchange current density on the nature of the metal are reflections of the different work functions for electron emission and in particular the extent to which the Fermi levels of the metallic electrons are affected by the presence of the metal-water interface.*

Two factors that cannot be entirely disregarded in the context of this discussion are, first, the analogy with liquid ammonia where an electron electrode (*1*) can be established involving solvated electrons (which in ammonia have indefinite stability with respect to the solvent) in reversible equilibrium with a platinum electrode and, second, the well-founded suspicion that in general when water is reduced by either physical or chemical means it prefers (perhaps for purely kinetic reasons) to yield e^-_{aq} rather than hydrogen atoms (*20*).

Ultimately there is the problem of the standard potential of e^-_{aq}. This is currently estimated to be -2.67 V (*35*)—it may increase slightly but not very much—which implies that e^-_{aq} is a very powerful reducing agent comparable to sodium metal. The hydrated electron is thus a stronger reducing agent than either the hydrogen atom, which has a standard potential of -2.1 V (*58*), or the solvated electron in ammonia where the value is -1.95 V (*1*).

It might be argued that this high value for $E°(e^-_{aq})$ indicates that e^-_{aq} is not involved in the hydrogen-evolution reaction because it implies that a potential of about -2 V would be required to generate it as an intermediate in this reaction at a cathode, whereas H_2 can easily be shown to be evolved at the so-called catalytic metals within 0.1 to 0.2 V from the reversible H_2 potential. However, this argument is not necessarily valid. Several factors would be overlooked in reaching such a conclusion.

* The reversible deposition of sodium on mercury electrodes and hydrogen overvoltage effects have also been interpreted as involving hydrated electrons (G. Dubpernell and D. J. Kenney, Abstract of paper presented at the 19th meeting of C.I.T.C.E. in Detroit, September, 1968).

(a) In the conventional manner the standard potential of e_{aq}^- should be equal to the standard free-energy change for reaction (12),

$$H_{aq}^+ + e_{aq}^- = \tfrac{1}{2}H_2(g), \Delta G_{12}^\circ = -61.5 \text{ kcal mole}^{-1} \quad (12)$$

for which Baxendale calculated a value of -61.5 kcal mole^{-1} (*35*). However, if we compare reaction (12) with the standard reference electrode reaction, (13),

$$H_{aq}^+ + e^-(Pt) = \tfrac{1}{2}H_2(g) \quad (13)$$

which is taken arbitrarily to have $\Delta G_{13}^\circ = O$, then for reaction (14),

$$e^-(Pt) = e_{aq}^- \quad (14)$$

we cannot conclude by difference that ΔG_{14}° equals $+61.5$ kcal mole^{-1} because ΔG_{12}° is on an absolute basis and not directly related to $\Delta G_{13}^\circ = 0$. (In principle nothing is implied about $e^-(Pt)$ in reaction (13) except that in order to be consistent with $\Delta G_{13}^\circ = 0$, it must have an arbitrary free energy of zero.)

(b) It may be misleading and invalid to presume that the free energy of reaction (14), whatever its value, constitutes an energy barrier in the hydrogen-evolution reaction simply because e_{aq}^- is an intermediate in the formation of H_2. After all, the potential required to form hydrogen atoms in reaction (15), $E_{15}^\circ = -2.1$ V,

$$H_{aq}^+ + e^-(Pt) = H \quad (15)$$

does not represent an energy barrier, because the hydrogen atom remains adsorbed on the surface. This not only lowers the energy barrier (presumably not the full 2 V) but also means that, since H_2 is formed at the surface, the electrode becomes reversible with respect to $H_2(g)$ not hydrogen atoms (adsorbed). Perhaps a similar detail of the mechanism involving e_{aq}^-, such as an association or pairing followed by reaction (4) at the surface, also makes that electrode reversible with respect to H_2. This is simply saying that in terms of equilibrium thermodynamics perhaps the nature of the intermediate is irrelevant for either mechanism and that the reversible potential is given by the activity of H_2 gas in equilibrium with the electrode.

(c) It is probably unreasonable to suppose that an electrode could ever in practice be made reversible with respect to a highly reactive, mobile, and negatively charged species such as e_{aq}^-. Perhaps in an electrolysis experiment the electron becomes hydrated some 3 to 10 nm from the surface and cannot again surmount the high field gradient of the double layer. Under these circumstances the only meaningful concentration at the surface that affects the potential at which e_{aq}^- is generated may be so small that there is essentially no thermodynamic barrier to its formation.

Baxendale (*35*) calculated E° (e_{aq}^-) from the standard free energy changes involved in the following reactions, starting with reaction (3),

$$e_{aq}^- + H_2O = H + OH_{aq}^-, \quad \Delta G_3^\circ = 6.3$$
$$H = \tfrac{1}{2}H_2(g), \quad \Delta G^\circ = -48.5$$
$$H_{aq}^+ + OH_{aq}^- = H_2O, \quad \Delta G^\circ = -19.1$$

so that we have reaction (12),

$$H_{aq}^+ + e_{aq}^- = \tfrac{1}{2}H_2(g), \Delta G_{12}^\circ = -61.5 \text{ kcal mole}^{-1} \tag{12}$$

and thus on the conventional (reduction potential) scale $E^\circ = -2.7$ V. Baxendale evaluated ΔG_3° from the equilibrium constant for reaction (3) obtained from the ratio of rate constants k_3/k_{-3}. Consequently ΔG_{12}° is absolute in that everything can be measured directly. Jortner and Noyes (*59*) have improved the estimation somewhat by using a more recent value for k_3, considering the standard states more exactly, and including a value for the solvation energy of the hydrogen atom, all of which results in the value -2.67 V.

It is worth noting that this value may be slightly too negative on two counts. First, the free energy of solvation of the odd electron species hydrogen may be more negative than the 4.5 kcal mole^{-1} assumed by Jortner and Noyes from an analogy with H_2 and He. Second, despite very great care being taken in obtaining this value (*31*) of $k_3 = 16\ M^{-1}s^{-1}$, it may be an upper limit only. Whereas ΔG_3° does not have a strong dependence on k_3, there are still uncertainties associated with reaction (3). For instance, it has not yet been proven that H is a product of that reaction.

III. THE FORMATION OF HYDRATED ELECTRONS IN VARIOUS SYSTEMS

A. The Ubiquitous Electron

Already in this chapter we have emphasized that many reactions previously attributed to hydrogen atoms (or nascent hydrogen) may have been initiated by e_{aq}^-, its conjugate base, under suitable conditions of pH. Hydrated electrons may be generated in various systems which fall into one of the following categories: (a) electrochemical: electric potential supplied externally (discussed already); (b) photochemical: photoinduced electron transfer, photoionization, or simply photoemission; (c) spontaneous chemical processes: the driving force being the chemical potential of the reactants; and (d) radiation chemical: ionization and excitation induced by the interaction with high-energy charged particles. Some of these methods of producing e_{aq}^- will give a homogeneous solution of species while others will not. For instance, in electrolysis, in the dissolution of active metals, and in photoinduced emission from metals the solvation of electrons close to the metal surfaces will give rise to nonuniform distributions that complicate studies on the reaction kinetics.

Extension of this topic into nonaqueous solvents can be followed through Refs. *(60)* and *(61)*. The role of solvated electrons in electrochemical studies of these systems has not yet been examined fully, but they are doubtless important, particularly in aprotic solvents where they may be fairly long lived.

B. Photolytic Processes

There are two main processes involved in photolytic systems—direct photoionization of a solute or charge-transfer-to-solvent excitation, and photoinduced emission from metals in aqueous solution. In each case an electron is released into the water in which it is presumed to become hydrated within about 10^{-11} sec, the dielectric relaxation time of the medium.

1. Photochemical Reaction

The dominant reducing species in γ-irradiated pure water is the hydrated electron and photolysis of water within its ionization continuum will probably follow the same mechanism. If suitable aqueous solutions are used, then the hydrated electron and its subsequent reactions can be readily observed. Both continuous illumination and microsecond flash photolysis techniques have been employed to generate e_{aq}^-, while spectrographs and spectrophotometric detection equipment have measured its lifetime and optical properties. Competitive scavenging techniques provide a means of calculating limiting quantum yields for e_{aq}^- (and the other photolytic by-products) in any system and collecting information on its chemical reactivity.

Below $\sim$200 nm water absorbs significantly. The extrapolated gas-phase value for the ionization potential of water is 12.6 eV, but recently it was proposed (*62*) that since molecules in the condensed phase often have substantially lower ionization potentials compared to the same discrete molecules in the gas phase, water might be ionized at about 7 eV. Nevertheless, various attempts to photoionize pure water with $\lambda \sim 147$ nm have been inconclusive (*63*)*, and severe technical difficulties with experimentation in the vacuum **uv** eclipsed any possible absorption signals in the only reported photolysis of water with λ as low as 58.4 nm (*20b*). The electronic properties of liquid water in the vacuum **uv** described recently (*64*) do not confirm the hypothesis that the diffuse nature and overlap of the water molecules in the condensed phase facilitate photoionization because no sign of an ionization continuum was observed as low as 105 nm; however, it was proposed that either the same processes were occurring in the liquid as in vapor phase but at different energies, or that additional processes were involved in the liquid state. This was the rationale behind the interpretation of two peaks in the absorption spectrum as an excitation transition at 8.3 eV and an interband transition at 9.6 eV.

The primary step in the photolysis of water at $\lambda \sim 185$ nm is

$$H_2O + h\nu \rightarrow H + OH \qquad \phi_{185} = 0.6$$

If e^-_{aq} were produced, the yield would be less than 0.1 of the yield of hydrogen atoms; of the latter, only 15% or so give H_2 gas whereas in irradiated water 50% react this way because of spurs,

$$H + H \rightarrow H_2$$

The photolysis of solutes in water at wavelengths greater than 200 nm has led to detailed investigations of many systems in which e^-_{aq} is an important precursor to the final products. One of the major problems associated with these experiments stems from the fact that often the products themselves have absorption bands in the same region of the spectrum as e^-_{aq} (a glance at Fig. 1 will show that any product absorbing strongly in the visible region falls into this category) thus interfering with both its identification and kinetic analyses based on the half-life of absorption signals. One such system is

$$I^-_{aq} + h\nu \rightarrow I_{aq} + e^-_{aq} \qquad \phi_{254} \sim 0.25$$
$$I_{aq} + I^-_{aq} \rightarrow I^-_{2aq}$$
$$I_{aq} + I^-_{2aq} \rightarrow I^-_{3aq}$$

* Note added in proof: Experiments have been reported which suggest that hydrated electrons may be formed by photolysis of H_2O at 195 nm (6.5 eV) [J. W. Boyle, J. A. Ghormley, C. J. Hochanadel, and J. F. Riley, *J. Phys. Chem.* **73**, 2886 (1969)].

The build up of I^-_{2aq} partially masks the absorption of e^-_{aq} as it has peaks at 290 nm, 350 nm, and 750 nm, while the I^-_{3aq} product causes a steady fall in the total transmitted light intensity. Cl^- and Br^- can also be photoexcited to release an electron and with $\lambda \sim 185$ nm the limiting quantum yields ϕ have been estimated to be $\sim$0.35 and $\sim$0.45, respectively, but their usefulness is considerably restricted by their "masking" tendencies. In addition there is a chance of back reaction between e^-_{aq} and some of the photochemical products that obviously would complicate the kinetics.

In these cases and many others a suitable charge-transfer-to-solvent band has provided the mechanism for solute photolysis. For example, the following series of sulfur salts has been investigated (*65*, *66a*), sulfate, sulfite, bisulfite, thiosulfate, and thiocyanate, and in each case the transition complex is thought to be a spectroscopically excited state of the solute ion; typically the flash photolysis of 10^{-4} to 10^{-2} *M* aqueous solutions gave

$$SO_3^{2-} \cdot H_2O \rightleftharpoons (SO_3^{2-} \cdot H_2O)^* \rightleftharpoons SO_3^- + e^-_{aq}$$
$$CNS^- \cdot H_2O \rightleftharpoons (CNS^- \cdot H_2O)^* \rightleftharpoons CNS + e^-_{aq}$$

It seems most probable that the e^- becomes solvated in the solvation sphere of the parent species immediately after photodetachment has occurred. The flash photolysis of phosphate anions in the vacuum **uv** also gives rise to transient optical spectra positively assigned to e^-_{aq} (*66b*).

If it is proposed that the short-lived intermediate is the hydrated electron, presumably the absorption bands of the transient in the different systems should have comparable characteristics. At first glance the λ maximum and width at half height of the absorption bands are not quite as similar as anticipated. Perhaps the reallocation of the molecules in forming a solvation sheath about the electron within the sphere of influence of the parent species contributes appreciably to the energetics; thus λ maximum and the broadness vary according to the nature and size of the ions involved. Certainly the degree of solvation of the ground and excited state has to be taken into consideration when evaluating the threshold photon energy for photodetachment processes in any polar medium. It has been observed that these quantum yields are generally temperature dependent and less than unity, which implies the participation of excited states in the primary photolytic act,

$$I^-_{aq} \rightleftharpoons (I^- \cdot H_2O)^* \rightarrow I_{aq} + e^-_{aq}$$

In this respect the spectrum of a photoejected electron that has become solvated should possibly resemble that of an *F* center rather than a free e^-_{aq}.

The most recent data published concerning variations in the features of modified spectrophotometric technique (*67*) measure the absorption spectrum in the stationary state after photoexcitation with a modulated source. The transmitted light is monitored by a phase-sensitive detector. The **uv** irradiation of all the negative species mentioned so far in this section gives the same spectrum identical with that established for e_{aq}^- in radiolyzed systems.

The photolysis of ferrocyanide at 254 nm produces e_{aq}^- with a limiting quantum yield higher than most other processes, $0.67 < \phi_{254} < 1.00$ (*11b*)

$$Fe(CN)_6^{4-} + h\nu \rightarrow Fe(CN)_6^{3-} + e_{aq}^-$$

The parallel reduction can be chemically induced. When the solvated ferrous cation is photoreduced, the electron is transferred from the solvation shell and the limiting quantum yield is low, $\phi_{254} = 0.06$ (*68*),

$$Fe_{aq}^{2+} + h\nu \rightarrow Fe_{aq}^{3+} + e_{aq}^-$$

There is, however, some controversy as to whether the intermediate in this reduction is e_{aq}^- or a hydrogen atom since recent kinetic salt effects support the latter mechanism.

Probably the least chemically complex photochemical system in which e_{aq}^- may be generated is the photolysis of H_2—saturated alkaline solutions. Here OH^- absorbs light of $\lambda < 210$ nm,

$$OH^- + h\nu \rightarrow OH + e_{aq}^-, \qquad \phi_{185} \sim 0.2$$

and the OH radical is converted via reactions (16) and (17),

$$OH + H_2 \rightarrow H + H_2O \tag{16}$$
$$H + OH_{aq}^- \rightarrow e_{aq}^- \tag{17}$$

to a second hydrated electron. Although somewhat inefficient in relation to some other systems (e.g., halide ions), none of the reaction products interfere with optical measurements and the predominant decay mechanism should be the same as reaction (4),

$$e_{aq}^- + e_{aq}^- \rightarrow (e_2^{2-})_{aq} \rightarrow H_2 + 2OH_{aq}^-$$

Consequently this reaction sequence results in no overall change. In this system the electron lasts for several hundred microseconds after a 25 μsec flash and it is thought that the $(e_2^{2-})_{aq}$ intermediate in the decay mechanism is stable possibly for milliseconds (*34*). However, the lifetime of the electron is extremely sensitive to submicromolar concentrations of oxygen or other impurities.

Organic substrates have been photolyzed to produce an e^- that is subsequently hydrated in the aqueous medium. For example,

$$Ph - NH_2 + h\nu \rightarrow Ph - NH_2^+ + e_{aq}^-, \qquad \phi_{254} = 0.025$$

The amines are merely one category of aromatic molecules having electron-donating groups that participate in photolytic reductions (*69*). In some organic glasses photoillumination of the solute gives rise to e_s^- via a different mechanism—that of biphotonic ionization. In this way electrons can be photoreleased from the solute molecules with light of a lower energy than is formally required by the ionization potential. Extensive work on a series of substituted aromatic amines (*70*) has provided a stimulating basis for further study of biphotonic processes from which unique information about the nature and lifetime of excited states of molecules or the behavior of ion pairs and e_s^- in an imposed electric field is made available. In a biphotonic ionization the molecule is first excited to an upper electronic state by a quantum of one energy and while in this state may absorb a second quantum of a different energy that excites it within its ionization continuum,

$$M \xrightarrow{h\nu_1} M^* \xrightarrow{h\nu_2} M^+ + e_s^-$$

The incident light beam is comprised of both wavelengths; for example, with the molecule tetra-methyl-paraphenylene diamine (TMPD) as a solute, 10^{-3} *M* in 3-methyl pentane at 77°K, a beam containing wavelengths of 335 nm and 380 nm selectively induced a biphotonic process. The lifetime of the excited state was ~4.5 sec, too long for a triplet state of TMPD and further experiments with different solutes in 3-methyl pentane indicated that the matrix itself was the excited state. Perhaps we might draw an analogy to liquid systems here and describe this as a charge-transfer-to-solvent in a condensed phase.

The production of e_{aq}^- in the reduction of biologically important molecules or as a reactive species in prebiological chemistry is currently receiving attention from workers in many disciplines—perhaps the most important biological reduction of all in the life cycle, the photosynthetic reduction of chlorophyll, may turn out to be another occasion on which e_s^- plays an aquiescent part.

2. Photoinduced Emission from Metals

Evidence pertaining to the production of e_{aq}^- following the **uv** illumination of a dropping mercury electrode in an aqueous solution has been discussed earlier in the context of electrode reactions. We merely repeat that the photocurrents observed are dependent on the wavelength of the illumination, the electrode potential, and the nature of the electrolyte. Again, the same sort of experiment has been carried out with a silver cathode embedded in glassy layers of 3-methyl pentane and illuminated with $\sim$185 nm at 77°K. Photocurrents were measured as a function of light intensity and distance from the electrode in an attempt to elucidate the mechanism for the decay of e_s^- and the possible role of correlated charge pairs in aqueous glassy solids (*71*).

C. Chemical Processes

1. The Reduction of Water

It is well known that metals with high reduction potentials, such as the alkali and most of the alkaline earths, reduce water to H_2. Hughes and Roach (*38*) and Shaede and Walker (*37, 72*) have reported experiments designed to identify the precursors of molecular hydrogen by competitive scavenging techniques. Both studies used sodium metal "diluted" with mercury to enable the scavengers to compete successfully with the bimolecular reaction occurring on the metal surface to form H_2 (a 10 ml sample of 1% sodium amalgam takes several hours to react completely). Hughes and Roach followed the H_2 production when Na—Hg reacted with acidified water containing a variety of additives—such as, Cu^{2+}, Zn^{2+}, Co^{2+}, $Co(NH_3)_6^{3+}$, and $Fe(CN)_6^{3-}$—and found the relative decrease in hydrogen production was consistent with the ratio of rate constants for the reaction of these additives with e_{aq}^-, as determined by pulse radiolysis, and not related to their rates of reaction with H atoms. Shaede and Walker studied the competition between N_2O and methanol (and isopropyl alcohol) for the species produced by Na-Hg in water at pH 1, 7, and 13. They followed the N_2 and H_2 production and from the known relative rate constants for N_2O and CH_3OH with e_{aq}^- and H concluded that hydrated electrons were produced at pH 7 and 13, but that N_2O could not significantly inhibit H_2 formation at pH 1, doubtless because of reaction (1).

Apparently then sodium reacts with water (even in fairly strong acidic solution) according to reaction (5),

$$Na + nH_2O \rightarrow Na^+_{aq} + e^-_{aq}$$

rather than reaction (7) as formerly supposed, and that combination of two e^-_{aq} leads to H_2. These observations are not particularly surprising when we consider the analogous reaction in liquid ammonia, various amines, and ethers, and doubtless any polar liquid (even the aprotic solvent hexamethylphosphoramide gives a stable blue solution when it reacts with sodium (*73*)).

It is often mentioned that films of alkali metals in contact with water appear blue just prior to their disappearance, and Jortner and Stein (*74*) reported measuring an absorption spectrum. Furthermore, Bennett, Mile, and Thomas find similar colorations when alkali metals are deposited in a low-temperature ice matrix (*75*). Although the structure and hence facility for solvating or trapping electrons are very different in water and ice, the **esr** studies of Bennett et al. certainly add credence to the notion that when sodium is placed in water, even ice, an electron is released into the solvent.

Cations having reduction potentials greater than 0.41 V, the potential necessary to reduce water at natural pH, spontaneously generate H_2 from water. Since the standard potential for U^{3+}/U^{4+} and Eu^{2+}/Eu^{3+} are 0.61 and 0.43 V, respectively, we have examined the intermediates in the reaction of crystalline UCl_3 and $EuCl_2$ with water using the N_2O/methanol competition. For the case of U^{3+} there is evidence that H atoms are not produced and the data are consistent with e^-_{aq} being formed by reaction (18) (*37*).

$$U^{3+}_{aq} \xrightarrow{H_2O} U^{4+}_{aq} + e^-_{aq} \qquad (18)$$

Perhaps also the reaction of $Ni(CN)_4^{4-}$, AsH_3, and TeH_2 with water or even reductions involving H_2 gas on a catalyst in alkaline solution have mechanisms involving the e^-_{aq}.

2. Intermediates in Electron Transfer Reactions

Considerable speculation is expended on the detailed mechanism of certain oxidation-reduction reactions involving a simple electron transfer, particularly on the question of whether an electron can be transferred over long distances through a suitably oriented solvent such as water held by hydrogen bridges. Interest is also converging on the question of whether in these processes the electron may be transferred to the solvent in which it becomes solvated and hence reacts as e_{aq}^- with the oxidizing agent. It is pertinent to follow an argument that has been presented (*76*) against the involvement of e_{aq}^- in the oxidation of Ru(II) by acidified water, but to apply it to the data on the oxidation of U^{3+} (*37*).

The stoichiometry is given by reaction (19),

$$U_{aq}^{3+} + H_{aq}^+ \rightarrow U_{aq}^{4+} + \tfrac{1}{2}H_2(g) \tag{19}$$

for which the standard potential is 0.61 V (*58*); but the proposed mechanism involves reaction (18),

$$U_{aq}^{3+} \rightleftharpoons U_{aq}^{4+} + e_{aq}^-$$

Since $\Delta E_{18}^\circ = 0.61 - 2.67 = -2.06$ V, it follows that the equilibrium constant $K_{18} = 10^{-33}$. This will equal the ratio of rate constants in forward and backward directions and because k_{-18} cannot exceed 10^{11} M^{-1} sec^{-1}, the diffusion-controlled limit, then k_{18} must be less than or equal to 10^{-24} sec^{-1}, which is much too slow to observe. This argument implies that the mechanism does not involve reaction (18).

However, let us consider the alternative mechanism in which hydrogen atoms, rather than e_{aq}^-, are intermediates in the formation of H_2. Reaction (20)

$$U_{aq}^{3+} + H_{aq}^+ \rightleftharpoons U_{aq}^{4+} + H \tag{20}$$

has $\Delta E_{20}^\circ = 0.61 - 2.1 = -1.49$ V, so by similar reasoning $k_{20} < 10^{-14}$ M^{-1} sec^{-1}. It is also true then that reaction (20) is too slow to be involved. Yet reaction (19) proceeds with a half-life of several minutes. Should we look for another mechanism (in which the free energy that will be released in the combination of two e_{aq}^- or two hydrogen atoms is available in the rate-controlling step) or should we question the inherent weakness of applying equilibrium thermodynamics to each step involving highly reactive species?

In a similar vein Sykes (*77*) has considered the possible involvement of e_{aq}^- as an intermediate in reaction (21),

$$Cr^{2+} + Fe^{3+} \rightleftharpoons Cr^{3+} + Fe^{2+} \tag{21}$$

through the steps (22) and (23),

$$Cr^{2+} \rightleftharpoons Cr^{3+} + e_{aq}^- \tag{22}$$

$$Fe^{3+} + e_{aq}^- \rightarrow Fe^{2+} \tag{23}$$

Making the stationary state assumption $d[e_{aq}^-]/dt = 0$, he obtains the relationship

$$\frac{-d[Fe^{3+}]}{dt} = \frac{k_{22}k_{23}[Fe^{3+}][Cr^{2+}]}{k_{-22}[Cr^{3+}] + k_{23}[Fe^{3+}]}$$

which is not obeyed by the experimental data. Sykes interprets this to mean that there is clearly no evidence for the intermediate formation of hydrated electrons in this electron-transfer reaction. There is another interpretation—that the steady-state approximation cannot be applied. It must be remembered that reactions (−22) and (23) proceed at the diffusion-contolled limit, thus there will mainly be a strong positional correlation between e_{aq}^- and its parent ion such that the system will be nonhomogeneous and ordinary rate constants will not apply. A simple competition between Cr^{3+} and Fe^{3+} for e_{aq}^- formed always in close proximity to Cr^{3+} is unrealistic for this type of system.

We respectfully suggest that it has not yet been satisfactorily demonstrated that hydrated electrons cannot be intermediates in electron-transfer reactions as typified by processes (19) and (21).

3. Base-Induced Conversions

A discharge through molecular hydrogen to produce hydrogen atoms has been used (*78*) as the basis of a technique to produce e_{aq}^- from reaction (2),

$$H + OH_{aq}^- \rightarrow e_{aq}^- + H_2O$$

Bases such as HPO_4^{2-} and $B_4O_7^{2-}$ have also been tried but found ineffective for this ionic dissociation. The hydrogen atoms are forced into water or aqueous solution and above pH 12 the conversion is extremely efficient. Distinction was made on the principle that Co_{aq}^{3+} ions react rapidly with e_{aq}^- regardless of the type of ligands associated with the transition metal, but hydrogen atoms show both a pH and ligand dependence in their reaction rates with the same complex. Competitive scavenging experiments using Co_{aq}^{3+} or more traditional scavengers such as N_2O or SF_6 produced a rate constant of $k_2 = 2.0 \times 10^7\ M^{-1}\ sec^{-1}$.

Since the OH radical can be converted to a hydrogen atom by reaction with H_2 according to (16), we also have a conversion of OH to e_{aq}^- available. The question then arises whether at higher pH under an atmosphere of H_2 this reaction proceeds in one direct step via reaction (24) (possibly forming $H + OH^-$ within a solvent cage),

$$O^- + H_2 \rightarrow e_{aq}^- \quad (24)$$

or is H produced homogeneously instead. Dainton (*11b*) has pointed out that ΔG_{24}° for the net reaction (24) is -23 kcal.

The reaction of OH_{aq}^- with several organic radicals also results in the appearance of a solvated electron as for instance in reaction (25)

$$RCHOH + OH_{aq}^- \rightarrow RCHO + e_{aq}^- \quad (25)$$

but in this case a hydrogen atom is abstracted from the parent radical (*11b*).

In pulse irradiated H_2-saturated fluoride solutions the following reaction gives rise to e_{aq}^- in addition to those produced radiolytically (*79*)

$$F_{aq}^- + H \rightarrow e_{aq}^- + HF_{aq},\ k_{26} = 1.0 \times 10^4\ M^{-1}\ sec^{-1} \quad (26)$$

However, the primary process seems to be a two-stage mechanism where the initial source of hydrogen atoms is from the OH via reaction (16)

$$OH + H_2 \rightarrow H + H_2O$$

The same results were obtained for the conversion of F_{aq}^- in a study utilising the ultrasonic excitation of water under a H_2 atmosphere to produce the necessary hydrogen atoms (*80*). From a consideration of the thermodynamics involved it has been suggested that fluorinated radicals such as CF_3 or C_6F_5 reacting with F_{aq}^- might generate e_{aq}^-.

4. Interconversion of Solvated Electrons

Another method currently receiving more attention is the feasibility of the transformation of one type of stable solvated electron into another (*81*). For example, by introducing the stable electrons produced in a solution of cesium in ethylenediamine into a dilute solution of water in ethylenediamine, the rate of loss of e_{ed}^- as a function of $[H_2O]$ may be followed spectroscopically. Different metals and solvents have been employed to generate solutions of e_s^- by this method in which the reactivity of e_s^- can be followed, in particular, for the ethylenediamine-water system the rate of reaction (27)

$$e_s^- + H_2O \rightarrow H + OH^- \quad (27)$$

was shown to have a second-order rate constant of 20 M^{-1} sec^{-1}, and the liquid ammonia analogue (28)

$$e^-_{amm} + H_2O \rightarrow H + OH^+ \tag{28}$$

was reported to be $k_{28} = 5 \times 10^{-3}$ M^{-1} sec^{-1}. One of the problems associated with this type of study is the difficulty in being certain about the real nature of the solvated electron that was reacting. For instance, do the steps $e_{ed} \rightarrow e_{aq}$ or $e_{amm} \rightarrow e_{aq}$ proceed at all, and if so, can we be certain they are not rate controlling?

D. The Hydrated Electron in Radiation Chemistry

In 1962 Hart and Boag published their now classic paper (*8*) describing an electron pulse-induced absorption, extending through the whole region of the spectrum, which on the basis of additional chemical evidence was positively attributed to a new species—the hydrated electron. This identification climaxed several years of experiments and predictions in radiation studies, although it is perhaps the predictions—in particular those at a meeting in 1953 (*82*)—that provide the most interesting background comparison for currently accepted ideas about the hydrated electron.

1. Speculation

The long apparent success of Weiss' free radical hypothesis (*83*) in accounting for the primary events in irradiated water seemingly eliminated any necessity or incentive to break from the established conceptual framework. One small discrepancy, however, led Platzman in 1953 to reconsider the fate of the thermalized electron which was written according to the Lea-Gray picture as

$$e^-_{thermal} + H_2O \rightarrow H_{aq} + OH^-_{aq}$$

He drew attention to these secondary electrons in the following words (*82*),

"Having attained thermal energy the electron finds it impossible to carry out the chemical reaction written in all the articles on radiation chemistry just because of the disparity between the time the actual reaction takes and the time which would be required to utilize the hydration energy which makes it possible. For this reason the electron becomes hydrated ... I mean that the electron polarizes the dielectric and is bound in a stable quantum state to it. Then ... the chemical reaction can proceed ... In between there is the time for hydration to take place which must, as Dr. Onsager says, be a minimum of the relaxation time, 10^{-11} sec." In short, he proposed the conversion of thermalized electrons into a hydrated species,

$$e^-_{thermal} \rightarrow e^-_{aq}$$

Later he added,

"If I am correct the trapped electrons will play a decisive role in constructing (the) chemical kinetics. It becomes very interesting to see whether these speculations can be substantiated or contradicted by experiments."

Further speculations were aired over the years; some, like the theories of electron escape and recapture, have proven more fruitful than others and still command the attention of different schools of thought (*84–86*); others, like problems of spatial nonhomogeneity in kinetic interpretations, have acquired new significance (*32*) as technology permits an examination of those events occurring within a nanosecond (*87*).

The first reported attempt to observe the absorption of the hydrated electron, assumed to be the visible region, was by Linschitz in pure water after a 1 μsec X-ray pulse. His failure to do so was attributed to the inadequate resolution (10 μsec) of the detection equipment, but in retrospect it was the sensitivity, not the resolution, that was the crucial limiting factor in this equipment.

By the late nineteen fifties the experimental response to these speculations had been sufficient to justify the existence of two different reducing species in irradiated pure water: there was no other way of interpreting the anomalous kinetics in the wide range of irradiated aqueous systems investigated. In 1960 Weiss (*88*) summed up many of the predictive feelings in a paper that discussed the probable existence and behavior of a negative Polaron, envisaged as H_2O^-. It will now be appropriate to look at some of the experimental work of this period.

2. Experimental Support

Some of the chemical evidence that eventually led to the acceptance of e^-_{aq} as a participant in the radiolysis of aqueous systems will be reviewed briefly.

Stein in 1952 while interpreting the kinetics of an irradiated aqueous methylene blue solution suggested that electrons may have interacted with the solvent; in this way he could explain the O_2 and CO_2 inhibition of the otherwise reversible decoloration of the dye as a competitive reaction between the electron and the hydrogen atom for the inhibitors, rather than the dye (*89*).

The proposal that the two reducing forms of "H-atoms" were an acid-base pair (H, e_{aq}^-) or (H, H_2^+) came from Barr and Allen (*6*) who studied the effects of irradiating hydrogen peroxide containing H_2 and O_2. The entity H_2^+ was thought by Weiss to be important in both radiation and photochemistry but conflicting data from different laboratories on the pH dependence of either pair led to the gradual adoption of (H, e_{aq}^-) as the most probable species.

Further work by Baxendale and Hughes (*4*) on the radiation-induced oxidation of methanol—cupric and ferric-sulphate systems in both H_2O and D_2O was concluded with the tentative proposal for an equilibrium at pH > 12

$$H_2O^- \rightleftarrows H + OH^-$$

Matheson later commenting on their work prophetically altered the symbol from H_2O^- to e_{aq}^-. Hayon and Weiss (*5*) and Hayon and Allen (*90*) outlined mechanisms to explain their pH-dependent yields of H_2 and Cl^- in X-ray and γ-irradiated aqueous solutions of monochloroacetic acid and other chlorinated organic molecules; these included reactions with solvated electrons. Czapski and Allen (*91*) showed that in neutral solution the reducing radicals are a basic species which in the H_2O_2—O_2 system could be represented by

$$e_{aq}^- + O_2 \rightarrow O_2^-$$
$$e_{aq}^- + H_2O_2 \rightarrow OH^- + OH$$

Finally the role of these electrons was no longer speculative. The dominant reducing species in these neutral solutions was demonstrated to have unit negative charge by Czapski and Schwarz (*7*) who used the Bronsted-Bjerrum theory of ionic reactions, a result confirmed by Collinson et al. (*92*), and Dainton and Watts (*93*). The initial experiments involved a study of the variation of the relative rate of reaction of the reducing species with hydrogen peroxide compared to other solutes such as H^+, O_2, and NO_2^- as a function of ionic strength. The Debye-Hückel theory of electrolytes and the Bronsted-Bjerrum theory indicate the rate constant k will depend on the ionic strength μ of the solution according to

$$\log_{10} k = \log_{10} k_0 + \frac{\alpha Z_A Z_B \mu^{1/2}}{1 + \mu^{1/2}}$$

where k_0 is the rate constant at infinite dilution of ions, α is a parameter, a function of dielectric (~ 1 for water), and $Z_{A,B}$ is the charge on the species in solution.

Thus the value of k depends on the sign of the charges on the reactants. If one reactant is uncharged, then the second term vanishes; if one of Z_A or Z_B is known, it is a simple matter to calculate the other once the dependence of k on μ has been determined.

Collinson et al. while investigating reaction rates for the reducing species with Ag^+ ions or acrylamide found that the reducing species at pH had zero charge but, as in the H_2O_2 experiments, a -1 value at pH 4.

One interesting question can be asked about the validity of the equation in circumstances when the species involved do not experience an ionic environment at all times. Should the electron be suddenly generated in so concentrated a solution that the electron reacts with the solute before acquiring an ionic atmosphere, then the complex will have an inbalance of charge with respect to the ionic atmosphere, which will influence the rate of subsequent reaction. Experimentally, then, it would not be possible to differentiate between an ion of -2 charge sharing an ionic atmosphere or an ion of -1 charge with its own atmosphere participating in the reaction. Coyle et al. (*94a*) have concluded that for the solvated electron the ion atmosphere is established within a few nanoseconds after the formation of the e^-, while electron scavengers do not remove the hydrated electron for about 100 nsec. However, in the case of diffusion controlled reactions—that is, $k \sim 10^{10}$ M^{-1} sec^{-1}—the equilibrium time may be of the order of the resolving time of the subnanosecond pulse radiolysis techniques and thus the treatment has to be modified to account for deviations in α.

Logan (*94b*) has shown that the Debye theory of diffusion-controlled reactions can lead to the same conclusions.

Keene in 1962 observed a transient absorption in the visible during and immediately after the microsecond pulse radiolysis of aqueous solutions; the signal disappeared in the presence of known electron scavengers (*96*).

3. Positive Identification

It was exactly at this time that Hart and Boag with their combination of two sensitive techniques, pulse radiolysis and flash spectroscopy, published the full details of their spectral and chemical analysis of irradiated aqueous solutions leaving little doubt but they had finally identified and characterized the hydrated electron (*8*). The apparatus they used is shown in schematic form in Fig. 6. In pulse radiolysis (pr), a very powerful tool that has emerged

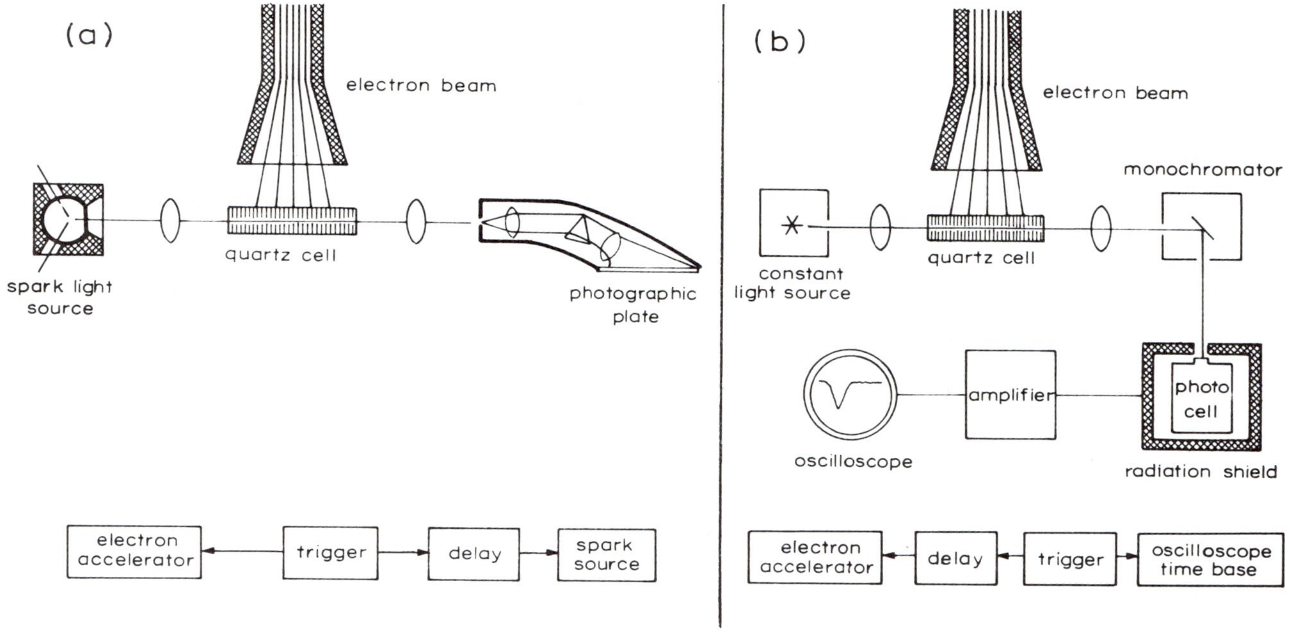

Fig. 6. Schematic diagrams of apparatus used in pulse radiolysis; (a) synchronized flash spectrographic technique; (b) kinetic spectrophotometric technique. These designs are applicable down to 1 nsec pulses.

this decade, the exciting radiation is a short, extremely intense burst of high-energy electrons. Since the combination of pulse radiolysis and competitive scavenging techniques also provides for the first time a method of measuring rate constants on an absolute basis, the bulk of information on the chemistry of e_{aq}^- has been collected this way, using pulses of several microseconds duration. In Hart and Boag's experiment, a 2 μsec pulse of 1.8 MeV electrons—giving a mean dose of between 4 and 12 krads per pulse—was used to generate transient species whose absorption characteristics were recorded photographically during a 4 μsec underwater spark between two uranium electrodes. The spark, which gave a good continuum from a ~400 nm to ~900 nm, could be triggered simultaneously or at a given delay after the pulse. The solutions they investigated were prepared from triply distilled water and were deaerated before use (the O_2 and CO_2 content were determined to be ~0.1 μm and ~0.4 μm, respectively). Absorption of the e_{aq}^- was observed in pure neutral water, in 0.05 *M* solutions of alkali metal carbonates, in the presence of electron scavengers such as O_2, CO_2, N_2O, and at specified time intervals after the electron pulse.

Later, more sophisticated electronic designs enabled pulses of a few nanoseconds to be available for experiments with still higher dose rates, thus opening up a new area of physiochemical studies. Recently the nanosecond time barrier has been broken (*87*) and now the 20-psec (0.02-nsec) fine structure pulses of a 30-nsec LINAC electron pulse are utilized in experiments that permit direct observations of the events that occur 10^{-11} to 10^{-9} sec after the passage of the ionizing particle through the medium. This elegant experimental design, shown in Fig. 7, delays the Čerenkov light flash emitted simultaneously with the electron pulse and then focuses it back into the reaction cell as the monitoring light for any absorbing species. However, if the latter have lifetimes in excess of 0.3 nsec, then the decay of the absorption signals will not be clearly defined as there is only an interval of 0.35 nsec between the pulses, and the species will still be actively absorbing when the subsequent pulse generates more of the same. Nevertheless this sort of time resolution promises to give rewarding information on the hydrated electron and the positive ion among the clusters of excited and ionized species and the fundamental processes of energy transfer within this very short time interval.

(The similarity between the chemical effects of silent electric discharge and high-energy radiation led to discharge experiments designed to reduce water. Carried out in an argon atmosphere, the electric discharge reduction nevertheless gave very low yields of e_{aq}^- (*95*).)

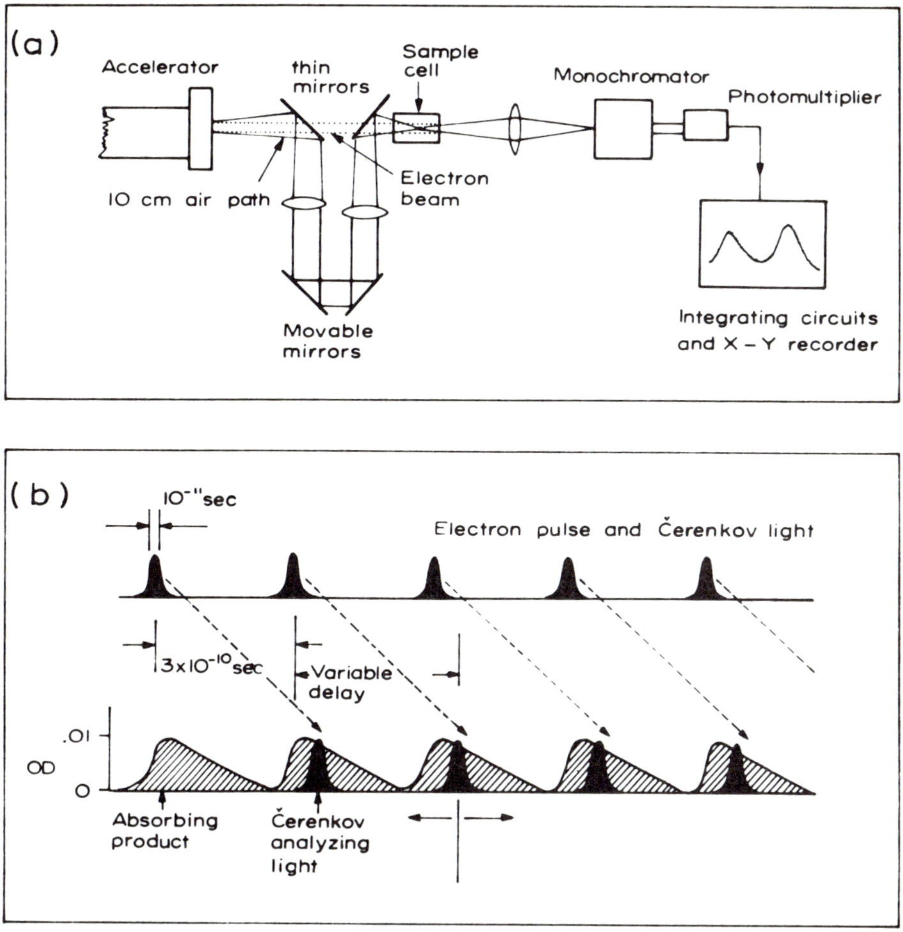

Fig. 7. Schematic diagram of apparatus and principle of subnanosecond pulse radiolysis; (a) components of a stroboscopic pulse radiolysis system in which the delayed Čerenkov emission acting as the analyzing light is subsequently focused through a monochromater and detected on a photomultiplier; (b) the fine structure pulses of the electron beam are selected as the ionizing radiation and the simultaneous pulses of Čerenkov emission are used to measure the transient absorbing species. (There is an interval of 0.3 nsec between consecutive pulses.)

In fact, the majority of aqueous systems presently under investigation with a view to elucidating the structure or thermodynamics and reactivity of e_{aq}^- generate this species radiolytically with either steady-state or pulsed radiation. In contrast to photochemical systems in which energy is selectively absorbed, both excitation and ionization occur extensively in radiolyzed systems without regard to the initial state of the molecules. Having absorbed the energy in a way characteristic of the type of radiation, the target material experiences a sequence of events that ultimately leads to the trapping of the electron. It will be pertinent at this stage to consider these events and some contemporary proposals for the trapping procedure.

4. Mechanisms of Radiolysis

As a fast charged particle travels through a target medium energy is transferred nonselectively from the incident particle to the medium, within 10^{-16} sec, as a number of events or spurs along its track. These spurs are considered to be an average deposition of 100 eV and at this stage are clusters of electronically excited species; the distance between the spurs is a function of the linear energy transfer (LET) characteristics of the particle. An elaborate model of tracks, spurs, and LET effects has been given by Mozumder and Magee (*86*). The excited and ionized species lose energy by collision, dissociation, and possibly solvent quenching within 10^{-12} secs and on an average one ion pair is produced for every 30 eV of energy absorbed. A typical spur may now consist of three such ion pairs or six radicals. The following processes are believed to represent the principal events leading to chemical species within spurs,

$$H_2O \rightarrow (H_2O^+ \cdot e^-) \qquad \text{possibly } H_2O^*$$

$$H_2O^+ + H_2O \rightarrow H_3O^+ + OH$$

$$H_2O^* \rightarrow H + OH$$

$$e^-_{thermal} \rightarrow e^-_{aq}$$

The actual distance traveled by a subexcitation electron (<10 eV) before thermalization is critical for recapture considerations. If the thermal energy of the electron does not exceed the attractive potential energy from the coulombic field of the parent positive ion, the electron will be recaptured. Two models have been proposed: one, the Samuel-Magee (*85*), assumes recapture within this field, the other, the Lea-Platzman (*82*), permits the electron to escape.

Within 10^{-11} sec (the dielectric relaxation time of water) any "free" thermalized electrons will be hydrated. The projection of the two theories, in which there are large differences in the estimate of the mean free path of the escaping electron, lead to values of the molecular product yields with which the experimentalist is primarily concerned. As the radicals diffuse the spur increases in size until overlap of adjacent spurs establishes a true homogenity of species in the medium. Intraspur and interspur reactions can occur, many at diffusion-controlled rates. A theoretical treatment of a one-radical one-solute problem has been given by Kupperman and extended to complex systems (*97*).

The precise sequence of events that lead to the formation of a solvent sheath about the electron and the nature of the stable quantum state in which the electron is trapped continues to invoke much discussion (see any review) as the experimental evidence is not unambiguous. Superficially we can picture the thermalized electron moving through the water relatively quickly, affecting polarization of water molecules along its track but not remaining long enough in the vicinity to be trapped itself. However, if there already existed a region of accidental polarization due to the random thermal motions of the molecules themselves, then the water dipoles will be polarized to a more significant degree in the field of the excess electron, which consequently would find itself trapped. In liquid water both electronic and orientational polarization are important, the significant time association with the latter corresponding to molecular rotation; the dielectric relaxation time of water is 10^{-11} sec. Schiller (*98*) regards the dielectric relaxation time as the decisive parameter in the trapping procedure. He assumes the track and spur model and considers the time-dependent dielectric properties of the medium and a nonconservative electric field in which the probability of the electron being trapped increases as the relaxation time (τ) increases. Experimentally the effect of a variation in τ on the electron escape probability may be investigated using scavengers in the liquid-water/supercooled water/ice system in which the chemical and physical properties as well as the static optical dielectric constants are similar but τ ranges from 10^{-11} sec (in water) to 10^{-5} sec (in ice). Schiller's predictions of total electron recapture in ice and considerable recapture in water were supported by the data collected.

Another model of electron capture is based on a time-independent dielectric constant and a static electric field. Freeman and Fayadh (*99*) use the cavities present in the liquid structure as the initial trapping centers, and suggest that the limiting factor of the mobility of the solvated electron is the physical migration of these cavities. They believe that the difference between electron binding in polar and nonpolar liquids is one of degree rather than kind arising from variations in the static dielectric constant. These conclusions were based on empirical observations; in a series of scavenger experiments the yield of electrons that escaped their parent spurs, G (free ion), was roughly proportional to the static dielectric constant ε of the liquid over a range $\varepsilon = 2.0$ for cyclohexane to $\varepsilon = 79$ for water, at 22°C. Since it is unlikely that any single factor governs electron-capture probabilities and the ultimate test of any theory lies in its predictive power and subsequent corroboration by experiment, the present state of affairs requires that these models be extended and interrelated with other facts from radiation chemical systems in order to provide a generally acceptable solution.

In all these calculations and correlations there are inevitably many assumptions which have to be made that often themselves warrant separate investigations, but one factor that consistently adds more uncertainty to the discussions than any other is the structure of liquid water, or the liquid solvent, in which the excess electron is "trapped." Water at room temperature is approximately 85% hydrogen bonded; if there were no regular structure, then all intermolecular distances would be equally probable beyond the distance of closest approach of two molecules. However, curves showing the relative electron intensity of neighboring water molecules as a function of distance from the center of any given molecule have two striking maxima at 2.9 Å and 4.5 Å. Data computed from X-ray measurements (*100*) indicate that these intermolecular distances predominate in the structure of liquid water, and that perhaps the regions of hydrogen bonding are tetrahedral arrangements as in ice where the distance between centers of oxygen atoms is 2.76 Å and the next nearest neighbors are 4.51 Å away. Regions of local icelike ordered structure have been proposed in water, but still little is known about the correlation, if any, between structural and dielectric properties of water or any other liquid.

IV. STRUCTURE OF THE HYDRATED ELECTRON

A. Origin of the Absorption Spectrum of e^-_{aq}

We have discussed some of the reactions of e^-_{aq} and the possible environmental interactions in the solvent that initiate the formation of a center which subsequently becomes a "trap." Once caught the electron digs into its potential well, causing the nearest dipoles to orientate about the trap to form an inner solvation sheath. There will also be some orientating influence felt by the next layer, possibly a third, but these will be more temporary arrangements susceptible to thermal effects. The nature of the potential well is not fully understood, but we hope that the following sections on the quantitative models constructed to describe the origins of the absorption spectrum of e_{aq}^- will succeed in communicating some intuition for the physical state of e_{aq}^-.

With the current emphasis on the elucidation of structural features via absorption and resonance spectra, it seems ironic that the simplest anion and radical of all, e_{aq}^-, should exhibit such a broad asymmetric absorption band lacking any hyperfine structure and—as yet—give a single **esr** signal devoid of hyperfine interactions. Nevertheless the spectral data associated with this electron transition have three possible origins (*20a*) and since the host medium is polar, any environmental influences on the energy of the transition can also be approximated in several ways. First, the observed λ_{max} may be regarded as the result of a single electronic excitation to an excited state in the potential well, the broadness and asymmetry of the spectrum suggesting the presence of different well sizes and vibrational effects. A second alternative interpretation sees the spectrum as comprising of an envelope of bands, indicating the excitation of the electron from a single-depth well into an ionization continuum. Finally, a third possibility interprets the spectrum as originating from the statistical distribution of solvation wells of various depths in which excitation is specific for one particular trap whether it be to an upper state or to a continuum.

The potential contours of the trap may be considered in at least two ways; either (a) the electron is in a natural physical vacancy in the liquid, which means that it resides largely in a hole of uniform potential and may be treated as a finite case of the electron-in-a-box problem, or (b) the medium, having been polarized in the presence of the excess electron, contains an artificial cavity with a central net unit positive charge with which the electron is associated. In the latter instance the potential field falls off coulombically towards the perimeter of the cavity and the system resembles the hydrogen atom problem.

When polarization of the medium is an important factor, the depth of neighboring molecules around the excess electron may be viewed as a continuous dielectric layer or as an atomic-structured medium. Further implications of these various ideas and the calculations pertaining to the vacancy and cavity potential well will be presented in the following sections.

B. Single Electronic Transitions

The electron we are considering is solvated in a polar medium and as such is trapped by a vacancy, a cavity, or perhaps in a potential well that is a combination of the two. The critical feature of the well is the potential field experienced by the electron, which varies according to the physical aspects of the trap.

(a) Within the vacancy created by the steric relationship of neighboring molecules in the liquid there is a uniform zero potential field, but the walls are of infinitely high potential. For a real situation the electron in the vacancy would see finite potential walls. The restriction of finite walls slightly distorts the wave functions used thus allowing a small probability that the electron will be found outside the box at any quantum level. As an electron-in-a-box problem the spacing between the quantum levels will be determined by the dimensions of the box, and it is conceivable that the first absorption ($n = 1$ to $n = 2$) also corresponds to the release of the electron from the vacancy. In Fig. 8(a) the vacancy is shown schematically. The presence of different sized vacancies (i.e., boxes) in the liquid must be considered the most reasonable cause for the broadening of the spectrum since the separation between quantum levels increases as n increases thus making several transitions within the box unlikely. The additional uncertainty in the location of the electron due to "tunneling" and perturbations of the vacancy by the vibrating molecular boundary also contribute to the shape of the spectrum.

The electron in liquid ammonia is known to be stabilized in a large hole but the situation in water is as yet unknown. It was on the above basis that the earliest model for an electron solvated in ammonia was formulated by Ogg (*101*) who assumed the electron to be trapped in a spherical vacancy with an infinite potential wall, surrounded by solvent molecules. The solvation energy and vacancy size derived from this model were not compatible with experimental measurements and so certain parameters such as surface tension and electronic striction were incorporated by other workers (*102*).

(b) If the electron finds itself caught in an area of favorable potential, which is then further polarized by the presence of an excess electron, a cavity would be created as a result of short-range repulsive forces. The potential of the cavity and associated electron now more clearly resembles the hydrogen atom case, which justifies the use of hydrogenic wave functions (i.e., s,p...) in assigning the spectrum to an optically allowed transition from the 1s ground state to the higher p levels. The first transition λ_{max} would be due to ($2p \leftarrow 1s$) excitation and presumably at higher energies the transition corresponding to electron escape is given by ($\infty p \leftarrow 1s$). A ($2p \leftarrow 1s$) transition would extend to three quarters of the ionization potential Ip, shown in Fig. 8(b). Because energy (Ep) is required to polarize the medium during the formation of the cavity, Ip will be greater than the heat of solvation ΔH_s by the same amount.

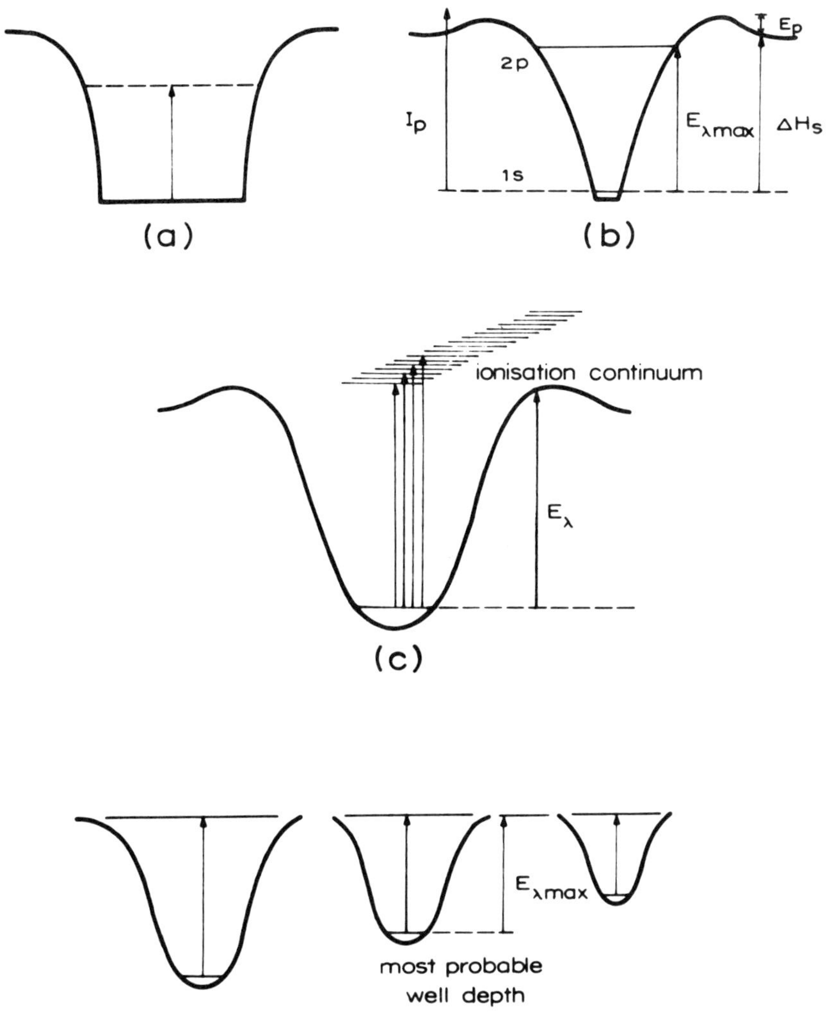

Fig. 8. Energy levels corresponding to possible absorption processes for e_{aq}^- (see text for details).

It transpires that the ΔH_s estimated by Baxendale has a value very similar to the observed $E\lambda_{max}$. (See p. 103 Ref. *20a.*)

(c) A combination of a vacancy and a cavity to give a single-depth potential well in which transitions occur into a continuum is the basis for a postulate that attributes the dominant photoabsorption processes in the visible region to the photorelease of the electron from its trap. The broadness of the spectrum would thus be accounted for, the continuum levels being analogous to exciton conduction bands. The absorption spectrum represents the envelope of the photoionization efficiency profile as shown in Fig. 8(c); it is fairly common for the photoionization cross section of vapor-phase molecules to exhibit this sort of profile.

(d) If there is a distribution of wells of different depths within the solvent, then a specific transition—be it electron-in-a-box, a ($2p \leftarrow 1s$), or to an ionization continuum—will take place at different energies. The result would be a broad structureless absorption spectrum that was an envelope characteristic of the solvent. Figure 8(d) shows a single electronic transition that varies in energy according to the well depths, although factors like interconversion through thermal equilibrium and other solvent temperature effects of necessity complicate the picture. Dainton, Salmon, and Zucker (*11b*) have photobleached e_t^- in a solid matrix and obtained a change of spectrum, which is consistent with this last interpretation.

C. Theoretical Aspects

All of these models employ hydrogenic wave functions as they have assumed a potential well with mainly cavity, and some vacancy, origins that give rise to a coulombic decrease from the center of the cavity to the outer edges. The solvent is treated either as a continuous dielectric medium or as possessing a critical local geometry of polar molecules as part of the detailed structure of the liquid.

The model of a polaron in a continuous dielectric medium has been developed by several workers (*103*, *104*); the essence of this approach lies in the difference between the role of the static and optical dielectric constants in the presence of rapid electronic motion. Jortner (*29*, *104*) based his self-consistent field calculations on this continuum in which he considered an excess electron caught in a self-induced polarization trap characterised by a decrease in coulombic potential at the perimeter of a spherical cavity. Having obtained a general expression for the ground state of the electron in a polar medium, he then performed a variational calculation until reasonable agreement was reached between calculated and observed values of certain parameters; these variational calculations will be outlined in more detail below. Recently, Natori and Watanabe (*106*) calculated values for the same range of parameters using a structural model of the hydrated electron, described as a quasitetrahedral arrangement of the oxygen atoms of the water molecules in which the electron is trapped. Iguchi (*107*) also departing from the cavity-continuum approach has proposed an atomic model for the hydrated electron differing only from the previous case in that the water dipoles are assumed to have spherical geometry about the electron and the degree of orientation of the molecules is made the variable that retrospectively accounts for anomalies unforseen by earlier calculations. Noda et al. (*108*) utilized the cavity-continuum model for the same purpose.

Weiss in his studies of irradiated frozen aqueous solutions has proposed a different conceptual approach to the problem of electron binding, and because of the similarity in λ_{max} for excess electrons in water and ice has moved away from the idea that D_s is the dominant parameter as is assumed in the polaron models (*109*). At low temperatures D_s becomes very small and the cavity continuum model predicts a red shift for λ_{max}, the opposite of what is actually observed. This discrepancy is sometimes avoided by using low frequency D_s, but in doing so the degree of orientational polarization is, by implication, very strong. In ice with a dielectric relaxation time 10^6 slower than that in water such an explanation is unacceptable, and the presence of pre-existing partially polarized trapping sites must be anticipated. Weiss, instead, proposes a bound electron-hole pair, which he refers to as an exciton of small radius, cleaved by appropriately selective electron or hole scavengers after which the usual properties of irradiated systems would be observed. The possibility of a triplet exciton is also entertained to explain the relatively long lifetime of the pairs in a scavenger-free system. The absorption is attributed to charge-transfer excitons; preliminary calculations using hydrogenic wave functions (really only applicable for excitons of large radius, hence this is an approximate treatment) with the optical dielectric constant and the reduced effective mass of the electron-hole pair as the critical parameters, estimate λ_{max} to be 1.6 eV, in reasonable agreement for observed values in both water and ice. The electron-hole pair can be represented as

$$(H_2O)^+ \cdots (H_2O)^-$$

but it is difficult to see how some of the typical features of the absorption spectra of e_s^- (namely, the broadness and asymmetry), **esr** data and the low hydration-free energy of -40 kcal for e_{aq}^- (which supports the idea of the electron, considerably delocalized over the solvent molecules, exhibiting an effective radius even larger than that of the iodine ion) fit into this model. Further work is to be published.

In interpreting data related to photobleaching and photoconductivity of electrons trapped in glassy matrices, or shifts in λ_{max} for electrons solvated in different media and the thermal dependence of λ_{max} in water and other solvents, many authors have made references to one or more of the above models. So we will conclude this section with an analysis of the variational calculations performed on the polaron basis in contrast to those on atomic models and compare some of the pertinent evidence now available in the literature with their predictions.

1. Cavity–Continuum Model

The interaction energy between an excess electron and the polarization field of a continuous dielectric medium, characterised by static (D_s) and optical (D_{op}) dielectric constants, can be estimated using two different approximations. The first, the electron adiabatic approximation, assumes that the binding energy of the excess electron is much less than that of the medium electrons; but whereas this might be valid when considering F centers, it is not so when treating the electron binding in a polar medium. Jortner, therefore, used the alternative self-consistent field (SCF) approximation in which electronic polarization effect of the medium electrons did contribute to the binding energy of the excess electrons (*29a*). The reference state is the polarized medium. The total energy of the system was then given by

$$E_i = \int \psi_i \left[-\frac{h}{8\pi^2 m} \nabla^2 - \frac{e}{2}\left(1 - \frac{1}{D_s}\right) f_i \right] \psi_i \, d\tau \qquad \text{(I)}$$

where ψ_i is a one-electron wave function, f_i is the electrostatic potential due to the charge distribution $e|\psi_i|^2$ which, in the general case, is

$$\nabla^2 f_i = 4\pi e |\psi_i|^2$$

He then assumes a cavity model for the solvated electron in which the solvent molecules in the immediate vicinity of the excess electron are repelled by short-range interactions due to the overlap of the orbitals of the excess electron with those of the medium electrons. The parameters affecting the size of the cavity will be the magnitude of the short-range repulsions relative to the energy required to form the cavity (given a value for the surface tension of the liquid) and the decrease in the self-energy of the trapped electron with increasing cavity dimensions. The excess electron is stabilized not by the formation of the cavity but by long-range interactions within the medium.

With this notion of a cavity the ground-state energy E and charge distribution (r) of a solvated electron can be evaluated using $1s$ hydrogenic wave functions in a variational calculation based on Eq. (I), with the cavity radius (R) as the variable.

Assuming the potential $f(r)$ to be continuous on the cavity boundary and constant within the cavity, the ground-state energy for the hydrated electron is given by

$$E_{1s} = \int_0^\infty \psi_{1s}\left[\frac{-h^2}{8\pi^2 m}\nabla^2\right]\psi_{1s}\,d\tau + \frac{e}{2}\left(1 - \frac{1}{D_s}\right)f_{1s}(R_0)\int_0^{R_0}\psi_{1s}^2\,d\tau$$

$$+ \frac{e}{2}\left(1 - \frac{1}{D_s}\right)\int_{R_0}^\infty f_{1s}(r)\psi_{1s}^2\,d\tau.$$

Jortner calculates the energy of the first excited state assuming that the absorption of e^-_{aq} corresponds to a $2p \leftarrow 1s$ transition. Here it is important to realise that the upper $2p$ state cannot be an equilibrium state since in accordance with the Franck-Condon principle the nuclei will still be in the same relative positions after the electronic transition. Therefore the orientational polarization for the $2p$ state is the same as the $1s$ state, but the electronic polarization will correspond to the new charge distribution. Again the variational method is used to give a best value for the energy of the $2p$ state with the nuclear configuration of the 1s state. The difference in total state energies corresponds to the transition energy

$$h\nu = E_{2p} - E_{1s}$$

which is written (according to Eq. (I) and its derivative for the $2p$ case) as

$$h\nu = \left[\int \psi_{2p}\left(\frac{-h^2}{8\pi^2 m}\nabla^2\right)\psi_{2p}\,d\tau - \int \psi_{1s}\left(\frac{-h^2}{8\pi^2 m}\nabla^2\right)\psi_{1s}\,d\tau\right]$$

$$+ \left[e\left(\frac{1}{D_{op}} - \frac{1}{D_s}\right)\int \psi_{2p}^2 f_{1s}\,d\tau - e\left(\frac{1}{D_{op}} - \frac{1}{D_s}\right)\int \psi_{1s}^2 f_{1s}\,d\tau\right]$$

$$+ \left[\frac{e}{2}\left(1 - \frac{1}{D_{op}}\right)\int \psi_{2p}^2 f_{2p}\,d\tau - \frac{e}{2}\left(1 - \frac{1}{D_{op}}\right)\int \psi_{1s}^2 f_{1s}\,d\tau\right] \qquad \text{(II)}$$

or, more explicitly,

$$h\nu = \Delta \text{ kinetic energy} + \Delta \text{ orientational polarization} + \Delta \text{ electronic polarization}$$

The energy of the ($2p \leftarrow 1s$) transition is thus determined to be 1.35 eV corresponding to a cavity radius of zero. This would have appeared to be a curious result had not the same parameters already been calculated in this way for the ammoniated electron where considerable agreement between theory and experiment was achieved. Indeed the conclusions are entirely consistent with the expectation that the cavity size in water would be much smaller than that in liquid ammonia when we recall the factors influencing the cavity dimensions. The simplest interpretation of $R = 0$ is that since the locus of the electron is not restricted to the formal notion of a cavity, the electron is delocalized over at least the inner solvent molecules. Thus it is more meaningful to consider the charge distribution (r) of the electron as the critical parameter, which on the above model is calculated to be 2.5 to 3 Å in the ground state and 4.9 Å in the excited state. Recent experimental work has determined the effective radius of e_{aq}^- in its ground electronic state to be 2.9 Å (*89*).

The actual values calculated for λ_{max}, the temperature dependence of λ_{max} $d(h\nu/dT)$, and the oscillator strength f calculated from the transition dipole moment $\int \psi_{1s} \mathbf{r}\, \psi_{2p}\, d\tau$ for cavity radius R are all compared to experimental values in Table I; included also are the calculations from other authors. If this hydrogenlike model is indeed the most accurate way of describing e_{aq}^-, then the first transition ($2p \leftarrow 1s$) should correspond to three quarters of the well depth given by ($\infty p \leftarrow 1s$) (see Figure 8). The large value of f for ($2p \leftarrow 1s$) suggests that higher transitions such as ($3p \leftarrow 1s$) will be very weak because in a one-electron system the sum of all the oscillator strengths must be unity. Furthermore, if the upper state is a $2p$ bound state, then we can make predictions about the outcome of photobleaching and photoconductivity experiments. In the case of the ammoniated electron no photoconductivity has been observed from which it is assumed that the upper state is a bound state. There is abundant evidence of photobleaching and photoconductivity exhibited by e^- trapped in glassy polar matrices, and a little for ice, but some of the interpretations give rise to ambiguities and no appropriate evidence to date is available for e_{aq}^-. The temperature dependence of λ_{max} has been measured (*110*) in Co^{60} γ-radiolyzed systems and fits reasonably well into the range Jortner predicted; the radius of e_{aq}^- seems to be increasing at higher temperatures. However, when the same measurements were made on a pulse radiolyzed system over the same temperature range, no dependence was observed. This is not the only anomaly since the temperature dependence of λ_{max} for other solvated electrons(*111*) appears outside the framework of that set by e_{aq}^- and e_{amm}^- using the parameters D_s and D_{op} to define the trap, although Dorfman (*112*) has found a qualitative correlation between λ_{max} and D_s for electrons solvated in alcohols.

Further series of calculations specifically on the electron solvated in various alcohols have been published by Noda, Fueki, and Kuri who use the cavity-continuum model with an electron adiabatic approximation. The potential function used contained the variables D_s and D_{op}; since they considered the interaction energy to be the result of long range interactions, the potential function is due only to the orientation polarization. The reference state is that of a nonpolarizing electron located in the medium at an infinite distance from the cavity; all the terms have the usual meaning,

$$V(r) = \left(\frac{1}{D_{op}} - \frac{1}{D_s}\right)\frac{e^2}{r} \qquad \text{for } r > R$$

$$V(r) = \left(\frac{1}{D_{op}} - \frac{1}{D_s}\right)\frac{e^2}{R} \qquad \text{for } r < R \tag{III}$$

With hydrogenic-type wave functions, variational calculations were performed for the energy of the ground state as a function of R, and contributions from electronic polarization in the $1s$ and $2p$ states added later. Noda et al. presented an interesting contour map of the transition energy $h\nu$ versus cavity radius as a function of the static dielectric constant (for those cases in which the binding energy of the solvated electron did not invalidate the approximation they had used). Then they matched the observed λ_{max} and D_s to obtain a best-fit cavity radius for the particular solvated electrons and found that this value decreased with increasing D_s for a series of seven alcohols with the exception of *n*-butanol. A similar curve of R against $h\nu$ exhibited approximately the same slope with no anomalies. One other correlation they tentatively proposed was between dR/dT (Å deg^{-1}) and the percentage of OH bonds (to the number of bonds in each alcohol) inferring that there is a relationship between the cavity dimensions and the strength of intermolecular bonds such as hydrogen bonds. Once again we are brought to a point where the structure of the liquid seems to be playing a significant but as yet unwritten part. The width of the absorption band for solvated electrons in alcohols at room temperature is related to the "normal" cavity radius and further correlations which they make indidate that the dimensions of the cavity increase with increasing temperature. The broadness of the band on the low-energy side is attributed to the presence of larger cavities in which the electronic transitions are taking place; thus they predict a blue shift in λ_{max}

when these electrons are photobleached with selective wavelengths. As yet such experiments have not been reported.

In a semi-empirical mood Jortner recently calculated a set of thermodynamic, optical, and kinetic data for e_{aq}^- by returning to the original polaron concept and Landau's self-trapping model (*29b*) expressed in the potential $V(r)$ in Eq. (III). Setting the cavity radius R to be 1.45 Å, about the size of a single solvent molecule, the agreement between computed and observed values was impressive (see Table I) and the indications were that this model could be fitted to any solvent data by selecting empirical values of R. However, any physical significance attached to the latter must still wait for a thorough understanding of the short-range molecular interactions in the liquid.

2. Atomic Models

In their calculations Natori and Watanabe (*106*) estimate the energy levels of an electron trapped in a tetrahedral arrangement of water molecules analogous to an electron in a lattice defect of an ice crystal. The potential field due to the nearest four (inside) hydrogen atoms, the four oxygen atoms, and the four (outside) remaining hydrogen atoms is computed for each case and the total potential field, $V(r)$, experienced by the electron (with the origin of the coordinate system at the center of the tetrahedron) is written as

$$V(r) = \sum_{s=1}^{12} V_s(\mathbf{r}_s) + V_0$$

where $\mathbf{r}_s$ is the position vector of the electron with respect to the position of the s nucleus, V_s is the potential due to s nuclei and V_0 is a potential induced by the molecules outside the immediate tetrahedral environment. Spherical symmetry is assumed for $V_s(r_s)$ which depends only on the value of r_s,

$$V_s(r_s) = -\frac{1}{\mathbf{r}_s} + \mathscr{J} \cdot \frac{|\psi_{1s}(\mathbf{r})|^2}{|\mathbf{r} - \mathbf{r}_s|}\, d\tau$$

once again a hydrogenlike wave function is used. If $\mathscr{J}$ is the fraction of the electron that occupies the 1s orbital, estimated from the dipole moments of water and the OH bond to be 0.674 atomic units, substituting a 1s wave function

$$\psi_{1s}(\mathbf{r}) = \pi^{-1/2} \exp(-\mathbf{r})$$

gives the general expression for $V_s(r_s)$,

$$V_s(r_s) = -0.326/r_s - 0.674\,(1 + 1/r_s)\exp(-2r_s)$$

In this way the total potential from the inside hydrogen atoms is calculated. Similar expressions are derived for the oxygen atoms, while the outside hydogen atoms are considered to be point charges.

Neglecting V_0, the hamiltonian $\mathscr{H}$ for the Schroedinger equation for a one-electron system becomes

$$\mathscr{H} = -\Delta/2 + \sum_{s=1}^{12} V_s(r_s)$$

from which the authors then proceed to carry out a variational calculation to estimate the eigenvalues E corresponding to the energy levels of the hydrated electron in both ground and excited states. Their results for λ_{max}, oscillator strength, and dimensions of the trap are reported in Table I for the purpose of comparison with other theoretical models and some experimental observations. There is a factor of 2 discrepancy between the calculated and observed λ_{max} which may be due to some extent to the neglect of V_0, and also to the fact some upper excited states may be contributing to the experimentally observed value; if the oscillator strength for their calculated transition is only 0.26, then such contributions are more feasible than in the previous model in which the oscillator strength was calculated to be 1.1. However, further evaluation of this approach becomes questionable in the light of recent **esr** experiments on trapped electrons in ice (*75*). Whereas this model would predict interactions between the unpaired trapped electron and four protons, the most recent **esr** indicates the interactions are between six equivalent protons. Since Natori and Watanabe's original assumption of an icelike structure for e_{aq}^- is no longer compatible in the tetrahedral order they propose, their calculations would have to be repeated on a new structure involving six water molecules. X-ray analysis for the structure of the ice used in the **esr** experiments showed that ice had a cubic structure at about 160°K, so the interactions of e^- with six equivalent protons could be best fitted to a model in which the e^- is at the center of a puckered hexagon of six oxygen atoms, the hydrogen atoms lying along the lines between adjacent oxygen atoms. Alternatively, three water molecules only need be involved and the e^- interact in an octahedral field of six protons. Although in ice the line width is narrow, indicating a fair degree of uniformity between the traps, the lack of hyperfine structure and relatively broad line widths observed for electrons trapped in solid alcohols at

77°K suggest that there is a wide range of trapping sites. The g value for e_s^- in ice ($g = 2.0008 \pm 0.0005$) was lower than the standard free-spin value ($g = 2.0023$), so the orbital of the excess electron does not possess full spherical symmetry and, as postulated earlier, overlaps with the molecular orbitals of its immediate environment. The situation for e_s^- in a range of alcohols is complicated by the appearance of **esr** signals due to other radicals, and the observed g values are higher than in ice but still less than the free-spin value. This negative shift is also characteristic of e_{amm}^- and F centers in alkali-halide crystals and was recently reported for e_{aq}^-. The **esr** signal from a hydrated electron has a remarkably narrow line width (less than 0.5 gauss) and a g value of 2.0002 ± 0.0002 (*25*), which would imply that the electron is delocalized over a few solvent molecules only and perhaps the cavities are structurally quite similar. If in liquid water these trapping sites exist prior to the arrival of the excess electron, then it is quite reasonable to suppose that certain transient molecular arrangements will be favored by the electron due to the minimal energy prerequisites for a "trap" and that these become stabilized by its presence. If the arrangement of dipoles is not appreciably distorted by the electron, then the traps may appear to be fairly uniform at the levels of signal sensitivity currently employed.

Even so, until the structure of liquid water is better understood we cannot avoid the suspicion that in matching theoretical models to empirical observations we are only getting a plausible idea of what the average, overall situation might be. The liquid environment is continually rearranging and local parameters (for instance, dielectric constants) may show marked variations in relation to bulk parameters used in the calculations.

Another theory of electron solvation has been put forward by Iguchi (*107*) who employs an atomic model comprised of oriented molecular dipoles in spherical geometry about the excess electron. He also used hydrogenic wave functions in the Schroedinger equation for a one-electron problem, computing the potential field part of the hamiltonian in the following way. First, the origin of the system is taken as the equilibrium point in the "trap" about which the electron moves at a radius **r** and toward which the solvent dipoles are oriented by their own static electric field. The orientation energy ε_γ of a molecule with a permanent dipole μ_0 at an angle γ to the field E is substituted into a Boltzmann-type density distribution to give n_γ, the density of molecules with ε_γ at T°K,

$$n_\gamma = \frac{n_0}{[1 + a(T - T_0]} \exp \frac{(-\varepsilon_\gamma)}{kT} \Big/ \int_0^\pi \exp \frac{(-\varepsilon_\gamma)}{kT} \sin \gamma \, d\gamma$$

where n_0 is the molecular density at T_0 (=273°K) and a is the volume expansion coefficient of liquid.

Immediately it is clear that this model incorporates the thermal dependency of λ_{max} not only as a function of γ but also a function of the thermal expansion properties of the liquid, which interpreted in another way would mean changes in the value of the effective radius of the solvated electron.

There are two components to the polarization: $P_{\perp}$ is normal to the radius **r** on the coordinate system, isotropic and, therefore, zero, while P_{11} parallel to the radius **r** contributes only for those dipoles outside the locus of r. Thus

$$P_{11}(\mathbf{r}) = \int_0^{\pi} n_{\gamma}\left(\mu_0 \cos\gamma - \frac{e\alpha}{r^2}\right)\sin\gamma\, d\gamma$$

$$= -n_{\mathrm{m}}\left\{\mu_0\left[\coth\left(\frac{e\mu_0}{kTr^2}\right) - \left(\frac{kTr^2}{e\mu_0}\right)\right] + \frac{e\alpha}{r^2}\right\}$$

where n_m is $n_0/[1 + \dot{a}(T - T_0)]$ and α is the molecular polarizability.

The first term represents the permanent dipole contribution and the second the induced polarization, which Iguchi computes separately. For all contributing dipoles the total potential $V(\mathbf{r})$ is

$$V(\mathbf{r}) = -4\pi e\mu_0\, n_{\mathrm{m}} \int_r^{\infty}\left[\coth\left(\frac{e\mu_0}{kTr^2}\right) - \left(\frac{kTr^2}{e\mu_0}\right)\right] dr$$

$V(\mathbf{r})$ is then used in the hamiltonian to calculate various energies for the ground and excited states of e_s^- at different temperatures. The induced polarization S_i is determined from

$$S_i = -2\pi e^2 n_m \alpha/\bar{r}_i$$

and added to the electronic energy E_i for each state, where $\bar{r}_i$ is the average radius of the electron orbital in a 1s or 2p state. The excitation energy is thus

$$h\nu = E_{2p} + S_{2\mathrm{p}} - (E_{1s} + S_{1s})$$

The main purpose of the calculations were to provide a more satisfactory model for electrons solvated in particular alcohols. As mentioned earlier the temperature dependence for these e_s^- in alcohol did not fit into the range predicted by Jortner's model. Values measured (*111*) for methanol and higher analogues were considered too large for the cavity model which implicitly required unusually high coefficients of thermal expansion. Some of these data are compared to those calculated values in Table I in which a volume coefficient of thermal expansion was assumed to be 1.0×10^{-3} Å deg^{-1} compared to the value of 3.0×10^{-3} Å deg^{-1} necessary in Jortner's model. The tabulated values show reasonable agreement with the experimental data, and since Iguchi is preparing an improved set of calculations incorporating a cutoff for the potential $V(\mathbf{r})$, further comments are perhaps superfluous.

In these correlations the transition corresponding to λ_{max} has been tacitly assumed to be a $2p \leftarrow 1s$ hydrogenlike excitation, with all the ramifications of an upper-bound state still unexplored. The absorption on the high-energy side of the λ_{max} is thought to be indicative of yet higher excited states, but no theory or evidence to date does more than make this another feasible proposition.

TABLE I

A Comparison between Observed and Predicted Values for Some Properties of Solvated Electrons According to Different Models

Parameter	Experimental values	Davydov	Jortner		Natori & Watanabe
In water			SCF approx.	Semi-empirical	
λ_{max} eV	1.75	2.1	1.35	1.65	0.80
$d(h\nu)/dT \cdot 10^3$ eV deg^{-1}	−2.9	—	−3.3 to −2.2	−3.3	—
R Å	0	—	0	1.45	—
r Å	2.9	—	2.5 to 3.0	2.8	—
fi	0.65	—	1.1		2.26
In ammonia					
λ_{max} eV	0.8	1.9	0.93		
$d(h\nu)/dT \cdot 10^3$ eV deg^{-1}	−1.1	—	−1.0		
R Å	3.5	—	3.3		
fi	0.7	—	—		
In ethanol			Noda, Fueki & Kuri		Iguchi
λ_{max} eV	1.77	1.7	1.77		1.69
$d(h\nu)dT \cdot 10^3$ eV deg^{-1}	−3.4	—	−2.99		−1.63
R Å	—	—	1.32		—
fi	0.87	—	—		—

Summary

The models and interpretations that have been discussed can explain in part the spectrum, its temperature dependence and photobleaching, but to what extent such explanations can give a true picture of the structure and quantum-mechanical description of the hydrated electron in particular, or solvated electrons in general, is still open to speculation.

ACKNOWLEDGMENT

We gratefully acknowledge the Defence Research Board and National Council of Canada for financial support during the writing of this chapter, which was completed in the spring of 1969.

APPENDIX

TABLE II

RATE CONSTANTS FOR REACTIONS BETWEEN HYDRATED ELECTRONS AND SOME INORGANIC SPECIES

Reactant	pH	Rate constant (M^{-1} sec^{-1})
Ag^{+}_{aq}	7	3.2×10^{10}
$Ag(NH_3)_2^+$	11.1	8×10^{10}
$Ag(CN)_2^-$	10	1.5×10^{9}
$Al(III)_{aq}$	6.8	2.0×10^{9}
$Al(OH)_4^-$	14	5.5×10^{6}
AsO_2^-	10.6	5.9×10^{8}
$HAsO_4^{2-}$	11	2.1×10^{8}
$Au(CN)_2^-$	10.6	4.2×10^{9}
BF_4^-	5.8	$<2.3 \times 10^{5}$
Br_2^-	7	1.3×10^{10}
BrO^-	13	2.3×10^{10}
BrO_3^-	7	7.8×10^{9}
CO	7	1×10^{9}
CO_2	7	7.7×10^{9}
HCO_3^-	—	$<10^{6}$
CO_3^{2-}	—	$<10^{6}$
CN^-	11.0	$<10^{6}$

TABLE II (*continued*)

Reactant	pH	Rate constant (M^{-1} sec^{-1})
CNO^-	11	1.3×10^6
CNS^-	7	$<10^6$
Cd^{2+}_{aq}	7	5.2×10^{10}
$Cd(H_2O)_4^{2+}$	6.5	4.8×10^{10}
$Cd(NH_3)_4^{2+}$	11.1	3.1×10^{10}
$Cd(CN)_4^{2-}$	10.0	1.4×10^8
$Ce(III)_{aq}$	—	$<10^9$
Cl^-	10	$<10^5$
$C1O^-$	10	7.2×10^9
$C1O_3^-$	10	3.5×10^8
$C1O_4^-$	—	$<10^5$
Co^{2+}_{aq}	—	1.2×10^{10}
$Co(NH_3)_6^{3+}$	11.1	9.0×10^{10}
$Co(CN)_6^{3-}$	10	4.1×10^9
$Co(NO_2)_6^{3-}$	—	5.8×10^{10}
Cr^{2+}_{aq}	6.9	4.2×10^{10}
$Cr(II)_{aq}$	11.2	1.9×10^{10}
$Cr(III)_{aq}$	10.9	4.6×10^{10}
$Cr(H_2O)_6^{3+}$	7.0	6.0×10^{10}
$Cr_2O_7^{2-}$	7	3.3×10^{10}
CrO_4^{2-}	13	5.4×10^{10}
Cu^{2+}_{aq}	7	3×10^{10}
$Cu(OH)_4^{2-}$	14	5.8×10^9
$Cu(NH_3)_4^{2+}$	11.1	1.8×10^{10}
$Eu(III)_{aq}$	5.5	6.1×10^{10}
F^-	7.2	$<2 \times 10^4$
HF	5	6×10^7
HF_2^-	5	3×10^7
$Fe(II)_{aq}$	12	1.0×10^8
$Fe(CN)_6^{3-}$	7,10.3	3.0×10^9
$Fe(CN)_6^{4-}$	—	$<10^5$
H^+	4.0-5.0	2.36×10^{10}
H_2	7	$<10^7$
H_2O	8.4	1.6×10^1
D_2O	10	$<2.2 \times 10^2$
H_2O_2	7	1.23×10^{10}
HO_2^-	13	3.5×10^9

TABLE II (*continued*)

Reactant	pH	Rate constant (M^{-1} sec^{-1})
$Hg(CN)_4^{2-}$	10	1.9×10^8
$Ho(III)_{aq}$	5.9	2.4×10^9
I_2	7	5.1×10^{10}
I_3^-	7	2×10^{10}
IO_3^-	7	7.7×10^9
IO_4^-	7	1.10×10^{10}
$Ir(NH_3)_6^{3+}$	7	1.3×10^{10}
K_{aq}^+	—	$<3 \times 10^4$
$La(III)_{aq}$	7.0	3.4×10^8
$Lu(III)_{aq}$	6.2	2.5×10^8
$Mn(II)_{aq}$	—	7.7×10^7
$Mn(CN)_6^{4-}$	—	2.5×10^{10}
MnO_4^-	7	2.2×10^{10}
N_3^-	11	$<5.6 \times 10^6$
NH_4^+	5.3	1.3×10^6
N_2H_4	—	$<10^8$
N_2O	7	5.6×10^9
NO	7	3.1×10^{10}
NO_2^-	7.0	4.6×10^9
NO_3^-	7.0	1.1×10^{10}
Na_{aq}^+	—	$<10^5$
$Nd(III)_{aq}$	4.7	5.9×10^8
Ni_{aq}^{2+}	7	2.9×10^{10}
$Ni(CN)_4^{2-}$	11	5.5×10^9
O_2	7	1.88×10^{10}
$Os(NH_3)_6^{3+}$	7	7.2×10^{10}
$Os(CN)_6^{-4}$	10.5	$<10^7$
$H_2PO_2^-$	6.8	$<1.0 \times 10^5$
$H_2PO_3^-$	—	1.7×10^7
$H_2PO_4^-$	7.9	1.5×10^7
$Pb(II)_{aq}$	7.0	3.9×10^{10}
PbO_2^{2-}	14	1.0×10^{10}
$PdCl_4^{2-}$	7.1	1.2×10^{10}
$PtCl_4^{2-}$	6.8	1.2×10^{10}
$PtCl_6^{2-}$	10	2.0×10^{10}
$Pr(III)_{aq}$	6.1	2.9×10^8
$Rh(NH_3)_6^{3+}$	7	7.9×10^{10}

TABLE II (*continued*)

Reactant	pH	Rate constant (M^{-1} sec^{-1})
$Ru(NH_3)_6^{3+}$	7	7.4×10^{10}
SO_3^{2-}	10	$<1.3 \times 10^{6}$
SO_4^{2-}	7	$<10^{6}$
$S_2O_3^{2-}$	11.9	$<10^{8}$
$S_2O_8^{2-}$	7	1.06×10^{10}
$Sb(V)_{aq}$	11	1.2×10^{10}
$Sm(III)_{aq}$	5.9	2.5×10^{10}
$Sn(II)_{aq}$	11	3.4×10^{9}
SnO_2^{2-}	14	6.2×10^{9}
SnO_3^{2-}	11	6.3×10^{8}
$Tb(III)_{aq}$	6.1	3.7×10^{8}
TeO_3^{2-}	7	6×10^{8}
TeO_4^{2-}	11	1.6×10^{10}
TiO_3^{2-}	11.5	5×10^{6}
$Tl_{aq}{}^{+}$	7	3×10^{10}
$Tm(III)_{aq}$	6.0	3×10^{9}
UO_2^{2+}	12	7.4×10^{10}
VO_3^{-}	11	4.9×10^{9}
Y^{3+}	—	2×10^{8}
Yb^{3+}	6.0	4.3×10^{10}
$Zn_{aq}{}^{2+}$	7	1.5×10^{9}
$Zn(OH)_4^{2-}$	3 M OH^-	7.5×10^{6}
$Zn(CN)_4^{2-}$	10.5	1.8×10^{8}

REFERENCES

1. H. A. Laitinen and C. J. Nyman, *J. Am. Chem. Soc.*, **70**, 3002 (1948).
2. G. Milazzo, *Electrochemistry*, Elsevier, Amsterdam 1963, p. 229.
3. A. T. Cameron and W. Ramsey, *Chem. Soc. J.*, *91*, 1266 (1907).
4. J. H. Baxendale and G. Hughes, *Z. Physik Chem.* (Frankfurt) **14**, 323 (1958).
5. E. Hayon and J. J. Weiss, *Proc. 2nd Int. Conf. Peaceful Uses Atomic Energy*, Geneva 1959, Vol. 29, p. 80.

6. N. F. Barr and A. O. Allen, *J. Phys. Chem.*, **64**, 928 (1959).
7. G. Czapski and H. A. Schwarz, *J. Phys. Chem.*, **66**, 471 (1962).
8. E. J. Hart and J. W. Boag, *J. Am. Chem. Soc.*, **84**, 4090 (1962).
9(a). E. J. Hart, *Science*, **146**, 19 (1964).
9(b). E. J. Hart, in *Actions Chimique et Biologiques des Radiations*, Ed. M. Haissinsky, Vol. 10, 1, (1966).
10. J. W. Boag, *Am. J. Roent, Radiation, and Nucl. Med.*, **15**, 896 (1963).
11(a). F. S. Dainton, *Gomberg Centenary Volume*, Butterworth, London, 1967.
11(b). F. S. Dainton, *Fast Reactions and Primary Processes in Chemical Kinetics*, ed. S. Claesson, Interscience, 1967, p. 185.
12. J. H. Baxendale, *Current Topics in Radiation Research*, Eds. M. Ebert and A. Howard, Vol. III, Wiley, 1967, p. 1.
13. M. S. Matheson, *Adv. Chem. Ser.*, **50**, 45 (1965).
14(a). M. Anbar, *Adv. Chem. Ser.*, **50**, 55 (1965).
14(b). M. Anbar, *Quart. Rev.*, **22**, 579 (1968).
15. J. K. Thomas, *Radiation Res. Rev.*, *1*, 183 (1968).
16. S. R. Logan, *J. Chem. Educ.*, **44**, 345 (1967).
17. H. A. Schwarz, *Ann. Rev. Phys. Chem.*, **16**, 347 (1965).
18. L. M. Dorfman and M. S. Matheson, *Progress in Reaction Kinetics*, ed. G. Porter, Vol. 3, Pergamon Press, 1965, p. 237.
19. U. Schindewolf, *Angew Chem.*, **80**, 165 (1968) (*Int. Ed.*, **7**, 1968).
20(a). D. C. Walker, *Quart. Rev.*, **21**, 79 (1967).
20(b). D. C. Walker, *Adv. Chem. Ser.*, **81**, 49 (1968).
21. M. Anbar and P. Neta, *Int. J. Appl. Radiation Isotopes*, **18**, 493 (1967).
22. J. P. Keene, *Radiation Res.*, **22**, 1 (1964).
23. J. J. Markham, *F Centers in Alkali Halides*, Academic Press, New York, 1966.
24. G. Lepoutre and M. J. Sienko, eds., *Solutions Métal-Ammoniac*, Benjamin, New York, 1964.
25. E. C. Avery, J. R. Remko, and B. Smaller, *J. Chem. Phys.*, **49**, 951 (1968).
26. K. H. Schmidt and W. L. Buck, *Science*, **151**, 70 (1966).
27. W. J. Moore, *Physical Chemistry*, Prentice-Hall, Englewood Cliffs, New Jersey, (1964).
28. P. J. Coyle, F. S. Dainton, and S. R. Logan, *Proc. Chem. Soc.*, 1964, 219.
29(a). J. Jortner, *Radiation Res. Suppl.*, **4**, 24 (1964).
29(b). J. Jortner, *Radiation Chemistry of Aqueous Solutions*, ed. G. Stein, Wiley (Interscience), New York (1969).
30. R. M. Noyes, *J. Am. Chem. Soc.* **86**, 971 (1964).
31. E. J. Hart, S. Gordon, and E. M. Fielden, *J. Phys. Chem.*, **70**, 150 (1966).
32. G. A. Kenney and D. C. Walker, *J. Chem. Phys.*, **50**, 4074 (1969).
33. L. M. Dorfman and I. A. Taub, *J. Am. Chem. Soc.*, **85**, 2370 (1963).
34. N. Basco, G. A. Kenney, and D. C. Walker, *Chem. Comm.*, 1969, 917.
35. J. H. Baxendale, *Radiation Res. Suppl.*, **4**, 139 (1964).
36. G. S. Barker, A. W. Gardner, and D. C. Sammon, *J. Electrochem. Soc.*, **113**, 1183 (1966).
37. D. C. Walker, *Can. J. Chem.*, **44**, 2226 (1966).

38. G. Hughes and R. J. Roach, *Chem. Comm.*, 1965, 600.
39. J. Rabani, W. A. Mulac and M. S. Matheson, *J. Phys. Chem.*, **69**, 53 (1965).
40. D. C. Walker, *Can. J. Chem.*, **45**, 807 (1967).
41. D. C. Walker, *Anal. Chem.*, **39**, 896 (1967).
42. N. Sato, private communication.
43. B. E. Conway, private communication.
44. G. J. Hills and D. R. Kinnibrugh, *J. Electrochem. Soc.*, **113**, 1111 (1966).
45. R. R. Hentz, Farhataziz, D. J. Milner and M. Burton, *J. Chem. Phys.*, **47**, 374 (1967).
46. W. L. Jolly, *Progress in Inorganic Chemistry*, Vol. 1, Wiley (Interscience), New York, 1959.
47. V. A. Yurkov, *Soviet Electrochemistry*, Vol. II, Consultants Bureau, New York, 1961, p. 85.
48. D. J. G. Ives and F. R. Smith, *Trans. Faraday Soc.*, **63**, 217 (1967).
49. M. Daniels, *Adv. Chem. Ser.*, **81**, 153 (1968).
50(a) H. H. Uhlig and R. C. Krutenat, *J. Electrochem. Soc.*, **III**, 1303 (1964).
50(b) R. C. Krutenat and H. H. Ulig, *Electrochim. Acta*, **11**, 469 (1966).
51. F. P. Bowden, *Trans. Faraday Soc.*, **28**, 505 (1931).
52. P. Delahay and V. S. Srinivasan, *J. Phys. Chem.*, **70**, 420 (1966).
53. M. Heyrovsky, *Nature*, **206**, 1356 (1965).
54. T. Kuwana, *Electroanalytical Chemistry*, ed. A. J. Bard, Vol. I, p. 197, Dekker, New York (1966).
55. T. Pyle and C. Roberts, *J. Electrochem. Soc.*, **115**, 247 (1968).
56. D. M. Brown, F. S. Dainton, J. P. Keene, and D. C. Walker, *Proc. Chem., Soc.*, 1964, 266.
57. E. J. Hart and E. W. Fielden, *J. Phys. Chem.*, **72**, 577 (1968).
58. W. M. Latimer, *Oxidation Potentials*, Prentice-Hall, Englewood Cliffs, New Jersey, 1952.
59. J. Jortner and R. M. Noyes, *J. Phys. Chem.*, **70**, 770 (1966).
60. " Solvated Electron," ed. E. J. Hart, *Adv. Chem. Ser.*, **50** (1965).
61. M. C. R. Symons, *Quart. Rev.*, **13**, 99 (1959).
62. F. Gutman and L. Lyons, *Organic Semiconductors*, Wiley, New York, 1967.
63. U. Solokov and G. Stein, *J. Chem. Phys.*, **44**, 2189 (1966).
64. L. Robinson Painter, R. N. Hamm, E. T. Arkawaa, R. D. Birkhoff, *Phys. Rev. Letters*, **21**, 282 (1968).
65. J. Barret, M. F. Fox, and A. L. Mansell, *J. Chem. Soc.*, A, 483 (1967).
66(a). L. Dogliotti and E. Hayon, *J. Phys. Chem.*, **72**, 1800 (1968).
66(b). J. R. Huber and H. Hayon, *J. Phys. Chem.*, **72**, 3820 (1968).
67. R. Devonshire and J. J. Weiss, *J. Phys. Chem.*, **72**, 3815 (1968).
68. J. J. Weiss, *The Chemistry of Excitation and Ionization*, ed. G. R. A. Johnson and G. Scholes, Taylor and Frances, London, 1967.
69. L. I. Grossweiner, Adv. *Radiation Biol.*, **2**, 183 (1966).
70(a). G. E. Johnson and A. C. Albrecht, *J. Chem. Phys.*, **44**, 3179 (1966).
70(b). W. M. McClain and A. C. Albrecht, *J. Chem. Phys.* **44**, 1594 (1966).
70(c). D. S. Kliger, J. D. Laposa and A. C. Albrecht, *J. Chem. Phys.* **48**, 4326 (1966).

71. W. Hamill, *J. Chem. Phys.* **49**, 2446 (1968).
72. E. A. Shaede and D. C. Walker, *The Alkali Metals*, special publication No. 22, The Chemical Society, 1967.
73. G. Fraenkel, S. H. Ellis, and D. T. Dix, *J. Am. Chem. Soc.*, **87**, 1406 (1965).
74. J. Jortner and G. Stein, *Nature*, **175**, 893 (1955).
75. J. Bennett, B. Mile, A. Thomas, *J. Chem. Soc.*, *A*, 1393 (1967).
76. F. Basolo and R. G. Pearson, *Mechanisms of Inorganic Reactions*, Wiley, 1967, p. 515.
77. A. G. Sykes, *Kinetics of Inorganic Reactions*, Pergamon Press, 1966, p. 222.
78(a). J. Jortner and J. Rabani, *J. Phys. Chem.*, **66**, 2081 (1962).
78(b). G. Czapski, J. Jortner and G. Stein, *J. Phys. Chem.*, **65**, 956 (1961).
79. M. Anbar and E. Hart, *J. Phys. Chem.*, **71**, 4163 (1967).
80. M. Anbar and I. Pecht, *J. Phys. Chem.*, **69**, 271 (1965).
81(a). R. R. Dewald, J. L. Dye, M. Eigen, and L. Demaeyer, *J. Chem. Phys.*, **39**, 2388 (1963).
81(b). L. H. Feldman, R. R. Dewald, and J. L. Dye, *Adv. Chem. Ser.*, **50**, 163 (1965).
81(c). R. R. Dewald, R. V. Tsina, *Chem. Comm.*, 1967, 647.
82. R. L. Platzman, *N.A.S.-N.R.C. Reports*, 1953, No. 305.
83. J. J. Weiss, *Nature*, **153**, 748 (1944).
84. L. H. Gray, *J. Chim. Phys.*, **48**, 172 (1951).
85. A. H. Samuel and J. L. Magee, *J. Chem. Phys.* **21**, 1080 (1953).
86. A. Mozumder and J. L. Magee, *Radiation Res.*, **28**, 203, 231 (1964).
87. M. J. Bronskill and J. W. Hunt, *J. Phys. Chem.*, **72**, 3762 (1968).
88. J. J. Weiss, *Nature*, **186**, 751 (1960).
89. G. Stein, *Discussions Faraday Soc.*, **12**, 227, 1952.
90. E. Hayon and A. O. Allen, *J. Phys. Chem.*, **65**, 2181 (1961).
91. G. Czapski, and A. O. Allen, *J. Phys. Chem.*, **66**, 262 (1962).
92. E. Collison, F. S. Dainton, D. R. Smith, and S. Tazuke, *Proc. Chem. Soc.*, 1962, 140.
93. F. S. Dainton and W. S. Watt, *Proc. Roy. Soc.*, **275A**, 447 (1963).
94(a). P. Coyle, F. S. Dainton, and S. Logan, *Proc. Chem. Soc.*, 1964, 219.
94(b). S. Logan, *Trans. Faraday Soc.*, **62**, 3416, 3423 (1966).
95. A. Yokohata, *Bull. Chem. Soc. Japan*, **42**, 658 (1969).
96. J. P. Keene, *Nature*, **197**, 47 (1963).
97. A. Kupperman, *Actions Chemiques et Biologique des Radiations*, ed. M. Haissinsky, Masson, Paris, 1961.
98. R. Schiller, *J. Chem. Phys.*, **47**, 2281 (1967).
99. G. R. Freeman and J. M. Fayadh, *J. Chem. Phys.*, **43**, 86 (1965).
100(a). J. Morgan and B. E. Warren, *J. Chem. Phys.*, **6**, 666 (1938).
100(b). G. W. Brady and W. J. Romanov, *J. Chem. Phys.*, **32**, 306 (1960).
101. R. A. Ogg, *Phys. Rev.*, **69**, 668 (1946).
102(a). W. N. Lipcomb, *J. Chem. Phys.*, **21**, 52 (1953).
102(b). R. A. Stairs, *J. Chem. Phys.*, **27**, 1431 (1957).
103. R. L. Platzman and J. Frank, *Z. Physik*, **138**, 411 (1954).
104(a). L. Landau, Physik, *Z. Soviet Union*, **3**, 664 (1933).
104(b). A. S. Davydov, *Zh. Eksperim Theor. F 1Z*, **18**, 913 (1948).

104(c). M. F. Deigen, *Zh. Eksperim Theor. F* 1*Z*, **26**, 300 (1954).
105. J. Jortner, *J. Chem. Phys.*, **30**, 839 (1959).
106(a). M. Natori, and T. Watanabe, *J. Phys. Soc. Japan*, **21**, 1573 (1966).
106(b). M. Natori, *J. Phys. Soc. Japan*, **24**, 913 (1968).
107. K. Iguchi, *J. Chem. Phys.*, **48**, 1735 (1968).
108. S. Noda, K. Fueki, Z. Kuri, *Bull. Chem. Soc. Japan*, **42**, 16 (1969).
109. J. J. Weiss, *Nature*, **215**, 151, (1967).
110. W. C. Gottschall and E. J. Hart, *J. Phys. Chem.*, **71**, 2102 (1967).
111. S. Arai and M. C. Sauer Jr., *J. Chem. Phys.*, **44**, 2297 (1966).
112. L. Dorfman, *Adv. Chem. Ser.*, **50**, 40 (1965).

The Fundamentals of Metal Deposition

J. A. Harrison

and

H. R. Thirsk

ELECTROCHEMISTRY RESEARCH LABORATORIES,
DEPARTMENT OF PHYSICAL CHEMISTRY,
UNIVERSITY OF NEWCASTLE UPON TYNE,
NEWCASTLE UPON TYNE, NE1 7RU, ENGLAND

I. INTRODUCTION

It must be stated at the outset that the primary object of this chapter is to delineate the areas where theory and experiment are most closely aligned. The experimental investigation of the kinetics of metal electrocrystallization from simple aqueous cations has proved difficult. The difficulty is that the state of the surface is not well defined before the measurement and also varies in a complex way throughout the electrolysis. Galvanostatic, a.c., potentiostatic measurements are not sufficient to arrive at a mechanism unless the surface can be controlled before electrolysis. At the moment this can only be achieved for two substrates, mercury and silver (see Section IV). This fact has allowed investigation of the variation of surface structure during electrodeposition, and much of this review will be concerned with progress in this area. Historically the subject of the mechanism of metal deposition has relied heavily on ideas derived from vapor deposition and these are reviewed briefly in Section II.

The overall process—that is, metal cation going to metal atom in the lattice—is complex, but considerable progress has been made in investigating systems that show extremes of behavior.

(a) The kinetics of the electrochemical step can be isolated from the crystal growth problem by depositing onto mercury, when the deposited metal is completely and rapidly soluble in mercury.

(b) The electrochemical step alone can be measured for solid metal substrates if the succeeding crystal growth step is fast. This can be achieved by ensuring that the substrate metal has a large number of imperfections—that is, growth sites. Prime examples of the technique are copper and indium deposition (to be discussed in Sec. IV) where the surface is prepared by electrodeposition.

(c) The growth and formation of nuclei can be simplified by depositing onto inert substrates with a low number of dislocations. Models for these processes are known from the deposition of nonmetallic lattices, primarily salts of mercury formed on a mercury surface. In comparison with metal lattices these have, because of their covalent nature, low rates of growth and the expected behavior can be tested without complications from diffusion.

Because of the relevance of certain basic studies of these anodic kinetics, a short account has, necessarily, been introduced in Sec. IV A (*3*). Indeed, it is clear that metal deposition is a special case of the more general subject of electrochemical phase changes or electrocrystallization. The older measurements (*1*, *2*) and some aspects of the electrochemical adatom model (*3*, *4*) have recently been reviewed. It is the purpose of this article to bring forward some of the newer theoretical results which incorporate bulk diffusion and discuss measurements of the type (a), (b), and (c) above. For clarity in exposition the material to follow is divided into two main sections, theoretical and experimental, with certain relaxations in this division in that experiments designed to justify the theory are included in Sec. III, which deals with models of the electrocrystallization process at an atomic level. The theory given in Secs. IIIA to IIIE will be mainly of academic interest until measurements with dislocation-free electrodes both on mercury and solids become widespread. The theory of Sec. IIIH has been used successfully several times. The review is weighted especially toward those aspects of the fundamental study of metal deposition that appear most promising. Some aspects that have been reviewed in detail recently—such as the formation of alloys (*5*), morphology of deposits (*6*), and development of plating processes (*7*)—have been deliberately omitted, as is the case with such topics as molten salts (*8*), electrodeposition in nonaqueous solvents (*9*), and electrophoretic deposition, which are of technological interest (for a general text see Ref. *8*). The fundamental processes of electrochemical discharge and lattice growth are probably similar in all these cases and even if this were not entirely true, they could best be studied by the methods reviewed here. Differences in bulk species in plating from complexes have a major effect on the final overall reaction and these are beyond the scope of the present review.

We hold quite firmly to the opinion that the subject of metal deposition is at the moment entering a more promising stage, with a deeper understanding of the areas in the theory of phase changes and of experiments that can be interpreted precisely.

II. VAPOR DEPOSITION OF METALS

Some of the more important results will be given briefly in this section. The phenomena closely parallel the electrochemical processes; in some areas the electrochemical theory and experiment are more developed, reflecting the ease with which currents can be measured and the surface energy controlled by potential. Also there is the fact that for many years an ideal dislocation-free metallic surface, mercury, has been available to electrochemists. Many of the ideas regarding nucleation and growth mechanisms can be tested on mercury, especially in the formation of anodic films, where diffusion effects are unimportant, the rate controlling step being demonstrably the growth of the crystalline phase.

Vapor-phase deposition is in principle simpler than the electrochemical case in that solvent is absent and uncharged atoms are deposited. There seems little doubt that under these conditions adatoms are formed on the surface which then diffuse to growth sites. The growth sites can be two- or three-dimensional nuclei or screw dislocations. All three types of growth mechanism are reported in the literature. For details of individual systems the reader is referred to existing reviews (*10*, *11*, *12*, *13*, *14*, *15*).

A crystal growing from its own vapor can develop by (a) two-dimensional nucleation and layer-by-layer growth and (b) a screw dislocation mechanism in which atoms diffuse to the growing edges and rotate the spiral.

Mechanism (b) was postulated by Burton, Cabrera, and Frank (*16*) to account for the fact that many crystals grow before the critical supersaturation for two-dimensional nucleation. The earlier theories of Kossel (*17*) and Stranski (*18*) had demanded mechanism (a); spirals are self-perpetuating structures that do not require nucleation. The dependence of growth rate on supersaturation has been measured for many systems and compared with theory. The theory is described in Sec. III for the electrochemical situation.

Two- and three-dimensional nucleation are commonly observed when metals are deposited on a foreign substrate. Two-dimensional films have been observed for gold on silver, platinum on gold, and lead on silver (*19*) by electron diffraction. Similar systems have been investigated electrochemically and are discussed in Sec. IVC.

Ultrahigh vacuum 10^{-10} torr is essential for the detailed investigation of vapor-phase kinetics, at least it must be possible to clean the surface and surroundings under these pressures. The condition is most nearly achieved in the field electron (FEM) and field ion (FIM) microscopes. The FEM has shown that when silver, nickel, zinc, cadmium and gold are deposited on to a single crystal tungsten tip the onset of crystallization can be detected; however, the resolution of 20°Å does not normally allow individual clusters to be detected. If crystallization occurs from a fixed critical concentration of adatoms, then the time for the appearance of nuclei is inversely proportional to the flux (Fig. 1); the electrochemical deposition of silver onto dislocation-free silver (Sec. IVA) shows a similar result. Vapor deposition of nickel, gold, and silver onto tungsten require layers of adatoms equivalent to 3.7, 2.8, 0.1 monolayers of adatoms before the appearance of crystals. The properties of such adsorbed layers have been examined in more detail (*20*) for nickel, (*21*), copper (*22*) on tungsten especially by work function

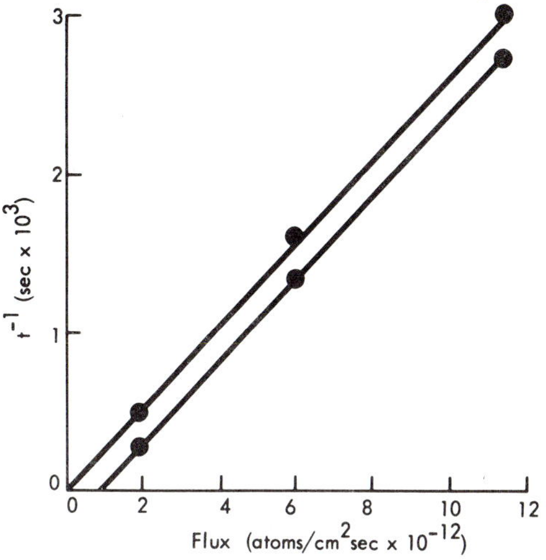

Fig. 1. Reciprocal time for nucleation versus flux for zinc nucleating on tungsten. Ref. (*13*).

measurement as a function of the number of monolayers. Other techniques—low-energy electron diffraction, electron diffraction—have also suggested that a highly ordered layer (see Ref. *20* for list of references) can be observed without the formation of crystallites. The critical nucleus size is small, in some cases less than 10 atoms. These two facts suggest that classical nucleation theory, which assumes the nuclei have bulk properties, may need reworking (*14*).

The field ion microscope is potentially a more powerful tool (*23*) for a very limited range of metals. It uses an imaging gas, usually helium or neon, and can in practice resolve adjacent tungsten atoms that are 2.74°Å apart. The diffusion coefficients for the movement of tungsten atoms over the various phases of tungsten have been measured (*24*).

III. MODELS OF THE ELECTROCRYSTALLIZATION PROCESS AT AN ATOMIC LEVEL

Introduction

The vapor-phase model shows that the surface features of Fig. 2 will probably be present on a metal surface. The step lines that are one atom high could be part of a single screw dislocation, a front formed by the interaction

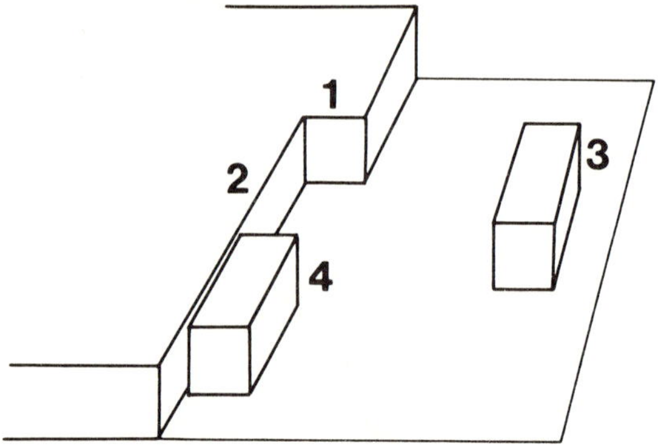

Fig. 2. Structural features probably present on a metal surface.

of two screw dislocations (Frank-Read source) or the edge of a two-dimensional nucleus formed on an originally flat low index crystal plane. Other properties of step lines are considered in Sec. IIIG. The positions on the surface are (1) kinks, (2) step lines, (3) adsorbed molecules, and (4) adsorbed molecules. It is thought that the number of kinks in step lines is high, so that at room temperature there are an equilibrium number of adsorbed molecules in the step line. The lattice grows at the step lines either from adatoms that diffuse along the surface or by direct deposition at the step line itself. The theory is now well worked out for the adatom model that is treated in Sec. IIIA to F. Direct deposition at the step lines treated in Sec. IIIG is less well known. However, it now seems likely that the earlier evidence for adatoms on real metals during electrodeposition can be completely explained by a direct deposition model; it is perhaps a little out of place here, but it is interesting to note Gerisher's comment on this matter (*46*) and his recent acceptance of this model.

A. Adatom Model. Potentiostatic Case; Complete Solution

Figure 3 shows the essential details of the adatom model. Adatoms arrive at the surface by diffusion and electrochemical reaction and then diffuse along the surface to the step line; the step lines do not advance. In a potentiostatic pulse experiment starting from the equilibrium potential the theory has been formulated (*25*) exactly for two cases:

(a) When the equilibrium concentration of adatoms is maintained at the edge and the rate of incorporation and the reverse reaction at the step line are both fast.

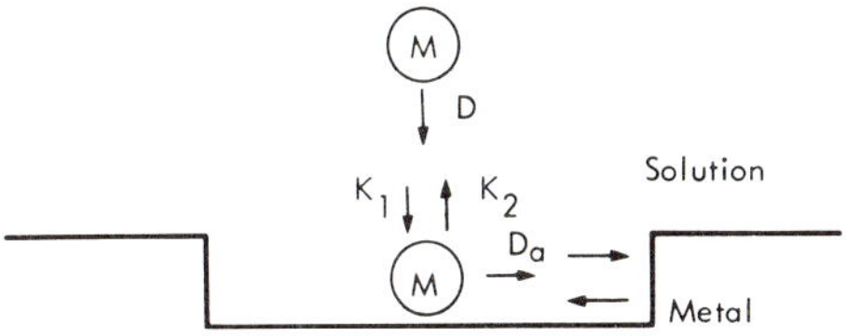

Fig. 3. Model for an electrochemical reaction with surface diffusion.

(b) When equilibrium is not maintained and the rate of removal of adatoms at the step line is slow.

The mathematical boundary conditions are shown in Fig. 4 for case (a). The solution, in the form of the Laplace transform, is

$$\bar{i} = zF(k_1 C^0 - k_2 C_a^0)\frac{\sqrt{pD}}{x_{00}(p)}\left\{1 + \frac{k_2\sqrt{pD}}{x_{00}(p)S_1(p)}\right\} \tag{1}$$

where

$$x_{00}(p) = (k_1 + \sqrt{pD})p + k_2\sqrt{pD}$$

$$S_1(p) = \sum_0^\infty \frac{\varepsilon_m(k_1 + \sqrt{(p + 1/\tau_m)D})p}{x_{m0}(p)}$$

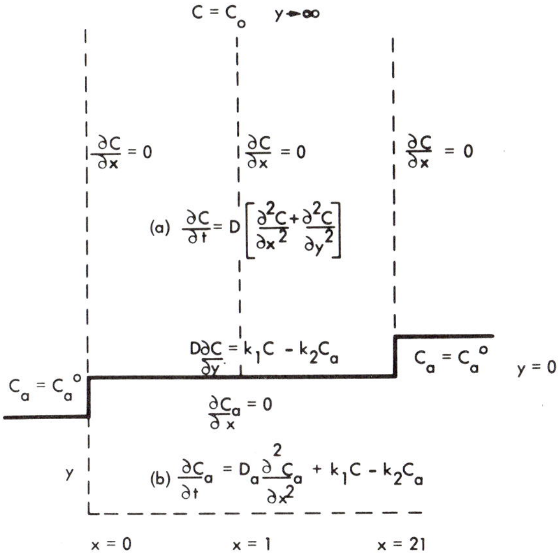

Fig. 4. Simultaneous partial differential equations (a), (b) with appropriate boundary conditions to be solved for the adatom model of Fig. 3. Ref. (*25*).

$$x_{m0}(p) = \left(k_1 + \sqrt{\left(p + \frac{1}{\tau_m}\right)D}\right)\left(p + \frac{1}{\tau_m^a}\right) + k_2\sqrt{\left(p + \frac{1}{\tau_m}\right)D}$$

$$\varepsilon = 1, \quad m = 0$$

$$\varepsilon = 2, \quad m = 1, 2 \ldots$$

$$k_1 = \frac{i_0}{zFC_0}\exp\left(\frac{\alpha zF\eta}{RT}\right); \quad k_2 = \frac{i_0}{zFC_a^0}\exp\left[\frac{(\alpha - 1)zF\eta}{RT}\right]$$

$$\frac{1}{\tau_m} = \frac{m^2\pi^2 D}{l^2} \qquad \frac{1}{\tau_m^a} = \frac{m^2\pi^2 D_a}{l^2}$$

where the Laplace transform is defined by

$$\bar{i}(p) = \int_0^\infty \exp(-pt)i(t)\,dt \tag{2}$$

and $i(t)$ is the current for a square centimeter electrode given by

$$i(t) = \frac{zF}{l}\int_0^l (k_1 C - k_2 C_a)_{y=0}\,dx \tag{3}$$

case (b) can similarly be solved by changing the boundary condition on the surface from

$$C_a = C_a^0 \qquad \text{at } x = 0 \tag{4}$$

to

$$D_a\left(\frac{\partial C_a}{\partial x}\right)_{x=0} = k_3(C_a - C_a^0) \tag{5}$$

Where k_3 is the rate of incorporation at the edge. The solution now becomes

$$\bar{i} = zF(k_1 C^0 - k_2 C_a^0)\frac{\sqrt{pD}}{x_{00}(p)}\left\{1 + \frac{k_2\sqrt{pD}}{x_{00}(p)(S_1(p) + pl/k_3)}\right\} \tag{6}$$

From these two general equations all the equations for this model that have appeared in the literature can be derived.

In case (a) the transform has been inverted numerically (*25a*) by the method of Weeks (*26*). Some results are given in Fig. 5, which shows the essential characteristics of the model for a limited set of data. When the distance between step lines (l) is small, adatoms are captured by the step before they can dissolve electrochemically and the model becomes identical to that for

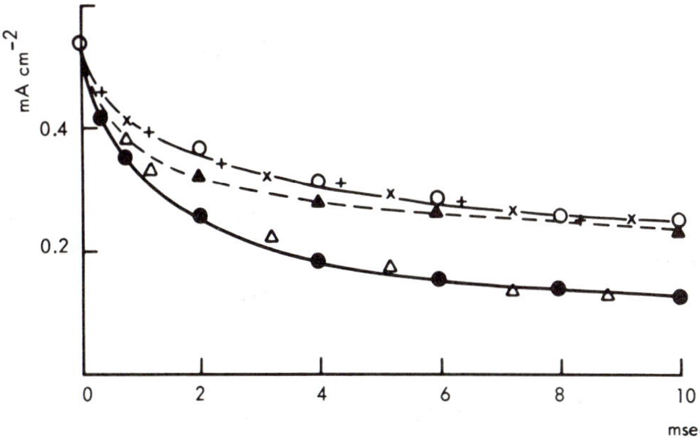

Fig. 5. Theoretical current-time transients for a potentiostatic experiment for the model of Figs. 3 and 4 with the parameters,

$\eta = 5.13$ mV	$\alpha = 0.5$
$k_1 = 3.16 \times 10^{-2}$ cm sec^{-1}	$i_0 = 2.76 \times 10^{-3}$ A cm^{-2}
$k_2 = 2.59 \times 10^2$ dec^{-1}	$C^0 = 10^{-6}$ moles cm^{-3}
$D = 10^{-5}$ cm^2 sec^{-1}	$C_a^0 = 10^{-10}$ moles cm^{-2}

$$(X),\ i = i_0 \left(\exp \frac{\alpha z F \eta}{RT} - \exp \frac{(\alpha - 1) z F \eta}{RT} \right) \exp \left(\frac{k_1^2 t}{D} \right) \mathit{erfc}\, (k_1 \sqrt{(t/D)})$$

(+), computed $D_a = 10$ cm^2 sec^{-1}, $1 = 10^{-2}$ cm, $m = 1$; (•), computed $D_a = 10^{-2}$ cm^2 sec^{-1}, $1 = 10^{-2}$ cm, $m = 9, 27, 256$; (O), computed $D_a = 10^{-2}$ cm^2 sec^{-1}, $1 = 10^{-4}$ cm, $m = 4$; (△), computed $D_a = 10^{-2}$ cm^2 sec^{-1}, $1 = 1.0$ cm, $m = 2$; (▲), computed $D_a = 10^{-1}$ cm^2 sec^{-1}, $1 = 10^{-2}$ cm, $m = 2$.
Ref. (*25a.*)

planar diffusion. If the surface diffusion coefficient D_a is small or l is large, then another extreme is reached when the electrochemical step and diffusion determine the current. Putting $D_a \to \infty$ in Eq. (1) leads to the expected expression.

$$i = i_0 \left(\exp \frac{\alpha z F \eta}{RT} - \exp \left[\frac{(\alpha - 1) z F \eta}{RT} \right] \right) \exp \left(\frac{k_1^2 t}{D} \right) \operatorname{erfc} \left(\frac{k_1 \sqrt{t}}{\sqrt{D}} \right) \qquad (7)$$

after inversion, which is the usual one for control by diffusion and electrochemical reaction; k_2 disappears as adatoms are captured by the step lines. Equation (7) is plotted in Fig. 5. It is clear that surface diffusion, as an observable rate-limiting process, will only be detectable experimentally within a narrowly defined range of the parameters i_0, D, D_a, l.

The equations have been written as a current but could equally well be formulated as a theoretical rate of advance of a step line v by the expression

$$v = \frac{Ib}{zFL} \tag{8}$$

where L is the length of growth step per cubic centimeter, b is the area per mole, I is the current in amperes. The rate of advance of an edge is an important quantity as it is independent of number of step lines or of the number of nuclei (in Sec. IIIH).

B. Adatom Model. Potentiostatic Case; Stationary State without Bulk Diffusion

Equations (1) through (6) can be inverted under some conditions. Neglecting diffusion in the solution, that is, $D \to \infty$, and putting $t \to \infty$ for the stationary state, gives

$$i = i_0\left(\exp\left(\frac{\alpha zF\eta}{RT}\right) - \exp\left[\frac{(\alpha - 1)zF\eta}{RT}\right]\right) \frac{1}{l}\sqrt{\frac{D_a}{k_2}} \tanh\left(\sqrt{\frac{k_2}{D_a}} \cdot l\right) \tag{9}$$

The properties of this equation have been thoroughly reviewed by Vetter (*4*). It is identical to equations derived directly by Lorenz (*27*), Vermilyea (*28*), Fleischmann and Thirsk (*29*), Damjanovic and Bockris (*30*), and equivalent to that derived by Burton, Cabrera, and Frank (*16*) for the vapor phase. Equation (9) can also be directly obtained by solving Eq. (10) by the Laplace method

$$\frac{\partial C_a}{\partial \varepsilon} = D\frac{\partial^2 C_a}{\partial x^2} + k_1 - k_2 C_a = 0 \tag{10}$$

with the conditions

$$\begin{aligned} x = 0, &\quad C = C_a^0 \\ x = l, &\quad \left(\frac{\partial C_a}{\partial x}\right) = 0 \end{aligned} \tag{11}$$

A simplified version of Eq. (9) can be obtained by calculating the steady current to an isolated step line by changing the boundary conditions to

$$\begin{aligned} x = 0, &\quad C = C_a^0 \\ x = \infty, &\quad C = {}_\infty C_a \end{aligned}$$

The solution is, for unit length of step line,

$$i = zF\sqrt{D_a k_2}(C_a^0 - \infty C_a) \tag{13}$$

A further calculation (*31*) has appeared in the literature which estimates the current into a single kink surrounded by a circularly symmetrical surface diffusion zone. The result assuming $r_0 \ll \sqrt{D/k_2}$ for the current into a kink radius r_0 is

$$i = 2\pi D(C_a^0 - \infty C_a)\frac{1}{\ln(1.13/r_0\sqrt{D/k_2})} \tag{14}$$

Equation (6) gives, similarly, the stationary value (*32*) when $D \to \infty$

$$i = \frac{zF}{l}(k_1 k_4 - k_2 k_3)\sqrt{\frac{D_a}{k_2}}\frac{\sinh(\sqrt{(k_2/D_a)}l)}{k_4\cosh(\sqrt{k_2/D}\,l) + \sqrt{D_a k_2}\sinh(\sqrt{k_2/D_a}\,l)} \tag{15}$$

Here the definition of the rate of incorporation given by Eq. (5) has been changed to retain the nomenclature of the original paper, namely,

$$D_a\left(\frac{\partial Ca}{\partial x}\right)_{x=0} = k_4 C_a - k_3$$

The Tafel slope can be formulated from Eqs. (9) and (15) for the two cases by substituting for k_2 and assuming, at low η, that $l\alpha^1/\eta$. The following limiting cases have been suggested: from Eq. (9), that is, for equilibrium (fast incorporation and removal) at the step line

(a) At high η

$$i = i_0 \exp\left(\frac{\alpha zF\eta}{RT}\right)\left[1 - \exp\left(-\frac{zF\eta}{RT}\right)\right]$$

(b) At low η

$$i = i_0^{1/2}\exp\left[\frac{(1-\alpha)Fz\eta}{2RT}\right]\left[1 - \exp\left(-\frac{zF\eta}{RT}\right)\right](zFDC_a^0)^{1/2}\frac{zF\eta}{2\pi V_1\sigma}$$

from Eq. (15), that is when incorporation at the step line is slow (*35a*).

(c) At low η and $(D_a k_2)^{1/2} > k_4$

$$i = \frac{zF}{l}\left[k_4 C_a^0 \exp\left(\frac{\eta F}{RT}\right) - k_3\right]$$

(d) At low η and $k_4 > (D_a k_2)^{1/2}$

$$i = \frac{zF}{l}\left(\frac{D}{k_2}\right)^{1/2}\left(k_1 - \frac{k_2 k_3}{k_4}\right)$$

(e) At high η, $(k_2/D_a)^{1/2} l$ small

$$i = zF\left(k_1 - \frac{k_2 k_3}{k_4}\right)$$

which is independent of step spacing. However, to use these relations it would be necessary to subtract bulk diffusion (probably in a rotating disk experiment) although in principle any technique with known diffusion control would be possible. To distinguish (a) to (e) and unambiguously separate them from direct deposition (Sec. IIIG) to step lines has not yet been investigated (see comment about stability of adatom model).

The stationary currents at a potential η in Eqs. (9) through (15) can be used as the initial condition to calculate the transients on switching to a potential η_2. This calculation has been attempted ignoring bulk diffusion (*33, 34*). This theory supports Gerischer's (*35*) experiment in which the overpotential η_2 is measured at the moment for which the instantaneous current directly at switching is zero; η_2 senses the average adatom concentration at η_1. After the subtraction of calculated η_{Nernst} when η_{R}, the reaction overpotential is small, Gerischer considered that for silver deposition the remainder was η_{ad}, from which C_{ad}^0 could be calculated. This type of measurement will only work for a fast process where the double-layer charging is negligible at the instant of switching from η_1 to η_2. It will be shown in Sec. IIIF that this is not unequivocal evidence for adatoms.

C. Adatom Model. Galvanostatic Case without Bulk Diffusion

The exact equation to be solved is

$$\frac{\partial C_a}{\partial t} = D_a\left(\frac{\partial^2 C_a}{\partial x^2}\right) + \frac{i_0}{zF}\left[\exp\left(\frac{\alpha_c F\eta(t)}{RT}\right) - \frac{C_a}{C_a^0}\exp\left(-\frac{\alpha_a F\eta(t)}{RT}\right)\right] \tag{16}$$

A simplified version of this equation has been solved (*36*), namely,

$$\frac{\partial C_a}{\partial t} = \frac{i}{zF} - v_0\left[\frac{C_a(\eta, t) - C_a^0}{C_a^0}\right] \tag{17}$$

where v_0 is the surface diffusion velocity. Equation (17) is equivalent to replacing $\partial^2 C_a/\partial x^2$ in Eq. (16) by

$$\frac{D_a}{l^2}(C_a - C_a^0) \tag{18}$$

and hence v_0 identified with $D_a C_a^0/l^2$. The result is

$$\eta(t) = \frac{RT}{zF}\frac{i}{i_0} + \frac{RT}{z^2F^2}\frac{i}{v_0}\left[1 - \exp\left(-\frac{v_0}{C_a^0}t\right)\right] \tag{19}$$

It was suggested by Bockris et al. (*36*) on the basis of Eq. (19) that

$$\eta_{t\to 0} = a + bt$$

$$a = \frac{RT_i}{zFi_0},$$

$$b = \frac{RT}{z^2F^2}\frac{i}{C_a^0} \tag{20}$$

$$\frac{\partial}{\partial t}[\ln(\eta_t - \eta_\infty)] = -\frac{v_0}{C_a^0} \tag{21}$$

$$\eta_{t\to\infty} = \frac{RT}{zF}\left[\frac{i}{i_0} + \frac{i}{zFv_0}\right] \tag{22}$$

so that i_0, C_a^0, D_a/l^2 could be calculated.

However, Rangarajan (*37*) has shown that the above conclusions are suspect. The stationary state, $\partial C_a/\partial t = 0$, can be solved directly from Eq. (16) with the conditions

$$\left(\frac{\partial C_a}{\partial x}\right)_{x=l} = 0; \qquad C_a = C_a^0 \qquad \text{at } x = 0 \tag{23}$$

The solution is

$$i = i_0\left(\exp\left(\frac{\alpha_c F\eta_\infty}{RT}\right) - \exp\left(-\frac{\alpha_a F\eta_\infty}{RT}\right)\right] \cdot \frac{1}{\Omega}$$

$$\cdot \exp\left(\frac{\alpha_a F\eta_\infty}{RT}\right)\tanh\left(\Omega\exp\left(-\frac{\alpha_a F\eta_\infty}{RT}\right)\right) \tag{24}$$

The nomenclature of the original paper is retained with

$$\Omega = l\left(\frac{i_0}{zFC_a^0 D_a}\right)^{1/2}$$

Equation (24) is identical with Eq. (9) as expected since the galvanostatic and potentiostatic stationary states must be the same. At low η when $i\Omega/i_0 \ll 1$

$$\eta_\infty = \frac{RT}{F(\alpha_c + \alpha_a)} \frac{i}{i_0} \Omega \tanh \Omega \tag{25}$$

This is quite different from the approximate Eq. (19). Rangarajan has similarly considered the nonstationary solution of Eq. (16). When it can be shown that under conditions favorable to surface diffusion, the rise time

$$\tau_g \approx 2l\sqrt{\frac{zFC_a^0}{i_0 D_a}} \tag{26}$$

when Ω is large. This is at variance with τ_g given by Bockris and Damjanovic

$$\tau_g = \frac{2C_a^0}{v_0} \tag{27}$$

It is possible to correct the theory by retaining Eqs. (27) and (22) and redefining

$$\frac{v_0}{C_a^0} \sim \frac{k_2}{\Omega} \tag{28}$$

The justification of ignoring bulk diffusion depends on the validity of the Sand equation

$$\frac{C_0}{C^*} = 1 - \frac{2t^{1/2}i}{\pi^{1/2}ZFC^*D^{1/2}} = 1 - \frac{t^{1/2}}{\tau_c^{1/2}} \tag{29}$$

that is, $\tau_g < \tau_c$. Unfortunately the theoretical interpretation of the rise time given above depends on Ω (that is, i_0) being large, so it is by no means certain that bulk diffusion does not play a role; unfortunately no reliable data are available for the magnitude of i_0, especially for the silver system which has been most investigated.

It might be possible to obtain i_0 data by depositing silver into mercury ($E°$ for Hg/Hg^+ is 10 mV cathodic to Ag/Ag^+) under some conditions but as far as we know this has not been attempted.

D. Surface Roughness

On a real electropolished metal surface the microscopic features with which the adatom model is concerned will be superimposed on a macroscopic structure. The literature on the current distribution problem to various macroscopic profiles has been thoroughly reviewed by de Levie (*38*). From the point of view of metal deposition it is important to establish when this will interfere. It can be roughly estimated from the dimensions of the surface irregularity for if it is supposed that the surface undulates with a characteristic height b then, in a potentiostatic diffusion-controlled experiment, surface structure will be important when $b \gg (Dt)^{1/2}$. Surface roughness will show ($t = b^2/D$) when $b = 0.1$, 1, 10 μ at times less than 10^{-5}, 10^{-3}, 10^{-1} sec in a potentiostatic experiment, or for frequencies greater than 10^5, 10^3, 10 Hz, in an impedance measurement. When $(Dt)^{1/2}$ is small, the current is proportional to the geometric area and when $(Dt)^{1/2}$ is large, the current is proportional to the apparent area.

E. Adatom Model. Layer-by-Layer Growth Controlled by Surface Diffusion

The two-dimensional model of Fig. 3, in which neighboring step lines compete for material is only realistic for short times of polarization, or for predicting the dependence of current on distance between the steps. If the steps are allowed to move, then the whole assembly of steps must be considered. Figure 6 shows a series of steps formed in principle by (a) the outward growth of a screw dislocation, (b) a progressive two-dimensional nucleation of layers on top of layers, and (c) growth of steps already present on a vicinal face. It is assumed in this section that the steps are of monomolecular

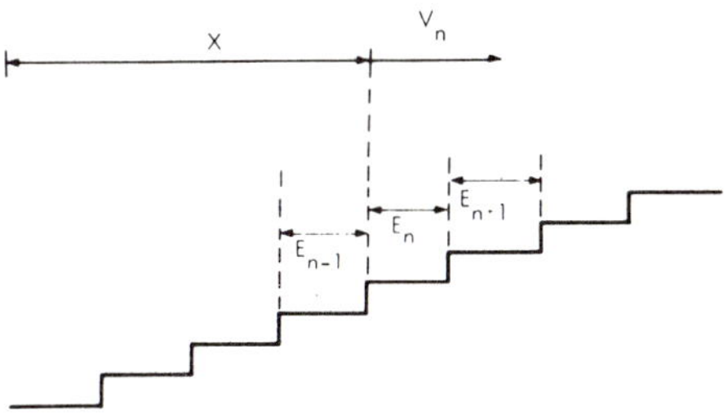

Fig. 6. Layer-by-layer growth controlled by adatom diffusion. E_n is a spacing between steps, V_n is a velocity. Ref. (*41*).

height, although clearly the model will also apply to larger steps introduced, for example, by deliberately misorientating the crystal near to a low index plane (see Sec. I). The problem of the stability of the steps and how they bunch together has been considered (*39*) by the method of Lighthill and Whitham (*40*); the theory treats a continuum model similar to that of the flow of traffic on highways. It is assumed that the surface can be represented by a step density versus distance profile and that dissolution only depends on the step density; the theory then describes the change in step density with time. An improvement of this theory to take account of the microscopic nature of the surface has been published by Mullins and Hirth (*41*) and will be described here. The velocity V_n of outward growth of an individual step will increase with the distance ε to either adjacent step.

Thus, in general,

$$\begin{aligned} V_1 &= f(\infty) + f(\varepsilon_1) \\ V_2 &= f(\varepsilon_1) + f(\varepsilon_2) \\ V_n &= f(\varepsilon_{n-1}) + f(\infty) \end{aligned} \tag{30}$$

In particular $f(\varepsilon)$ could have the form of Eq. (9); that is,

$$f(\varepsilon) = A \tanh \frac{\varepsilon}{\delta_{\mathrm{m}}} \tag{31}$$

where

$$A = V_{\mathrm{max}}/2 \qquad \text{and} \qquad V_{\mathrm{max}} \tag{32}$$

is the velocity of an isolated step, δ_{M} the root mean square distance for surface diffusion of an adatom. The authors assume only that $f(\varepsilon)$ is a monotonically increasing function of ε. They show that an evenly spaced, infinite train of steps will start pairwise grouping from the rear and spreading from the front; the effect of adsorbed impurities has also been considered. A more complicated model in which random two-dimensional nucleation and step motion combine has been investigated by Bertocci (*42*) in a digital computer simulation. The use of this method is described in Sec. IVD for the dissolution of copper.

F. Adatom Model. AC Method with Bulk Diffusion

The complete impedance can be directly calculated from Eqs. (1) and (6) by multiplying by p, substituting $j\omega$ for p, and assuming that the current potential characteristic can be linearized about the equilibrium potential.

If a potential $\Delta\eta = \Delta\eta \cos \omega t$ is applied to the cell, then the impedance Z_f is given, from Eq. (1), by

$$\frac{1}{Z_f} = \frac{1}{R_t}\left\{\frac{j\omega\sqrt{j\omega D}}{x'_{00}(j\omega)}\left[1 + \frac{k\sqrt{j\omega D}}{x'_{00}(j\omega)S'_1(j\omega)}\right]\right\} \tag{33}$$

where

$$j = \sqrt{-1}; \qquad R_t = \frac{RT}{zFi_0}$$

$$k'_1 = \frac{i_0}{zFC_0}; \qquad k'_2 = \frac{i_0}{zFC^0_a}$$

and the definitions of X'_{00}, S'_1 are the same as Eqs. (1) and (4) with k_1, k_2 replaced by k'_1, k'_2. A similar expression can be written from Eq. (6) when lattice growth is slow,

$$\frac{1}{Z_f} = \frac{1}{R_t}\left\{\frac{j\omega\sqrt{j\omega D}}{x'_{00}(j\omega)}\left[1 + \frac{k_2\sqrt{j\omega D}}{x'_{00}(j\omega)(S'_1(j\omega) + j\omega l/k3)}\right]\right\} \tag{34}$$

The frequency dependence of Eq. (34) has been calculated numerically by Rangarajan (*43*) and the results compared with the less exact treatment of Lorenz (*44*), the latter treatment being of special interest in that it also included bulk diffusion. The main features of Eqs. (33) and (34) will be discussed here; the reader is referred to the original papers for the details of the calculation. It can be shown by substituting for X'_{00} and splitting off the first term of the series S'_1 that when adatom diffusion is important, Eq. (34) can be represented by a model shown in Fig. 7.

The double-layer capacity would be in parallel with this where

(a) A transfer resistance, $R_t = \dfrac{RT}{zFi_0}$

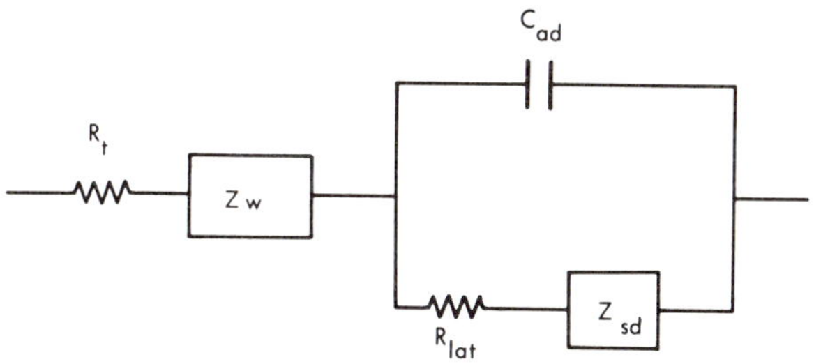

Fig. 7. Impedance Z_f of interface corresponding to Eq. (34). Ref. (*43*).

(b) A Warburg impedance, $Z_w = \dfrac{k_1}{z^2F^2C_0\sqrt{jwD}}$

(c) A pseudo capacity, $C_{ad} = RTz^2F^2C_a^0$

(d) A lattice transfer resistance, $R_{lat} = \dfrac{k_2}{k_3} l R_t$

(e) An adatom diffusion impedance,

$$Z_{Sd} = 2Rt \sum_1^\infty \frac{(k_1 + \sqrt{(p + 1/\tau m)D})k_2}{[(k_1 + \sqrt{(p + 1/\tau m)D})(p + 1/\tau_m^a) + k_2\sqrt{(p + 1/\tau m)D}]}$$

The behavior of the series elements of this circuit and the phase angle ϕ has been calculated by putting trial parameters into the above. The reader is referred to the original paper for details (*43*).

Where bulk diffusion is important

$$D_a/l^2 \gg k_2;\ \omega < D_a/l^2$$

the equations become the usual Warburg impedance in series with the transfer resistance (Fig. 8).

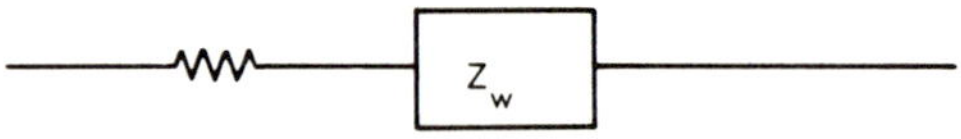

Fig. 8. Usual Warburg impedance for comparison with Fig. 7.

A classical Randles linear plot (*45*) (R versus $\omega^{-1/2}$) may still be obtained under some conditions even when surface diffusion and lattice incorporation effects are present, but in this case the intercept is altered and is not equal to R_t but is

$$\approx R_t\sqrt{\frac{k_2' l}{D_a}} \tag{35}$$

It can also be shown from Eqs. (33 and 34) that under many conditions it is impossible to decide if the slow step is directly at the edge or is a preceding surface diffusion step with fast incorporation at the edge.

G. Direct Deposition Model. Bulk Diffusion Only

Other types of mechanism besides the adatom route can be envisaged and indeed must be if a really satisfactory examination of the overall problem is to be adequately presented. Assuming that the reaction is entirely diffusion controlled, then depending on the distribution of surface features the lattice could grow by (a) diffusion from hemispherical zones to kinks in the step lines, (b) diffusion from hemicylindrical zones surrounding straight (to a first approximation) step lines when the number of kinks is large, or (c) diffusion from a planar layer when the number of step lines and kinks are large.

The equations for steady-state bulk diffusion to isolated sites in cases (a), (b) and (c) are well known, namely,

(a) $$i = nF2\pi r_0(C^* - C_0) \tag{36}$$

(b) $$i = \frac{nF\pi D(C^* - C_0)}{ln\ 1/r_0} \tag{37}$$

(c) $$i = \frac{nFD(C^* - C_0)}{\delta} \tag{38}$$

Eqs. (14) and (36), (13) and (37) have been compared numerically (*46*). With $D_a = 10^{-6}$ cm^2 sec^{-1}, $D = 10^{-5}$ cm^2 sec^{-1}, $k_2 = 10^6$ sec^{-1}, $L = 10^{-5}$ cm, $r_0 = 2°$Å, the calculation suggests about an order of magnitude in favor of direct deposition. The uncertainty in the parameters (*49*), however, and the unrealistic assumption of totally isolated sites makes the basis of this calculation rather indefinite.

More detailed investigations have been given by other authors. Vermilyea (*28*) has calculated, by the method of Burton, Cabrera, and Frank (*16*), the stationary behavior assuming that steps (a), (b), and (c) occur in series. Equating the fluxes at the boundaries between the different types of diffusion leads to

$$i = \frac{zFDC^0[1 - \exp(-zF\eta/RT)]}{[\delta - y_0/2 + y_0/\pi \ln y_0/x_0 + x_0 y_0/\pi(1/r_0 - 2/x_0)]} \tag{39}$$

This equation could, in principle, be compared with experimental data. However, a more exact calculation, which could be used, is given below.

Fleischmann and Thirsk (*29*) have compared the steady-state adatom model of Sec. IIIB with the cylindrical diffusion model taking a finite electrochemical rate. This analysis suggested that when

$$\left(\frac{k}{D}\right)^{1/2} l < 100 \tag{40}$$

surface diffusion is possible. Absence of data for i_0, D_a, C_a^0, however, make it impossible to apply this criterion to metal systems. Control by hemicylindrical diffusion implies that the linear Nernst diffusion layer, which exists further out in the solution, must be thin. By equating the fluxes due to hemicylindrical and linear diffusion, complete control by linear diffusion is expected when $l \sim \delta$. Figure 9 describes in more detail (*33*) a model of the

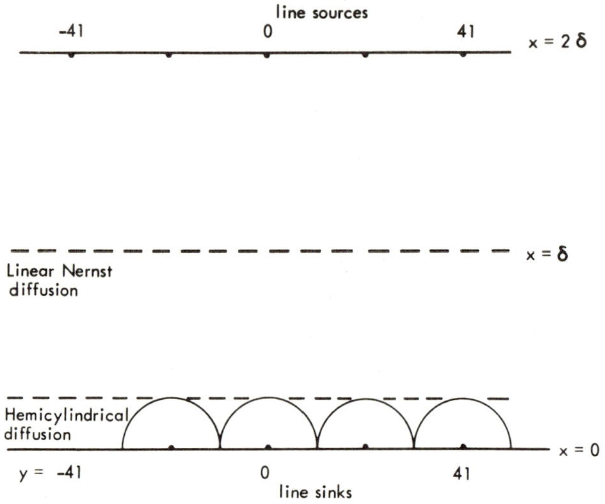

Fig. 9. Model of the discrete nature of the electrode surface at $x = 0$. The sinks are separated by impervious material. The associated row of sources serves to keep the concentration contant at $x = \delta$, the outer boundary of the Nernst diffusion layer. Ref. (*33*).

surface when direct deposition is taking place. An equivalent mathematical model is shown in Fig. 10. If the behavior of one sink is known, then by summing the sinks by the method of Casper (*47*)—that is, by Fig. 10—the behavior of Fig. 9 can be calculated. Considering a single-line sink the concentration distribution, generated at time *t*, after the sink is formed is

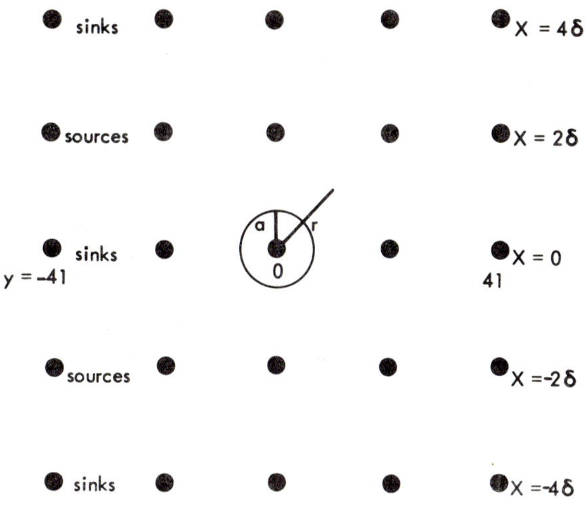

Fig. 10. Mathematical model equivalent to the situation of Fig. 9. The system extends to infinity along the x and y axes. Ref. (*33*).

given by

$$C = C_1 - \frac{Q}{4\pi D} \int_{r^2/4Dt}^{0} \exp(-u)\, du \tag{41}$$

where Q is the strength of the sink and C_1 is the uniform concentration at $t = 0$. By definition of the exponential integral $E_i(x)$,

$$C = C_1 + \frac{Q}{4\pi D} E_i\left(-\frac{r^2}{Dt}\right) \tag{42}$$

at long times and small x,

$$E_i(-x) = \gamma + \ln x \tag{43}$$

where γ is Eulers constant, then

$$C = C_1 + \frac{Q}{4\pi D} \ln \frac{r^2}{4Dt} + \frac{\gamma Q}{4\pi D} \tag{44}$$

Equations (42) or (44) could be used to calculate the time dependence by summing over an appropriate number of sinks in Fig. 10 and this could be compared with Fig. 5, the exact adatom model. The calculation has only been attempted in the stationary state and the changes in concentration in the hemicylindrical and linear diffusion regions calculated. The result is

$$\left(\frac{\Delta\eta}{\Delta\eta_{\mathrm{Nernst}}}\right)_{\eta_{\mathrm{small}}} = \frac{\Delta C}{\Delta C_{\mathrm{Nernst}}} \tag{45}$$

$$\frac{\Delta C}{\Delta C_{\mathrm{Nernst}}} = \frac{l}{\delta\pi}\ln\frac{16\,\delta^2}{\pi^2 r_0} + \frac{2l}{\delta\pi}\sum_1^{\infty}\ln\left\{\frac{1+\cosh \pi nl/\delta}{\cosh \pi nl/\delta - 1}\right\} \tag{46}$$

where ΔC is the total concentration change in the solution Equation (46) is shown in Fig. 11. The value observed in Gerischer's silver experiment de-

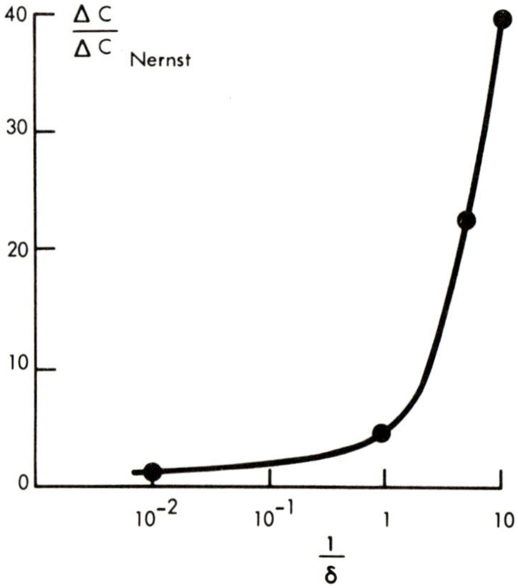

Fig. 11. The relative concentration change as a function of $1/\delta$ at constant apparent current density. Ref. (*33*).

scribed in Section IIIB is $\eta_{\mathrm{adatom}}/\eta_{\mathrm{Nernst}} = 1.5$. (The galvanostatic work of Bockris and Mehl (*48*) and the ac measurements of Lorenz (*44*) are essentially equivalent but not so convenient for comparison with theory.)

Figure 11 shows that a value $\Delta C/\Delta C_{\mathrm{Nernst}} = 1.5$ is equally compatible with a direct deposition model for reasonable values of the parameters δ, l. The exact features on the surface that determine the current are not known.

The attempts in this section to calculate the distribution of material about sites are rather crude. Intuitively we might expect the adatom model of Sec. IIIA always to degenerate, after first setting up a stationary adatom distribution, into some form of direct diffusion to the step. This proposition has not yet been investigated theoretically, but would be possible by extending the theory given in Secs. IIIG and IIIA.

H. Nucleation and Growth Models

1. Potentiostatic; Slow Step at the Edge without Diffusion

In the preceding sections it has been assumed that the metals grow at step lines in dislocations. Although this is true for real metal surfaces, there are conditions under which a metal phase will build up by nucleation and growth processes. Not many of these equations have been applied to metals, although there are many situations that demand this type of analysis for example, (a) the early stages of deposition of metals on inert substrates, (b) the growth of metals on substrates of the same metal containing few dislocations, and (c) the growth of the same metals, as the substrate, at specific sites on the surface (dendrite formation).

If the kinetics are controlled by nucleation and growth, the information that can be obtained about the slow step is substantial. The subject has been reviewed in detail fairly recently (*3*). In this section only a brief account will be given together with some more recent results. The equations apply generally to the formation of a metal or nonmetal phase. It is convenient to consider two limiting cases of growth two-dimensional (cylinders) and three dimensional (hemispheres). The free energy of formation of one surface is clearly less than that required to form two new surfaces. On this ground it is to be expected that three-dimensional lattice growth occurs layer by layer. The consequences of this are analyzed below. On this argument only the various two-dimensional models would have any fundamental significance. However, it is convenient to be able to interpret experimental results in terms of three-dimensional models where growth of this type occurs. The rates under these conditions must be such that the detailed atomic arrangement of the three-dimensional surface can be ignored.

a. Two-Dimensional Nucleation and Growth. The calculation of characteristic potentiostatic current-time curves in the early stages of growth before the nuclei have overlapped, based on simple geometrical arguments, is as follows. For one nucleus

$$i = zFkS$$

$$= \frac{\rho z F}{M} \frac{\partial V}{\partial r} \frac{\partial r}{\partial t} \tag{47}$$

where V is the volume of a nucleus, S is the area of a nucleus, k is the rate constant (moles cm^{-2} sec^{-1}). Equation (47) gives r as a function of t, (as $\partial V/\partial r$ and S are both known functions of the geometry) and consequently gives $i_{(t)}$. Multiplication by N_0, the number of nucleation sites, gives $i_{(t)}$ for instantaneous nucleation. The expression for progressive nucleation of the nucleation sites in the form

$$N = N_0[1 - \exp(-At)] \tag{48}$$

can be calculated from

$$i = \int_0^t i(u)\left(\frac{dN}{dt}\right)_{t=t-u} du \tag{49}$$

where $i(u)$ is given by Eq. (47) and dN/dt by Eq. (48) with the substitution $t = (t - u)$. A is a nucleation rate constant; for small values of A Eq. (48) can be approximated to $N = N_0 At$.

Table I gives a summary of expressions for an unbounded electrode, calculated on this basis, unless otherwise stated using the simplified nucleation law.

The dependencies are all of the form $i \propto t^{\beta}$ and are rising transients in which β is a constant depending primarily on the geometry and type of nucleation. From the slope a function containing k and N_0 for instantaneous nucleation and k and A for progressive nucleation can be calculated. It has been suggested that the potential dependence of k alone can be investigated by nucleating with a large square pulse and growing at a small overpotential where nucleation is minimal. A better method is to study the relative potential dependence of k by instantaneously pulsing the potential at a fixed stage of growth of the layer to a new potential: A can in principle be measured electron optically.

Relations of the type, given in Table I, are also known in the growth of precipitates from aqueous solution, aerosols, and the vapor phase and phase

TABLE I

CURRENT INTO ISOLATED NUCLEI (WITHOUT OVERLAP)

t	$t^{1/2}$	$t^{3/2}$	t^{0}
(1) $i = \frac{2zF\pi M}{\rho} N_0 h k^2 t$	(4) $i = \frac{zF\rho}{M} N_0 \pi \theta'^3 D^{3/2} t^{1/2}$	(5) $i = \frac{zF\rho}{M} (\pi\theta'^3 D A^{3/2}) \frac{2}{3} t^{3/2}$	(6) $i = \frac{zF\rho}{M} \pi\theta^2 D$
(a) Instantaneous	Instantaneous	Progressive	Instantaneous
(b) $2D$	$3D$	$3D$	$2D$
(c) Periphery	Periphery	Periphery	Periphery
(d) Fast	Slow	Slow	Slow
(2) $i = zFAL^2 kt$			
(a) Progressive			
(b) $1D$ needle			
(c) End (cross section L^2)			
(d) Fast			
(3) $i = \frac{zFh\rho}{M} \pi\theta^2 DAt$			
(a) Progressive			
(b) $2D$			
(c) Periphery			
(d) Slow			

TABLE I—*continued*

t^3	t^2	$\exp(t)$
(7) $i = \dfrac{2zF\pi M^2}{3\rho^2} Ak^3t^3$	(8) $i = \dfrac{2zF\pi M^2}{\rho^2} N_0k^3t^2$	(10) $i = zFk\dfrac{4LM}{\rho}\left[\exp\left(\dfrac{4kM}{L\rho}(t - t_0) + \ln t_0\right)\right]$
(a) Progressive	Instantaneous	Instantaneous
(b) $3D$	$3D$	$1D$ (needle cross section L^2)
(c) Periphery	Periphery	Round long axis (r_0 ht. at t_0)
(d) Fast	Fast	Fast
	(9) $i = \dfrac{zF\pi M}{\rho} hAk^2t^2$	(11) $i = \dfrac{zFkM}{\rho} 2\pi x_0 \left[\exp\left(\dfrac{kM}{\rho} x_0(t - t_0) + \ln r_0\right)\right]$
(a)	Progressive	Instantaneous
(b)	$2D$	$3D$
(c)	Periphery	Base periphery into section x_0 high (r_0 is radius of hemisphere at t_0, $r_0 > x_0$)
(d)	Fast	Fast

(a) Nucleation type.
(b) Growth type.
(c) Site of slow step.
(d) Diffusion.

θ, θ', see Eq. (79), (92)

changes during solid reactions. Electrochemically these equations have been investigated mainly for the formation of nonmetallic films where the slow step appears to be at the edge of the growing nucleus. This seems to be because a complex lattice that involves anions and cations requires a rearrangement at the kink sites before development.

When nuclei overlap, the form of the current-time curve can still be calculated exactly provided that overlapping is random, the growth outwards is uniform and the boundaries of the electrode surface do not interfere. In this situation the Avrami Eq. (50a) allows the microscopic growth parameter for a single nucleus (S_{ex}) to be related to the system as a whole. The equation is full discussed in Ref. (*3*). Using the Avrami (*49*) equation

$$S = 1 - \exp(-S_{ex}) \tag{50a}$$

$$i_{(t)} = zFkS \tag{50}$$

together with Eq. (47), where S_{ex} is the total area of nuclei without overlap, $i_{(t)}$ can easily be calculated. Some values are shown in Table 2 for cylinders.

TABLE II

CURRENT CALCULATED WITH OVERLAP

(1) $i = \dfrac{zF\pi M}{\rho} hAk^2t^2 \exp - \dfrac{\pi M^2Ak^2t^3}{3\rho^2}$

(a) Progressive
(b) 2*D*
(c) Periphery
(d) Fast

(2) $i = \dfrac{zF\pi M}{\rho} N_0k^2t \exp - \dfrac{\pi M^2N_0k^2t^2}{\rho^2}$

(a) Instantaneous
(b) 2*D*
(c) Periphery
(d) Fast

(a) Nucleation type.
(b) Growth type.
(c) Site of slow step.
(d) Diffusion.

Two commonly occurring expressions (Eqs. (1) and (2), of Table II) are plotted in Figs. 12 and 13 in terms of nondimensional quantities; that is,

$$\frac{i}{i_m} = \frac{t}{t_m} \exp\left[-\frac{1}{2}\left(\frac{t^2 - t_m^2}{t_m^2}\right)\right]$$

$$\frac{i}{i_m} = \left(\frac{t}{t_m}\right)^2 \exp\left[-\frac{2}{3}\left(\frac{t^3 - t_m^3}{t_m^3}\right)\right]$$

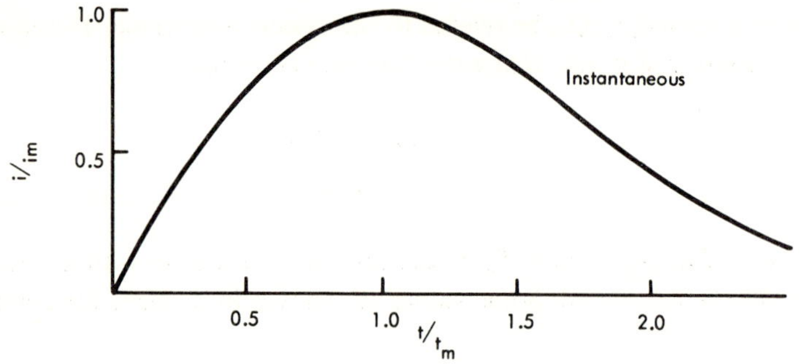

Fig. 12. Theoretical nondimensional curve for instantaneous nucleation of a two-dimensional layer.

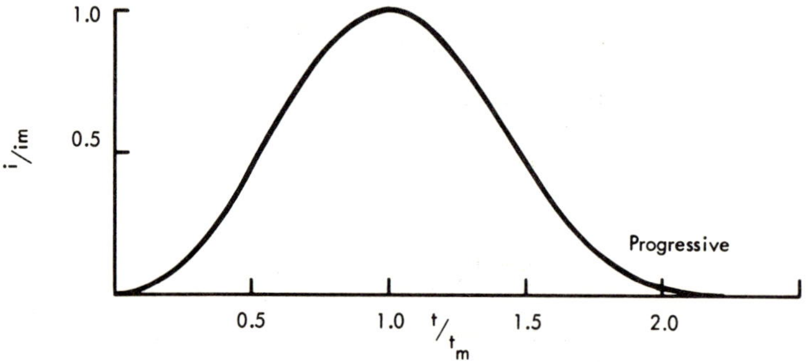

Fig. 13. Theoretical nondimensional curve for the progressive nucleation of a two-dimensional layer.

for instantaneous and progressive two-dimensional nucleation, respectively. The experimental $i - t$ plots will have a similar shape. A nondimensional plot is a convenient way of comparing theory and experiment. The variation of k^2A with potential and concentration can be investigated by plotting log i_m or log t_m against potential. For progressive nuclei it can be shown by calculating i_m and t_m from Eqs. (1), (2), of Table II, that

$$\frac{\partial \log i_{\mathrm{m}}}{\partial \eta} = \frac{1}{3} \frac{\partial \log Ak^2}{\partial \eta}$$

$$\frac{\partial \log t_{\mathrm{m}}}{\partial \eta} = -\frac{1}{3} \frac{\partial \log Ak^2}{\partial \eta}$$

The expected value of $i_m t_m$ (which is invariant with potential for this model) serves as a diagnostic test for a two-dimensional mechanism and is given in Fig. 25 and referred to again in Sec. IIIH 4a.

b. Two-Dimensional Layer-by-Layer Growth. The equations in Table II are for the formation of a single layer. The more general problem of layer-by-layer growth can be attempted in a similar manner (*50*). Consider a patch of the nth layer (dS_n) formed at time u which will generate a current due to the succeeding layer at time t given by

$$di_{n+1} = q_{\mathrm{mon}}\, 3\beta_{n+1}(t-u)^2 \exp[-\beta_{n+1}(t-u)^3]\, dS_n \tag{51}$$

so that the total current due to the formation of the $(n + 1)$ layer

$$i_{n+1} = q_{\mathrm{mon}} \int_0^t 3\beta_{n+1}(t-u)^2 \exp[-\beta_{n+1}(t-u)^3] \frac{dS_n}{du}\, du \tag{52}$$

$$i_{n+1} = \int_0^t 3\beta_{n+1}(t-u)^2 \exp[-\beta_{n+1}(t-u)^3] i_n\, du \tag{53}$$

since i is known from Table II, Eq. (1), then the current contribution of the successive layers and hence the overall current can be calculated by numerical desk calculator integration. Under the condition that is expected for metal deposition, Eq. (53) gives the curve shown in Fig. 14. The steady state, which is constant, is reached after the oscillations have died away. It is clear that evidence for two-dimensional growth can only be obtained in a transient experiment by pulsing from a potential where no reaction occurs into the deposition region. The results of Fig. 14 extend the earlier computer simulation (*51*) which was not taken for sufficiently long times. The present calculation does not substantiate the earlier supposition that the curves depend independently on V, A that is on the rate of advance of the edge and the nucleation rate.

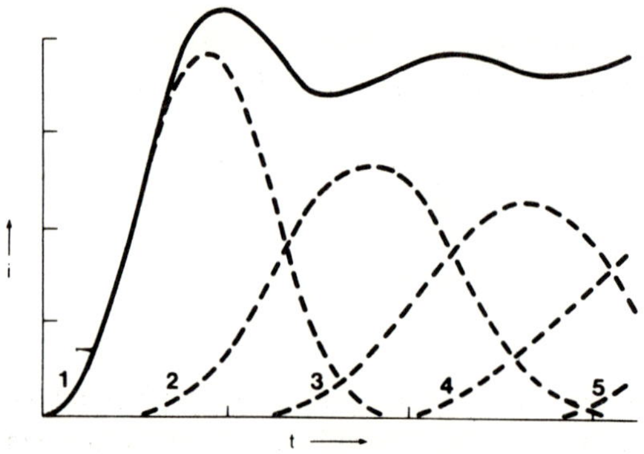

Fig. 14. Expected current-time curve (full line) for two-dimensional layer by layer growth. The numbered dashed lines refer to the contribution from successive monolayers. Ref. (*50*).

c. Three-Dimensional Nucleation and Growth. The problem can be attacked (*52*) in the manner of Fig. 15 where the rates of advance V_1, V_2, V_3 are defined. A slice x from the surface and height dx will grow out horizontally as Eq. (2) of Table II, the assumption being that the nuclei are distributed at random on the surfaces and the interaction of slices at a height x can be described by the Avrami equation. A particular slice will be created at a time x/V_2. Therefore for the slice dx and a constant number of nuclei N_0,

$$di = \frac{2nF\pi}{\rho} N_0 M k_1^2\left(t - \frac{x}{V_2}\right)\exp\left[-\frac{\pi}{\rho^2} N_0 k_1^2 M^2\left(t - \frac{x}{V_2}\right)^2\right] dx \tag{54}$$

and on integration,

$$i = \frac{nF\rho}{M} V_2\left[1 - \exp\left(-\frac{\pi}{\rho^2} N_0 k_1^2 M^2 t^2\right)\right] \tag{55}$$

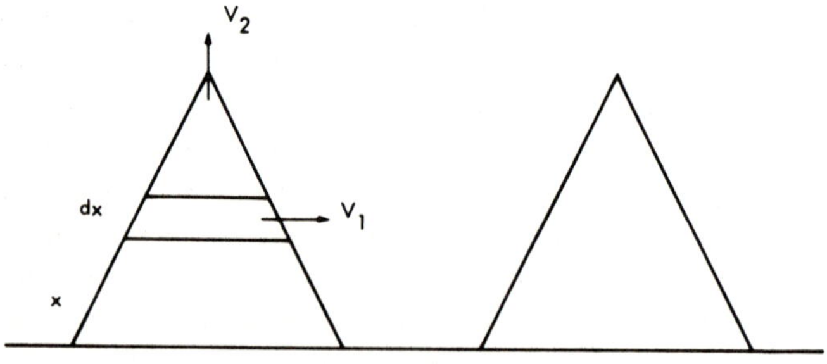

Fig. 15. Model for three-dimensional nucleation and growth of centers. Base plane is fixed.

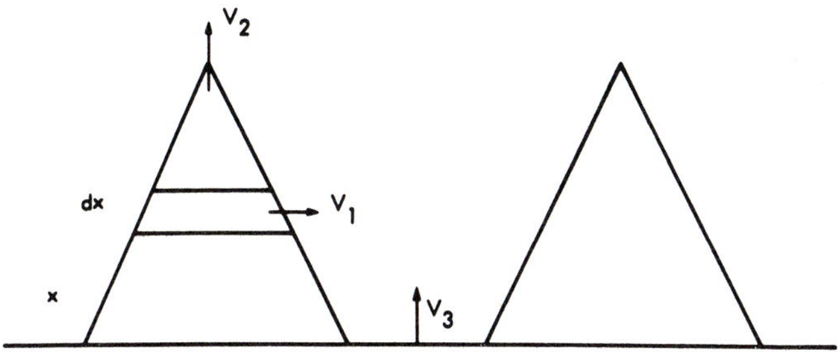

Fig. 16. Model for three-dimensional nucleation and growth. Base plane grows at a different rate to the cones.

similarly progressive nucleation by $N = At$ gives

$$i = \frac{nF\rho}{M} V_2\left[1 - \exp\left(-\frac{\pi M^2 k_1^2 A t^3}{3\rho^2}\right)\right] \tag{56}$$

Equation (56) is shown in Fig. 17. A similar calculation can be effected for hemispheres.

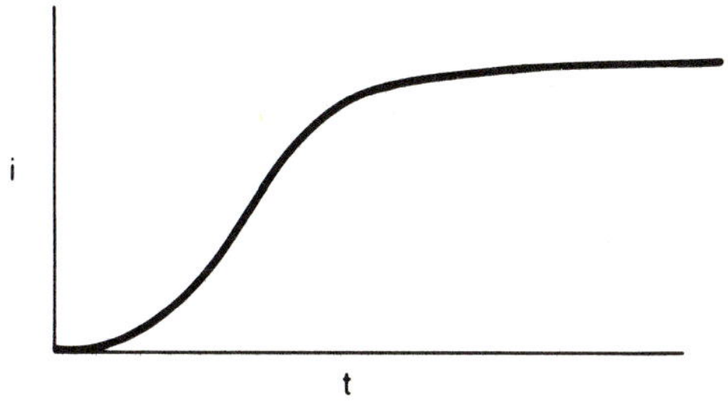

Fig. 17. Expected current-time transient for the model of Fig. 15 and Eq. (56). (Diagrammatic.)

At $t = 0$, $i = 0$ the current in Eq. (56) rises with a characteristic time dependence $i \propto t^3$, then it reaches a stationary stage. Figure 17 is to be compared with Fig. 14 in which the three-dimensional lattice is formed by layer growth. The value of the steady state is the same in both cases.

Equation (56) describes the growth at particular sites and can be extended by assuming that deposition also occurs on the base plane as in Fig. 16. This would be expected when some surface feature causes a cone to form. The rest of the surface has sufficient growth sites for an essentially smooth growth.

Consider the current in two parts, i_s into the smooth surface and i_p into the cone. The effective increase in extended area of the base of the cone is

$$S_{ex} = N_0 \pi y^2 = N_0 \pi v_1 \left(1 - \frac{V_3}{V_2}\right)^2 t^2 \tag{57}$$

using Avrami's method to obtain the real area at any time

$$S = 1 - \exp(-S_{ex}) \tag{58}$$

the current i_s is given by

$$i_s = \frac{nFV_3 \rho}{M}(1 - S) \tag{59}$$

$$= \frac{nFV_3 \rho}{M} \exp\left[-\pi V_1^2 \left(1 - \frac{V_3}{V_2}\right)^2 t^2\right] \tag{60}$$

i_p is calculated as before Eq. (54) except this time the slice at height x measured from the surface at $t = 0$, only has a life between $x = V_2 t$ when it is formed and $V_3 t$ when it dies.

$$i_p = \int_{V_3 t}^{x} 2nF \frac{\pi}{\rho} N_0 M k_1^2 \left(t - \frac{x}{V_2}\right) \exp\left[-\frac{\pi}{\rho^2} N_0 k_1^2 M^2 \left(t - \frac{x}{V_2}\right)^2\right] dx \tag{61}$$

$$i_p = \frac{V_2 nF\rho}{M} \left\{1 - \exp\left[-\pi N_0 V_1^2 \left(1 - \frac{V_3}{V_2}\right)^2 t^2\right]\right\} \tag{62}$$

The total current i is given by the sum of Eqs. (60) and (62)

$$i = i_p + i_s \tag{63}$$

Figure 18 shows the expected time dependence. At $t = 0$, $i = nFV_3 \rho/M$ and as $t \to \infty$, $i = nFV_2 \rho/M$.

d. Potentiostatic; Slow Step at the Edge Limiting Growth. Two types of limitation have been considered: (a) that nuclei only grow to a certain size and (b) that the value of k, the growth constant, can be limited by a preceding chemical reaction. A further possibility, that of limitation of k by diffusion, is considered in Sec. IIIH4. Case (a) might be observed in practice if adsorption of organics is in competition with crystal growth, or if some physical process such as sites of limited size curtails growth. Assuming an unbounded

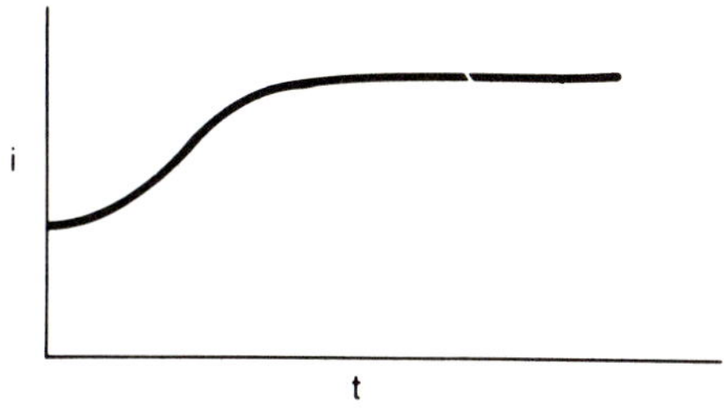

Fig. 18. Behavior of current as a function of time according to the model of Fig. (16) and Eqs. (63), (62) and (60). (Diagrammatic.)

electrode, the current can be calculated as follows (*53*) for two-dimensional progressive nucleation,

$$i_{t \leqslant t'} = \frac{zF\pi Mhk^2At^2}{\rho} \exp\left(-\frac{\pi M^2k^2At^3}{3\rho^2}\right) \tag{64}$$

If t' is the time for the first nucleus to reach its maximum size; that is, for any nucleus $u = t'$ is limiting age. For times $t > t'$ then a single center without overlap, Eq. (3) of Table I with Eq. (49), gives

$$i = \int_0^{t'} \frac{2zFM\pi hk^2A}{\rho} u\, du + \int_{t'}^{t} \frac{2zF\pi Mhk^2A}{\rho} u\, du \tag{65}$$

as, $t > t'$, $i = 0$

$$i = \frac{zF\pi Mhk^2At'^2}{\rho} \tag{66}$$

Using the Avrami theorem, Eq. (50) leads to

$$i_{t > t'} = \frac{zF\pi Mhk^2At'^2}{\rho} \exp\left[\frac{\pi M^2k^2A}{\rho^2}\left(\frac{2}{3}t'^3 - t'^2t\right)\right] \tag{67}$$

At $t = t'$ Eqs. (67) and (64) are equal and i reaches a maximum. Equations (67) and (64) are plotted in Figs 19–21. Instantaneous nucleation would obey

$$i = \frac{2N_0 zFM\pi hk^2t}{\rho}$$

until

$$t' = \frac{d\rho}{Mk}$$

then the current would fall to zero. It has been suggested (*53*) that Eqs. (64) and (67) would represent approximately the behavior of a finite electrode in which interaction of the two-dimensional centers with the boundary would be significant. A test, by simulation on graph paper in the manner of Sec. IIIH 4a, shows a reasonable correspondence (shown in Figs. 19–21 for a variety of data) provided $t' < t_m$, where t_m is the position of the maximum current for unrestricted growth. The calculation of current time curves for one nucleus interacting with a boundary is described in Sec. IVA1.

Case (b) in which k is limited by preceding chemical reaction can be calculated (*53*) by considering the reaction occurring in the solution

$$MX^+ \underset{k_{-1}}{\overset{k_1}{\rightleftharpoons}} M^+ + X \tag{68}$$

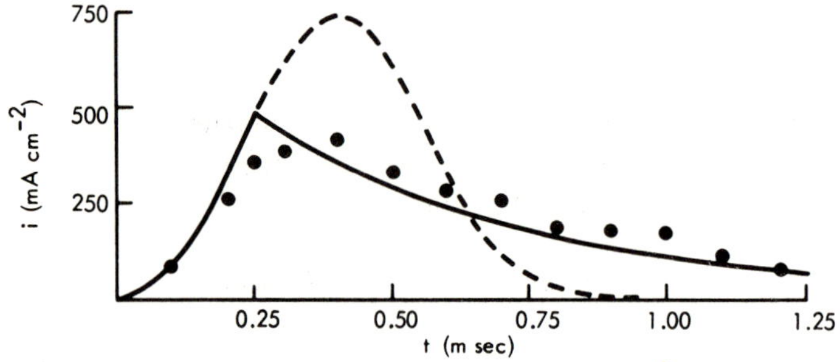

Fig. 19. Current-time transients for limitation of growth;
——— theoretical according to Eqs. (64), (67).
- - - - progressive nucleation Eq. (1) of Table II.
● simulated. Ref. (*53*).

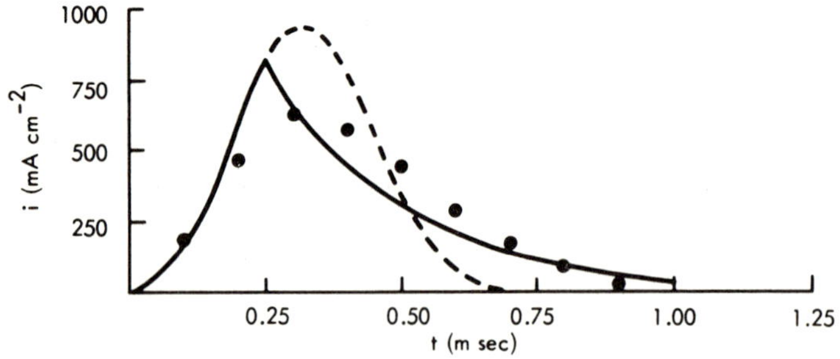

Fig. 20. Current-time transient for limitation of growth of a two-dimensional layer. Symbols as in Fig. 19. Ref. (*53*).

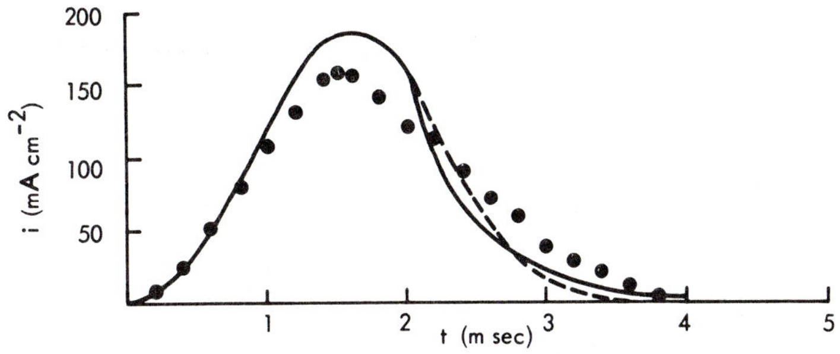

Fig. 21. Current-time transient for limitation of growth of two-dimensional layer. Symbols as in Fig. 19. Ref. (*53*).

The metal lattice is formed by reduction of M^+. The reaction layer x is given by

$$x = \sqrt{D\tau} \tag{69}$$

$$x = \left(\frac{D}{k_{-1}C_{M^+}}\right)^{1/2} \tag{70}$$

and the current i, for a linear reaction zone, is given by

$$i_{m'} = Fk_1 C^0_{MX^+}\, x \tag{71}$$

$$= FD^{1/2}K(k_{-1})^{1/2}\frac{C_{MX^+}}{(C_{M^+})^{1/2}} \tag{72}$$

where $K = k_1/k_{-1}$.

Similar equations can be written for hemispherical and hemicylindrical reaction zones. Experimentally a limiting value of $i_{m'}$ would be observed provided that A also reaches a limit. Equation (72) is to some extent unrealistic in that diffusive transport should also be considered (see Section 4a). The possibility of Eq. (68) occurring as a heterogeneous reaction on the surface has not been treated.

2. Galvanostatic Conditions; Slow Step at the Edge

Typical transients, which are well known for the nonmetallic film case, are shown in Fig. 22. These have been discussed by Vermilyea (*54*).

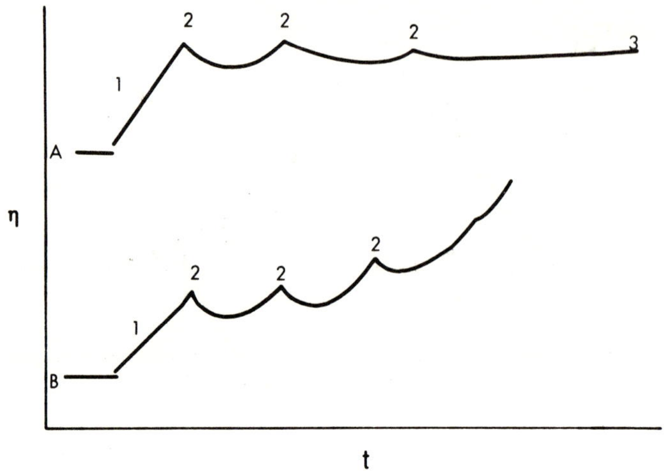

Fig. 22. The form of galvanostatic transients for the nucleation and layer by layer growth of (A) a metal on an electrode, (B) an anodic oxide film. 1, double layer charging; 2, high nucleation rate region; 3, steady state.

In the early stages Eq. (47) gives the form of the potential time curve directly

$$V(t) = \frac{Mk(t)}{\rho} = \frac{1}{2}\frac{I}{q_{\text{mon}}}t^{-1/2} = \frac{dr}{dt} \tag{73}$$

Some assumption must be made about the potential dependence of k to compare experiment and theory; the easiest would be to linearize the expected dependence.

$$k = k_0\left\{\exp\left(\frac{\alpha zF\eta}{RT}\right) - \exp\left[\frac{(\alpha - 1)zF\eta}{RT}\right]\right\} \tag{74}$$

The complete curve accounting for overlap can be calculated provided a constant number of nuclei is found; that is, following Avrami,

$$\frac{It}{q_{\text{mon}}} = 1 - \exp(-S_{\text{ex}}) \tag{75}$$

therefore

$$S_{\text{ex}} = -\ln\left(1 - \frac{It}{q_{\text{mon}}}\right) \tag{76}$$

also

$$S_{\text{ex}} = N_0\pi\left(\int_0^t \frac{Mk(t)}{\rho}\,dt\right)^2 \tag{77}$$

solving Eq. (77) gives

$$\frac{Mk(t)}{\rho} = \frac{1}{2(\pi N_0)^{1/2}} \frac{I}{q_{\mathrm{mon}} - It} \left[\ln \frac{q_{\mathrm{mon}}}{q_{\mathrm{mon}} - It} \right]^{-1/2} \tag{78}$$

This plot is shown in Fig. 23. The form of the experimental data of Fig. 22

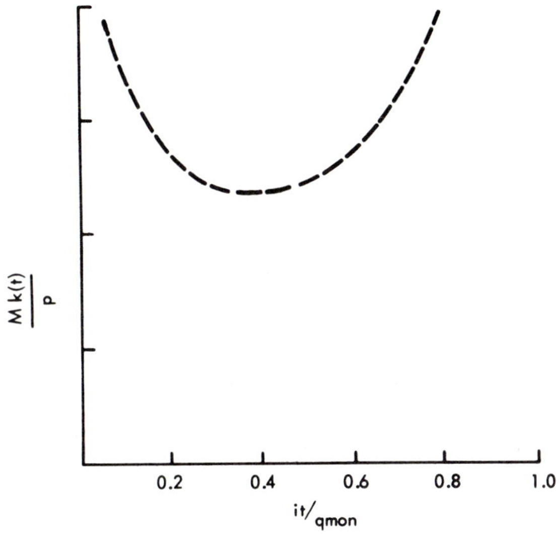

Fig. 23. Plot of Eq. (78). The theoretical rate of advance of an edge is plotted vertically against time. This would correspond to the growth of a single layer. Ref. (*50*).

can be understood by assuming that initially N_0 nuclei are formed instantaneously when η and, therefore k, is large. The steady state will be the same as that discussed in the section on layer-by-layer growth (Sec. 1b), so here also experiments must be used on an unbounded electrode and pulse. It is apparent that exact kinetic data is most easily obtained by a potentiostatic method. Curves of this shape have been observed for metal deposition (see Sec. IV2).

3. Sweep; Slow Step at the Edge

This method is in general very sensitive and probably the most effective way of picking out the qualitative features of a system. The advantage lies in being able to detect progressive, as opposed to instantaneous, nucleation and to test for bulk diffusion. However, the theory has not been worked out satisfactorily.

4. Potentiostatic Conditions with Diffusion

Considering two-dimensional growth, two cases can be calculated as follows:

(a) The two-dimensional center is surrounded by a fixed diffusion zone and the rate of advance of the edge is also fixed.

(b) At fixed potential the diffusion zone increases with time and the rate of advance of the edge is governed by the flux. These two cases will now be considered separately.

a. Fixed Diffusion Zone. A fixed diffusion zone could arise either by adatom diffusion, where the diffusion zone is set up by the electrochemical reaction on the bare surface, or by a preceding chemical reaction in the solution. This corresponds to control by the rate at the edge and by preceding processes; it is not possible to differentiate between the two, experimentally, on the evidence of current-time curves alone without further evidence. A diffusion zone increasing with time, described below, is properly due to two-dimensional metal deposition by diffusion in solution. This situation (*55*) has been simulated on large sheets of graph paper by advancing the nuclei and their surrounding diffusion zone (assuming two-dimensional square nuclei and diffusion zones) at a fixed rate and nucleating at random only in the completely uncovered area. The current at any stage of growth is proportional to the total peripheral length of nuclei. Some examples are shown in Fig. 24; when the diffusion layer is large, the curves fall with time; when the diffusion layer is thin, the current-time curves approach those predicted for slow step at the edge curves (Eqs. (1) and (2) or Table 2). The equations have the important property that $i_m t_m/Q_m$ is a constant, independent of potential, which has a calculated value 1.02 for progressive and 0.61 for instantaneous nuclei. However, when the stationary adatom diffusion zone thickness is larger than zero, $i_m t_m/Q_m$ deviates progressively from the values given above, as shown in Fig. 25. A metal system that shows these characteristics—that is, $i_m t_m/Q_m \ll 1$—and grows to a monolayer (of unit cell height) is nickel on mercury. The experimental conditions are described in Sec. IV.

b. Diffusion Zone Increasing with Time. The problem without overlap has been considered by Frank (*56*) for the growth of one two-dimensional nucleus expanding at a rate determined by the flux. The diffusion zone is concentric to the nucleus. The radius at time t is considered to obey

$$R(t) = \theta\sqrt{Dt} \tag{79}$$

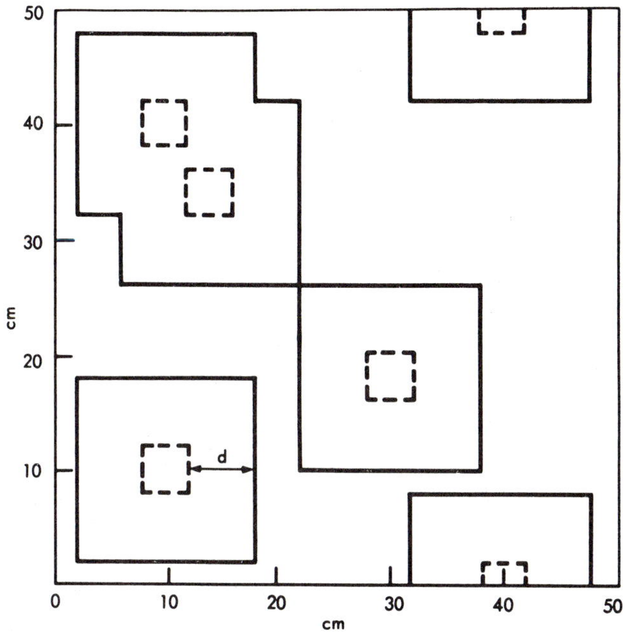

Fig. 24. A typical simulation diagram for instantaneous nucleation $N = 6$, $d = 3.0$ cm, $a = 2.5 \times 10^3$ cm². The dashed lines enclose the growing nucleus. The area between the full and dashed lines contains the diffusion zone. Ref. (*55*).

where θ is a constant given by the expression,

$$(C^* - C^0)\frac{M}{\rho} = \frac{1}{4}\theta^2 \exp\left(\frac{1}{4}\theta^2\right)E_i\left(-\frac{1}{4}\theta^2\right)$$

where E_i is the exponential integral that is a standard function; calibration graphs are given in the original paper. A more rapid but less exact determination of θ can be made by assuming the radius of curvature is large and linear diffusion parallel to the surface holds.

From Eq. (47),

$$R(t) = \frac{2M}{\rho}\frac{c^*}{(\pi)^{1/2}}\sqrt{Dt} \tag{81}$$

For instantaneous nucleation the Avrami treatment gives

$$i = q_{\text{mon}}\pi\theta^2 D \exp(-\pi\theta^2 D N_0 t) \tag{82}$$

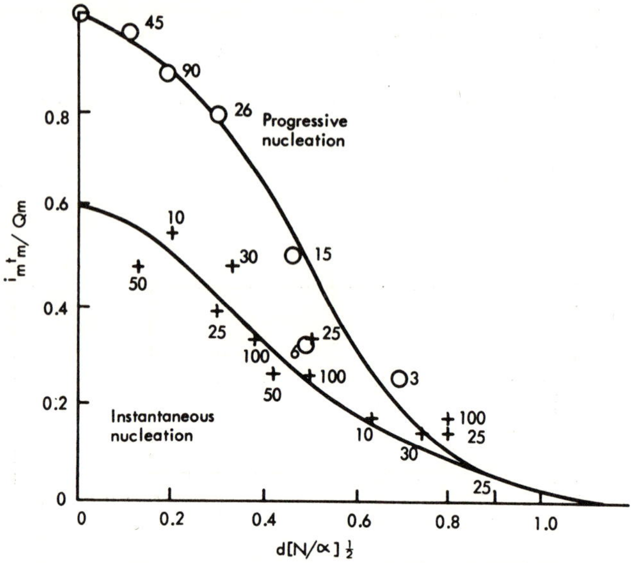

Fig. 25. $i_m t_m / Q_m$ as a function of dN/a. N the number of nuclei is given on the graph. $a = 2.5 \times 10^3$ cm^2. Ref. (*55*).

and for progressive nucleation it gives

$$i = q_{\text{mon}} \pi \theta^2 DAt \exp\left(\frac{-\pi\theta^2 DAt^2}{2}\right) \tag{83}$$

the forms of Eqs. (82) and (83) are shown in Figs. 26 and 27. Figure 26 shows a continuous fall in current compared with Fig. 27 which is similar to Fig. 12 and, experimentally, indistinguishable from it.

c. Diffusion Zones Overlapping. An alternative model to that for Eq. (79) might hold when a large number of very small nuclei are formed initially if it is assumed that the individual diffusion zones soon disappear and the rate of growth is controlled by planar diffusion perpendicular to the surface. The problem (*57, 58*) is similar to the adsorption of organics. The equations to be solved are

$$\frac{\partial c}{\partial t} = D \frac{\partial^2 c}{\partial x^2} \tag{84}$$

$$\frac{\partial a}{\partial t} = \left(\frac{A'D}{M}\right)\left(\frac{\partial c}{\partial x}\right)_{x=0} \tag{85}$$

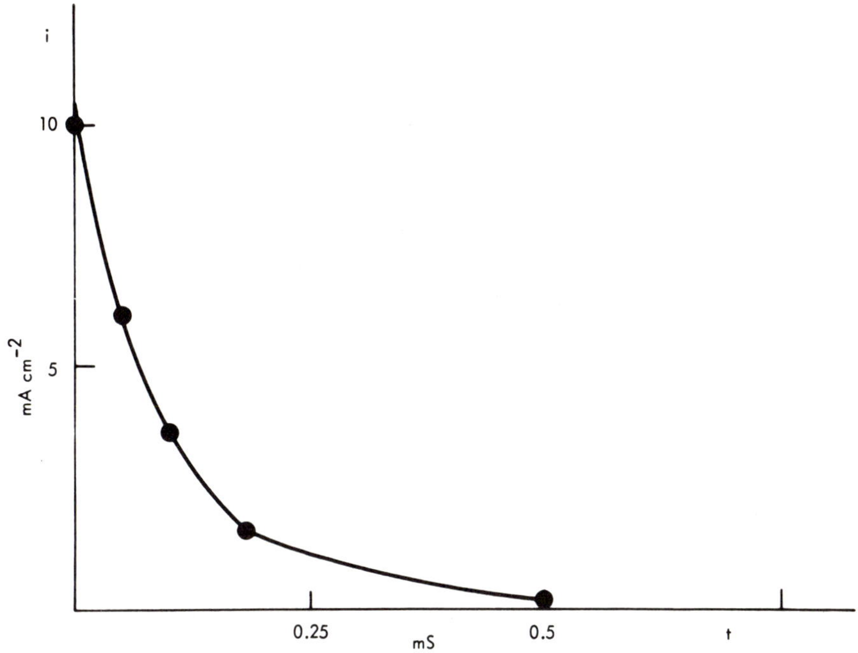

Fig. 26. Calculated current-time transient for two-dimensional growth under diffusion control according to Eq. (82), instantaneous nucleation. Ref. (*50*).

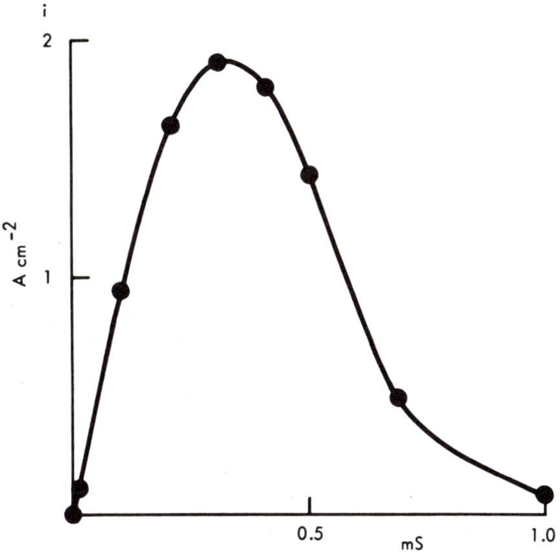

Fig. 27. Calculated current-time transient for two-dimensional growth under diffusion control according to Eq. (83), progressive nucleation. Ref. (*50*).

$$\frac{C_{x=0}}{C^*} = a(t)\exp\left(\frac{-nF\eta}{RT}\right) \qquad \text{at } x = 0 \tag{86}$$

$$\frac{C \to C^*}{C = C^*} \qquad \frac{x \to \infty}{t = 0} \tag{87}$$

where a is the activity of the deposit and m is the number of moles in the deposit at full coverage. The equations can be solved by the Laplace transform method to give

$$i = nFA'C^*\sqrt{\frac{D}{\pi t}} - \left[\left(\frac{nFA'^2C^{*2}D}{m}\right)\exp\left(-\frac{nF\eta}{RT}\right)\right.$$
$$\left.\exp\left\{\left[\frac{c^{*2}DA'^2}{m^2}\exp-\frac{2nF\eta}{RT}\right]t\right\}\operatorname{erfc}\left\{\left[\frac{c^*D^{1/2}A'}{m}\exp-\frac{nF\eta}{RT}\right]t^{1/2}\right\}\right] \tag{88}$$

which is shown graphically in Fig. 28. A' is the electrode area. Equation (88) does not distinguish between adsorbed adatoms and adatoms that have crystallized into two-dimensional nuclei. Experimental evidence for Eq. (88) is given in Sec. IV.

d. Three-Dimensional Growth. In a similar way the growth of hemispherical nuclei can be formulated; only the growth in the early stages has been calculated. If the radius is assumed to increase as before,

$$R = \theta'\sqrt{Dt}$$

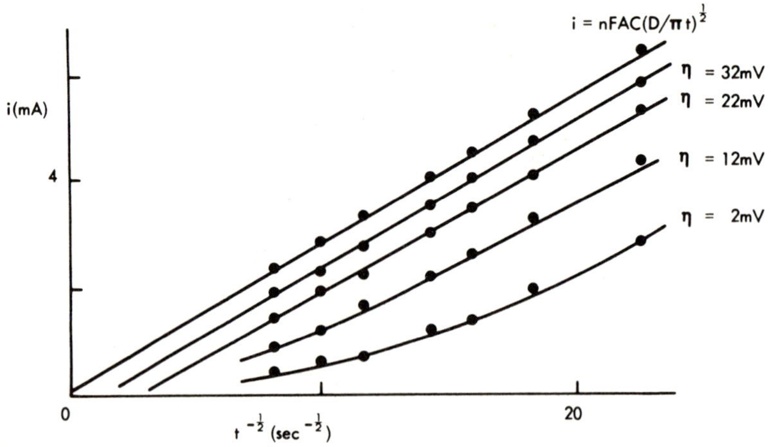

Fig. 28. Plot of $i - t^{-1/2}$ predicted from Eq. (88) for the formation of a layer from a large number of nuclei with overlapping diffusion zones. Ref. (*59*).

and θ' is a constant given by

$$(C^* - C^0)\frac{M}{\rho} = \theta'^3\left(\exp\frac{1}{4}\theta'^2\right)\left\{\frac{1}{2}\pi^{1/2}\left[1 - \operatorname{erf}\left(\frac{1}{2}\theta'\right)\right] - \frac{1}{\theta'}\exp\left(-\frac{1}{4}\theta'^2\right)\right\} \tag{89}$$

the transients before overlap for progressive and instantaneous nucleation are therefore given by

$$i = \frac{zF\rho}{M}(\pi\theta'^3 DA^{3/2})\frac{2}{3}t^{3/2} \tag{90}$$

$$i = \frac{zF\rho}{M}N_0\pi\theta'^3 D^{3/2}t^{1/2} \tag{91}$$

An approximate value of θ' assuming linear diffusion instead of spherical diffusion is given by

$$\theta' = \frac{2MC^*}{\rho\pi^{1/2}} \tag{92}$$

Equation (92) in Eq. (91) has been used to estimate the number of nuclei formed in the early stages of growth for mercury (*59*) on pyrolitic graphite. The complete problem with overlap would be complicated; it would, however, be possible to produce a solution with the same limitations as Eq. (82). Although this has not yet been calculated it would be important to test the amount of charge contained in a transient for various values of N_0. A completely realistic solution would clearly also have to include overlapping of the diffusion zones.

5. Linear Sweep with Diffusion in the Solution

The calculation (*60*) can be carried out as in the last section by solving the diffusion equation

$$\frac{\partial c}{\partial t} = D\frac{\partial_2 C}{\partial x^2}$$

under sweep conditions

$$\begin{aligned} C_{0,t} &= C^0\exp(\beta^* t) \\ C_{x,0} &= C^0 \\ C_{x,t} &= C^0 \qquad \text{as } x \to \infty \end{aligned} \tag{93}$$

where $\beta^* = nFU/RT$, is the sweep rate $\sec^{-1}$. The solutions of these equations for the flux $D(dC_{0t}/dx)$ are well known. The calculation has only been attempted for linear diffusion although hemispherical diffusion could be considered as in Sec. 4d. Setting the flux given by Eq. (47) equal to that from Eq. (93) gives, for growth without overlap,

$$i = 3F\gamma N_0^{1/3} C^0 (\beta^* D)^{1/2} \exp(\beta^* t) \operatorname{erf}[(\beta^* t)^{1/2}]$$

$$\times \left\{ QC^{1/3} - F\gamma N_0^{1/3} C^0 D^{1/2} \beta^{*-1/2} \left[\exp(\beta^* t) \operatorname{erf}(\beta^* t)^{1/2} - \frac{2(\beta^* t)^{1/2}}{\pi^{1/2}} \right] \right\}^2 \quad (94)$$

where $QC =$ the initial quantity of deposit and $\gamma = 2\pi(3M/2\pi)^{2/3}$.

Some numerical graphs of this expression have been given in the original paper.

I. Model for the Interpretation of A, the Nucleation Rate Constant

The rate of nucleation is determined by the free energy necessary to form a critical size of nucleus. Nuclei smaller than the critical size will dissolve and those of greater size will grow. The critical free energy can be calculated by taking a cell, one plate of which is an infinite flat electrode the other a two-dimensional hemicylindrical nucleus (height h) on an infinite flat electrode of the same material.

$$\Delta G = -nzF\eta + \sigma_3 2\pi rh \quad (95)$$

where n is the number of moles in a two-dimensional nucleus given by

$$n = \frac{\pi r^2 h \rho}{M}$$

The first term in Eq. (95) is the free energy change for unit volume of layer. Substituting for n in Eq. (95) and differentiating to find the maximum radius r^* and maximum free energy ΔG^* gives

$$r^* = \frac{\sigma_3 M}{zF\eta\rho} \quad (96)$$

$$\Delta G^* = \frac{\pi M h \sigma_3^2}{zF\eta\rho} \quad (97)$$

The assumption of a stationary (*61*) distribution of nuclei gives

$$A = Z \exp\left(-\frac{\Delta G^* \mathcal{N}}{RT}\right) \tag{98}$$

Z is a frequency factor. Attempts to calculate the absolute magnitude of Z are reviewed in Ref. (*3*). Equation (97) can be substituted in Eq. (98) and the rate of nucleation calculated as

$$A = Z \exp\left(-\frac{\pi M h \sigma_3^2 \mathcal{N}}{zF\eta\rho RT}\right) \tag{99}$$

Equation (99), however, is inexact, as σ_3 is also a function of potential. Taking into account the potential dependence of σ_3 when differentiating Eq. (95) gives

$$r^* = \frac{M\sigma_3}{2zF\eta\rho} + \frac{M}{\rho zF}\frac{d\sigma_3}{d\eta} \tag{100}$$

which corresponds to Eq. (96). This could be inserted in Eq. (95) to enumerate ΔG^* and hence the rate of nucleation from Eq. (98). σ_3 is, however, difficult to measure for a solid metal.

The situation is somewhat easier to evaluate when a film is deposited on an inert substrate—that is, of a different material. In fact for mercury the interfacial tension with or without a film is measurable (*62*). Arguments similar to the derivation of Eq. (100) and the nomenclature of Fig. 29 leads to

$$A = Z \exp\left\{\frac{-\pi\mathcal{N}h^2\sigma_4^2}{[h\rho zF\eta/M + (\sigma_2 + \sigma_3 - \sigma_1)]RT}\right\} \tag{101}$$

Equation (101), which has been discussed by Vermilyea (*54*), shows that if $(\sigma_2 + \sigma_3 - \sigma_1)$ is negative a stable monolayer can be formed before the equilibrium potential for the bulk phase.

Equation (101) can be written in identical form to Eq. (99) if η is redefined with respect to the monolayer equilibrium potential. The previous objections

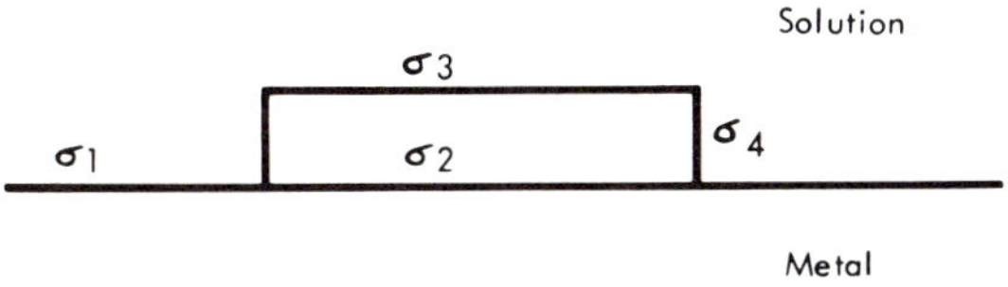

Fig. 29. The formation of a metal film on a substrate. σ is a surface energy.

also apply to Eq. (101); namely, the interfacial tensions are probably potential dependent and only the overall interfacial tension is a measurable quantity. In spite of these difficulties, the prediction of Eq. (101) is observed in practice and a number of stable monolayers, which cannot thicken until $\eta = 0$, have been observed on mercury. In some cases this is not observed, as a nucleation overpotential is needed to form the layer. The evidence is collected in Table III. Metal layers on inert solid substrates in some cases, also form monolayers. The potentials involved are large compared to mercury and it is possible that a surface compound is involved. An attempt to investigate the kinetics of metal monolayer formation is described in Sec. IV.

The general case for two-dimensional nuclei of other shapes has been calculated by Pangarov (*77*) by an alternative method. The treatment is based on a treatment by Stranski and Kaischew in which the length of edges *Li* and their specific edge energy χ_i are calculated, for the equilibrium form of the nucleus, from thermodynamic and structural arguments. Brandes' equation gives the work of formation

$$W_{hkl} = \tfrac{1}{2} \sum \chi_i L_i$$

In general for an inert substrate

$$W_{hkl} = \frac{B_{hkl}}{ze_0 \eta + \psi_0 - C_{hkl}} \tag{102}$$

where

$$B_{hkl} = b\psi_1^2 \text{ and } C_{hkl} = a\psi_1;$$

ψ_1 is the work for breaking a bond between nearest neighbors; a, b, are constants depending on the lattice; and ψ_0 is the work to separate an atom from the substrate. The corresponding expression for the formation of a nucleus on substrate of the same metal i is

$$W_{hkl}^{k} = \frac{B_{hkl}}{e_0 \eta} \tag{103}$$

Figure 30 shows this expression graphically for various equilibrium surfaces of an fcc crystal. Similar curves have been drawn for bcc, the hexagonal close packed tetragonal lattice of tin (*77*) and the diamond-type lattice (*78*). W_{hkl}^{k} depends on the overpotential. At each value of η the nucleus with the

TABLE III

THE REVERSIBLE POTENTIALS OF MONOLAYER PHASES AT 25°C

Electrolyte	Electrolyte (aqueous)	Phase formed	Difference between monolayer and bulk potential	Charge of thin film $\mu C\ cm^{-2}$ (10%)	Hysteresis between formation and reduction	Reference to previous work
Hg	0.1 *M* KCl	Hg_2Cl_2 (110) orientation	Probably +18 mV	90	Not measurable	63, 64
Hg	1 *M* NaOH	HgO (010)	+15 mV	200	1 mV	65, 66
	0.1 *M* Na_2HPO_4					
	0.1 *M* Na_2HPO_4		−43 mV	100	3 mV	67
Hg	+0.1 *M* NaH_2PO_4 +0.6 *M* KNO_3	Hg_2HPO_4	−16 mV	125	15 mV	
Hg	1 *M* $HClO_4$	Hg oxalate	+24 mV	80	3 mV	68
Hg	1 *M* $NaHCO_3$ +0.5 *M* Na_2S	HgS Random	+4 mV	180	7 mV	69
Tl(Hg) 1%	1 *M* KCl	TlCl (100) orientation	+10 mV +40 mV	80 20	1 mV 0.1 mV	70
Cd(Hg) 1%	1 *M* NaOH	$Cd(OH)_2$ (0001) orientation	+15 mV	260	9 mV	71, 72
Solid Ag	1 *M* NaOH	Ag_2O				73
Solid Ag (100) face perfect crystal	6 *M* $AgNO_3$	Ag	0	190	14 mV	86–88
Hg	1 M NaOH +1 *M* Pyridine	Pyridine ad layer	Not defined	9	0.1 mV	74, 75
Pb on Ag	1 *M* KCl	Pb	+60 mV		100 mV	76
Tl on Ag	1 *M* KCl	Tl	+100 mV		20 mV	76

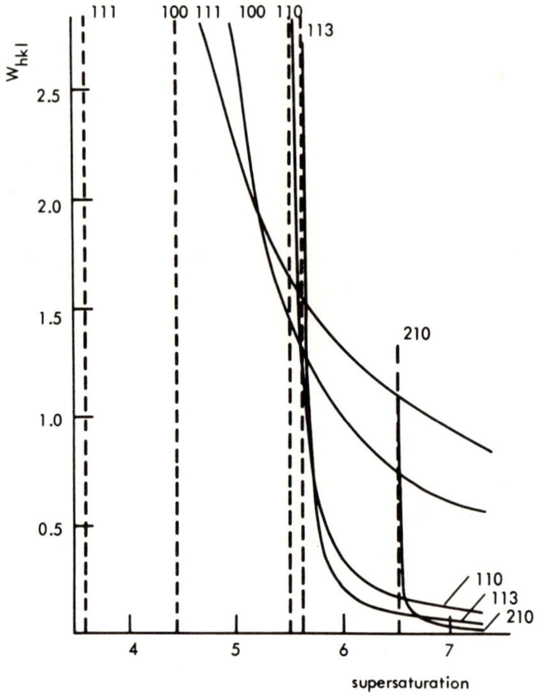

Fig. 30. Relative values of the work W_{hkl} as a function of supersaturation for the fcc lattice. Ref. (*77*).

lowest value of W_{hkl}^k will be preferred. Assuming that the formation of a single two-dimensional nucleus determines the subsequent orientation of three-dimensional growth, Pangarov has shown that for the metals iron on platinum (*79*), cobalt on platinum (*80*), and nickel on platinum (*81*) the orientation of the deposit changes progressively with the overpotential in the correct sequence as given by a diagram similar to that of Fig. 30. These systems were, however, investigated galvanostatically and are not amenable to exact analysis because of hydrogen evolution. A metal that has been investigated in more detail is that of silver onto platinum (*82*). The experiments were performed by applying a potentiostatic pulse (*ms*) to nucleate a single nucleus and growing this at a lower potential to microscopic size; the observed orientations as a function of potential closely followed the sequence of Fig. 30. It is possible to identify experimentally (*83*) the overpotentials η_1 and η_2 at which the 111 and 100 planes and 100 and 110 planes respectively, appeared with equal probability—that is, the crossing points of the curves in Fig. 30 given by

$$\psi_1 = \frac{ze_0(\eta_2 - \eta_1)}{q_1 - q_2} \tag{104}$$

Thus ψ_1 and hence B_{hkl}, C_{hkl}, W^k_{hkl}, and n^k_{hkl} can be calculated directly.

The appearance of twins for silver on platinum has been investigated experimentally (*84*) and theoretically (*85*). W^h_{hkl} can be calculated for a normal nucleus on its own substrate and this can be compared with a nucleus formed in the twinning position W^t_{hhl}. The probable proportions of twinned to normal nuclei as a function of potential can then be estimated as

$$\rho = \frac{A(t)}{A(t) + A(n)} \tag{105}$$

where A is given by the Volmer equation $A \propto \exp - W_{hkl}/RT$, the equation is plotted in Fig. 31. The twinning does increase in practice with potential but the details are uncertain.

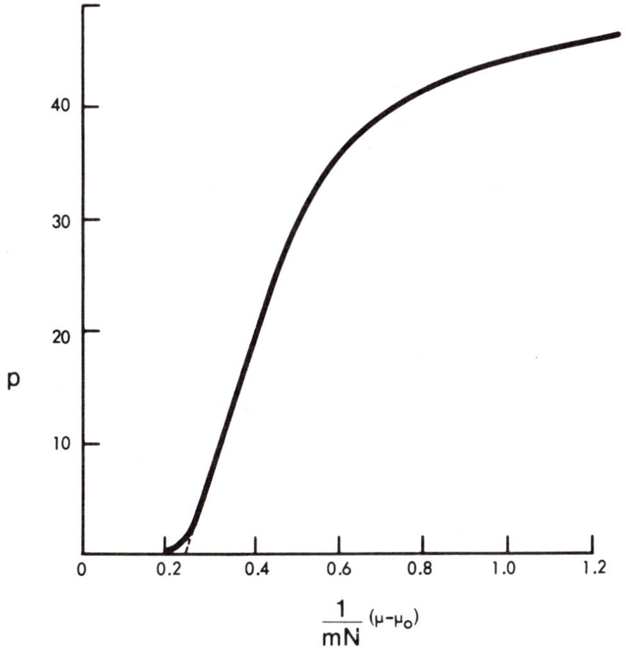

Fig. 31. Probability P as a percentage for formation of nuclei in twinning position as a function of overpotential $Ze_0\eta$ (expressed in units of Ψ_1). Ref. (*85*).

J. Model for K, the Rate of Lattice Growth

It is expected that k will vary with potential in the same way as for an electrochemical reaction—that is,

$$k = k_0\left(\exp\frac{\alpha nF\eta}{RT} - \exp\left[\frac{(\alpha - 1)nF\eta}{RT}\right]\right. \tag{106}$$

However all attempts to verify this experimentally have failed. The nonmetallic systems for which the potential variation has been measured show a much higher potential variation than that of Eq. (106). The earlier measurements (*63*), (*71*), (*75*) were obscured by ohmic effects in the solution and the reported potential dependence must be discounted. The reason for this enhancement is not yet known.

IV. EXPERIMENTAL PROCEDURES

It has been suggested that on a solid metal, which contains a large number of dislocations, a large number of growth sites are available without the need for nucleation; thus the measurements (about 1959) described in Sec. IIIG can probably be interpreted, as has been discussed in some detail above, as a complex solution diffusion problem rather than by adatom diffusion. More recent work has, however, proved helpful in the development of experimental methods and three satisfactory routes seem to be available for investigating the kinetics of metal deposition from simple aquo ions: (a) deposition onto metal surfaces with few dislocations in them; (b) deposition onto inert substrates; and (c) comparison of behavior of amalgams with solid metals.

The measurements have been limited to very few inorganic cations by the following simplifying conditions: (i) little or no complexing of the cations should occur with the anions in solution; (ii) the anion should not be specifically adsorbed; and (iii) hydrogen evolution should be absent.

These conditions are most nearly attained for Ag^+ ions; substantial progress has been made for each of these experimental routes, and the experimental results that fall under these headings will be discussed.

A. Dislocation-Free Metal Surfaces and Absence of Hydrogen Evolution

1. Silver on Silver

A most important series of experiments by Budewski et al. (*86*, *87*. *88*) has shown that it is possible to produce single crystal surfaces of silver that behave as if they have few or no dislocations. The electrodes are made by growing a seed single crystal with a small ac signal superimposed on dc into a tapering capillary. The dislocations are progressively "killed" at the glass walls as the crystal grows into the capillary. Detailed descriptions of the apparatus, which incorporates a microscope to view a small area of the electrode (0.1 mms), are given in the literature and the whole experiment has been most elegantly filmed (*88a*).

Surfaces of silver, in 6 *N* purified $AgNO_3$ have remarkable properties. They behave as ideal polarisable electrodes up to an overpotential of 10 mV. At potentials greater than this, classical two-dimensional nucleation and growth occurs. This corresponds to an adatom build up of about 50% over the equilibrium value that is the expected value for a solid as predicted from Eq. (98). Unfortunately, other substrates of platinum and carbon also show about a 10-mV overpotential for nucleation, so this evidence is not unambiguous.

Figure 32 shows a galvanostatic experiment in which the oscillations contain monolayer amounts of charge. Figure 33 shows a potentiostatic experiment in which the potential was maintained constant by a simple low-resistance potentiometer circuit; the characteristic shapes of the curves can be reproduced by arguments similar to those of Sec. IIIH with the instantaneous formation of a single nucleus assumed to occur at a preferred site on the edge of the crystal. At low η, assuming Eq. (97) can be linearized,

$$i = k\eta L(r) \tag{107}$$

where k is a constant ($\text{ohm}^{-1}\ \text{cm}^{-1}$), L is the periphery of a growing layer, f is the surface per atom, Ze_0 is the charge per atom, and r is the radius of the growing nucleus, $A = fk\eta/Ze_0$. By a change of variable

$$\rho' = \frac{A}{k} = \frac{r}{R}; \qquad \lambda = \frac{L}{R}; \qquad r = At$$

therefore

$$i = Rk\eta\lambda(\rho')$$

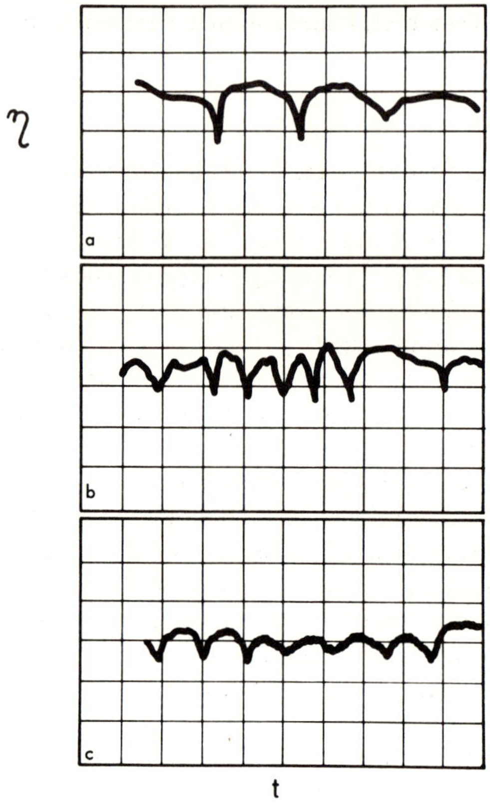

Fig. 32. Overvoltage oscillations on a dislocation free [100] surface at constant current of
(a) 1.9×10^{-4} amp cm^{-2}
(b) 3.8×10^{-4} amp cm^{-2}
(c) 7.6×10^{-4} amp cm^{-2}
(a) x axis; (b) 0.5 sec cm^{-1}; c, 0.2 sec cm^{-1}; y axis, 5.2 mV cm^{-1}. Ref. (*87*).

$\lambda(\rho')$ has been calculated for three cases shown in Fig. 34 where the non-dimensional variable q/R is introduced. The results are shown in Figs. 35 and 36. These agree qualitatively with the experimental results of Fig. 33. The nucleation rate of a two-dimensional nucleus has also been investigated and compared satisfactorily with an amended version of Eq. (99) as shown in Fig. 37 (see also Fig. 1 for equivalent vapor-phase experiment).

Silver surfaces parallel to the 111 and 110 planes with few dislocations in them show a behavior that has been attributed to the growth of screw dislocations; thus a maximum is observed on a smooth surface in a galvanostatic experiment that could be explained by the winding of the originally straight

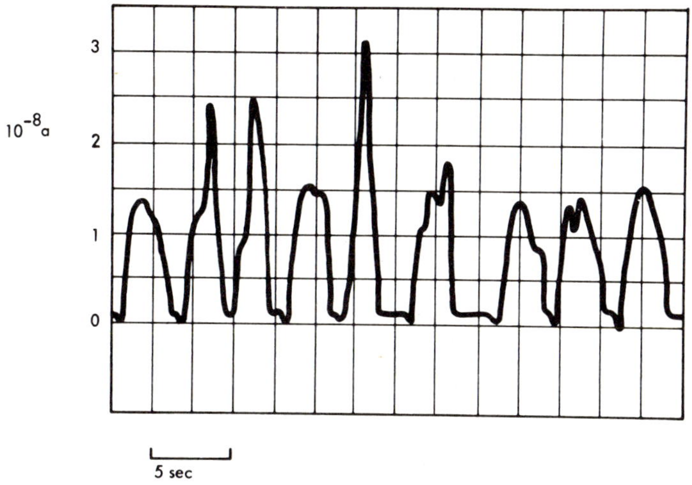

Fig. 33. Current-time curves at $\eta = 6$ mV after $\eta = 13$ mV for 1 μsec. A single two-dimensional nucleus is responsible for the transient. Ref. (*87*).

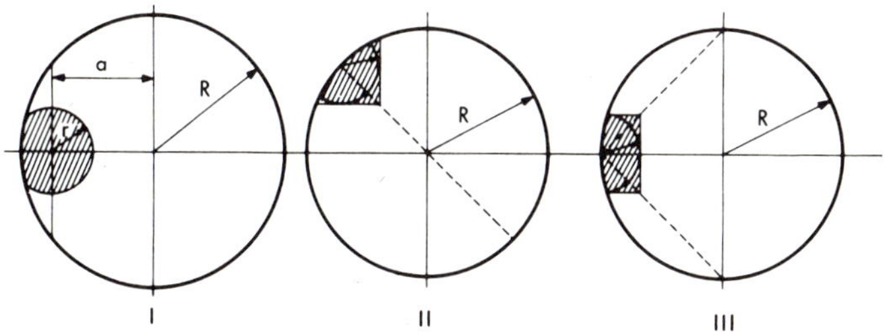

Fig. 34. Models for the growth of a single two-dimensional nucleus and its interaction with the boundary. Ref. (*87*).

step into a spiral. The maximum disappears on subsequent current pulses but reappears on standing when the spirals "smooth" and revert to their equilibrium shape.

2. Metals on Mercury

Measurements have also been attempted on a mercury substrate. The experiments are then limited to metals that are essentially insoluble in the silver. Deposition of nickel (*55*), from the thiocyanate complex, onto mercury shows the potentiostatic current-time curves shown in Fig. 38; the charge under the curves corresponds to a monolayer of unit cell height. Directly after double layer charging, the initial current behaves like an electrochemical

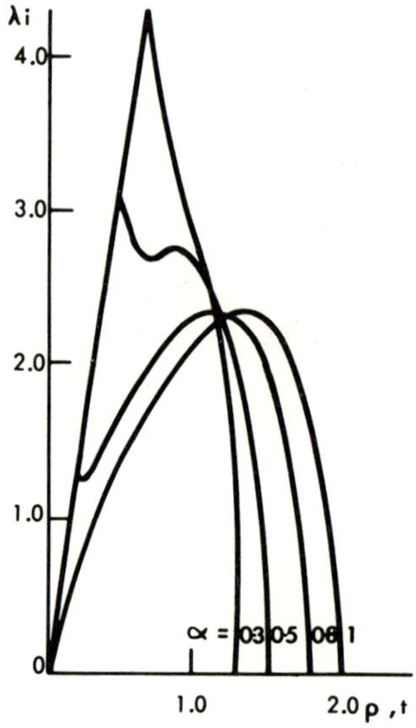

Fig. 35. Predicted current-time curves case *I* of Fig. 34 for various values of α. Ref. (*87*).

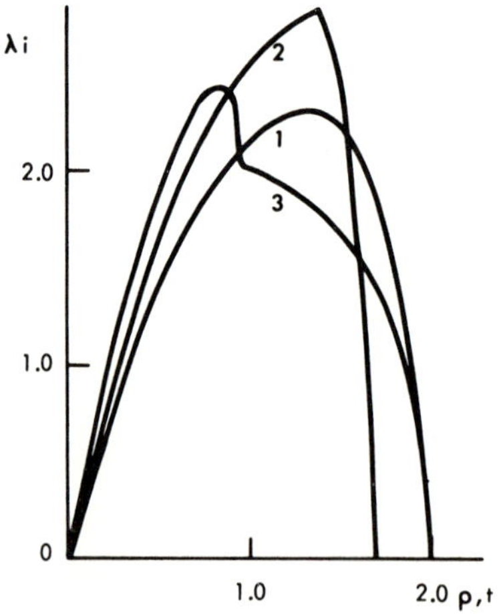

Fig. 36. Predicted current-time curves; (1) case 1, $\alpha = 1$; (2) case II; (3) case III of Fig. 35. Ref. (*87*).

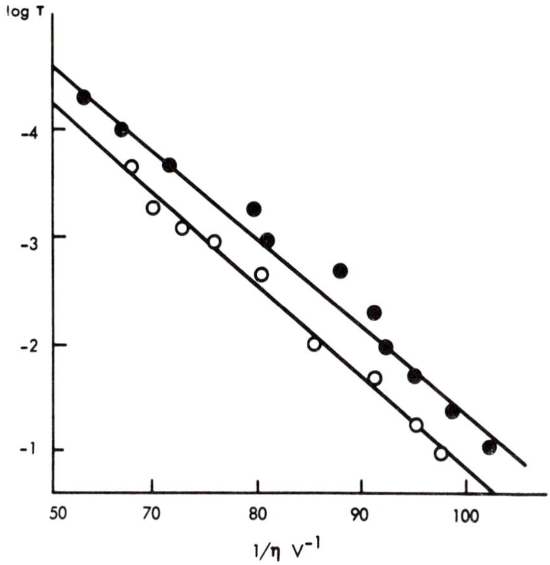

Fig. 37. Nucleation time τ for the first nucleus on dislocation free (100) as a function of reciprocal of η. Ref. (*87*).

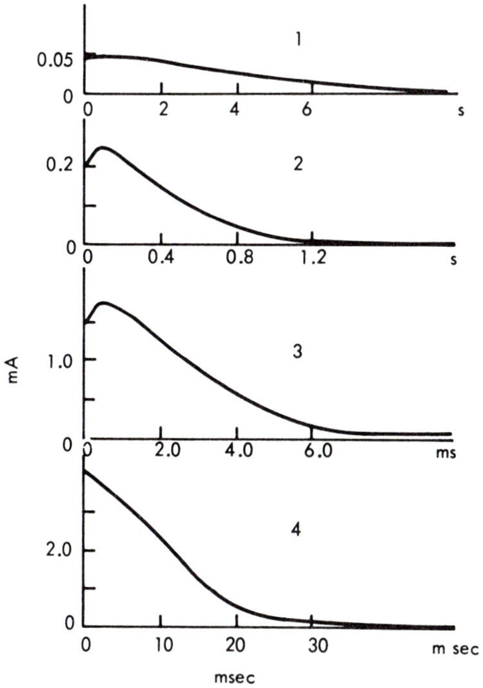

Fig. 38. Cathodic current-time transients for 0.1 *M* $Ni(NO_3)_2$ + 0.1 *M* KSCN + 2.2 M KNO_3. Area of mercury drop 0.055 cm^2. (1) −570; (2) −600; (3) −640; (4) −670 mV. Ref. (*55*).

reaction on to the whole of the mercury surface. At later times crystal growth takes effect and the curve goes through a maximum. This suggests that adatom diffusion, which must link the electrochemical discharge with the crystal growth step, is a major part of the mechanism. Diffusion in the solution is probably ruled out as the electrochemical reaction is slow.

The experimental time curves show features similar to the simulated curves described in Sec. IIH4a—that is $i_m t_m/Q_m \ll 1$.

3. Anodic Films on Mercury

Perhaps the most exact information about layer growth has been obtained for anodic films on mercury and amalgams. A brief review is given in Table IV. In these systems the Eqs. (1) and (2) of Table II for the growth of the first

TABLE IV

Electrode	Electrolyte	Phase	Mechanism
Hg[63,64]	KCl HCl $^*HgCl_4^{2-}$	Hg_2Cl_2	Low 1 multilayer high succession of mono-layers
Hg[65,66]	NaOH $^*Hg(OH)_2$	HgO	1 monolayer + 1 multilayer
Hg[67]	HPO_4^{2-}	Hg_2HPO_4	2 monolayers +1 multilayer
Hg[68]	Oxalate	HgOX	1 monomonolayer 1 multimulilayer
Hg[53]	Barbituric acids	Hg barbiturate	1 monolayer
Hg[69]	S^{2-} $2HgS_2^{2-}$	HgS	2 monolayers
Hg[89]	KSCN $^*Hg(SCN)_2^-$		
Tl (1%)[70]	KCl $^*Tl^+$	TlCl	2 monolayers +1 multilayer
Cd (1%)[71,72]	IMNaOH $^*Cd(OH)_4^{2-}$	$Cd(OH)_2$	3 monomonolayers
Ag (single crystal)[73]	IM NaOH $^*Ag(OH)_2^-$	Ag_2O	1 monolayer +1 multilayer

monolayer, without complication from diffusion, have been substantiated many times. The earlier i versus t measurements, however, were quantitatively in error due to IR drop and the kinetic schemes given in these papers have not been substantiated (*63*), (*70*), (*71*).

As a complementary experimental procedure to transient measurements, ac impedance studies show that in many cases a solution-soluble species is also formed in parallel with the layer formation. The nature of the complex has been indentified by the potential and concentration dependence of the faradaic impedance contribution and is also given in Table IV.

B. Metals on Mercury with Hydrogen Evolution

Investigation of fuel cell reactions has led to an interest in the properties of thin films of catalytic metals. For example, it has been shown that ruthenium deposition follows the nucleation and growth kinetics except that hydrogen is codeposited (*73*). The system may be regarded as a model system, and many metals that codeposit hydrogen can reasonably be expected to have features in common with it. Further investigation (*91*) has, however, shown that some of the conclusions in Ref. (*90*) need to be modified as follows. Ruthenium and platinum have a low hydrogen overvoltage and when the metal is codeposited with hydrogen onto mercury, the hydrogen reaction is confined to sites on the platinum and ruthenium nuclei. Alternating-current measurements in 10^{-4} M $RuCl_3$ in 10^{-2}—HCl solution show that a metal monolayer (i.e., corresponding to a layer of cell unit height requiring 1580 μC) is formed more negative than -0.3 V (SCE) whereas codeposition of hydrogen starts at about -0.6 V (SCE). Similar conclusions apply to platinum, except that the metal is deposited at all potentials more negative than anodic calomel formation. The thermodynamic reversible hydrogen potential in these solutions would be approximately 0.26 V against SCE. Measurements are limited to metal ion concentrations of 10^{-4} M because of the chemical reaction between mercury and the group VIII metal ions in chloride solutions to form metal and calomel. If the metal is grown outside the hydrogen region with a well-defined potentiostatic pulse and the potential jumped instantaneously into the hydrogen region $i^0{}_{H_2}$, the instantaneous hydrogen current has the behavior shown in Fig. 39. Two limiting types of behavior might be expected: (a) with hydrogen evolved only on the edges of the two-dimensional cylinders, in which case the current would go through a maximum at about half coverage and then decrease to zero; (b) with hydrogen evolved on the

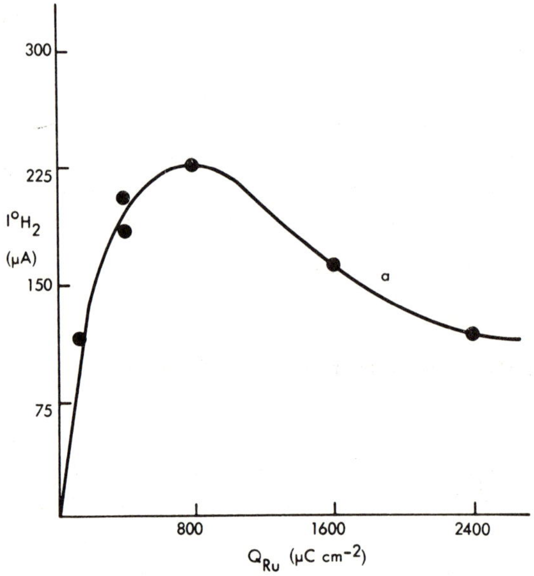

Fig. 39. Instantaneous hydrogen current (I^oH_2) as a function of ruthenium deposit ($\mu C\ cm^{-2}$) formed at -0.4 V (S.C.E.) in 10^{-4} *M* Ru + 10^{-2} *M* H^+. A fresh mercury drop for each experiment. Ref. (*91*).

top surface when the current would rise to a constant value. Figure 39 shows a behavior intermediate between these two, the distribution of current density between top and edge being about 100 : 1. This assumes that the nuclei at full coverage lock together and the residual current is not due to some mismatch in the layer; the fact that thickening the layer further does not decrease the residual current supports this. $i^0_{H_2}$, as a function of potential, compared with the behavior of platinum and ruthenium on a carbon substrate and plotted as a current density (in Fig. 40) shows that the metal layers on mercury behave quite differently from three-dimensional nuclei of the bare metals on carbon. The ruthenium and platinum layer on mercury have 120 mV^{-1} tafel slope intermediates in potential range between the hydrogen evolution reaction on pure mercury and on bare platinum and ruthenium. Bare platinum has the usual 30 mV^{-1} slope due to the diffusing away of molecular hydrogen. Bare ruthenium has a 40 MV^{-1} slope and a concentration dependence that supports.

$$H^+ + e \rightleftharpoons H\bullet$$
$$H\bullet + H^+ + e \rightarrow H_2$$

The modification of the catalytic properties of ruthenium and platinum

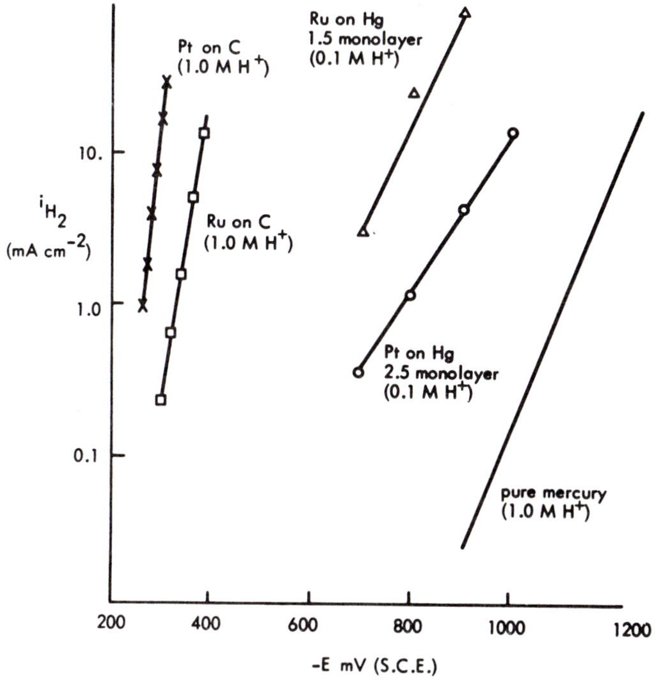

Fig. 40. Potential dependence of the hydrogen evolution reaction on different surfaces. Ref. (*91*).

on mercury is possibly due to (a) the nuclei being covered with a thin layer of mercury and having a type of amalgam structure, or (b) these films being less effective as catalysts than bulk phase. (a) is by far the most likely. However, it has not as yet proved possible to form small or monolayer nuclei on an inert substrate that is noble and flat enough. These measurements open up the possibility of following the kinetics of metals in which the codeposition is unavoidable. The extension of these ideas to the initial stages of the plating situation is clearly desirable.

C. Metals on Inert Metal Substrates without Hydrogen Evolution

Most work in this field has been directed toward stripping voltammetry (see Barendrecht, this series, Vol. 2). However, a few papers have appeared that show the initial stages of metal deposition. A table of suitable combinations of depolariser and substrate that do not react together has been given by Schmidt and Gygax (*92*). The reason for using an inert substrate is twofold: (a) it is of great advantage to be able to take this metal layer on and off without affecting the substrate; and (b) the early stages of nucleation and growth can be investigated before overlap. These measurements move one step nearer than the last section to the situation of practical plating.

In potentiostatic linear sweep experiments essentially two cases arise (other cases are possible and these will be discussed later). These are with metals that give a prepeak and a main peak during deposition and with metals that give a main peak only. The rising portion of the main peak starts at the bulk equilibrium potential, whereas the prepeak starts before; typical examples are shown in Fig. 41. Two theories are available to account for this phenomenon. The first is due to Schmidt and Gygax (*92*) and interprets the prepeak as a metal adsorption wave and reproduces the experimental curves theoretically by coupling an appropriate isotherm to the bulk diffusion conditions. Written for a thin layer cell,

$$\frac{\partial C_{M^+}}{\partial t} = \frac{D\,\partial^2 C_{M^+}}{\partial x^2}$$

$$C_{M^+} = m \qquad \text{at } t = 0$$

$$\frac{\partial C_{M^+}}{\partial x} = o \qquad \text{as } x \to \delta$$

$$\frac{\partial C_{M^+}}{\partial x} = \frac{i(t)}{zFAD} \qquad \text{at } x = 0 \tag{108}$$

where δ is the electrolyte thickness in the cell. It can be shown that to an approximation the solution of Eq. (108) is the obvious one.

$$(C_{M^+})_{x=0} = m - \frac{y}{\delta} \tag{109}$$

where y is the amount (mole cm^{-2}) of material that is removed from the solution to the surface. For a reversible reaction,

$$E = E^0 + \frac{RT}{zF}\ln\frac{(C_{M^+})_{x=0}}{a_m} \tag{110}$$

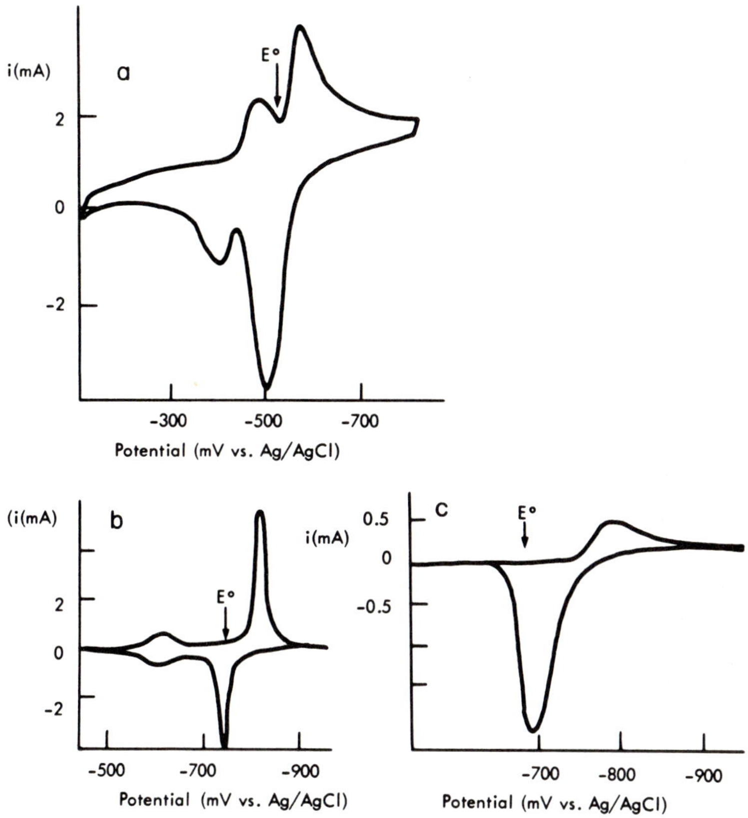

Fig. 41. Sweep curves for some metal systems: (a) lead into a (111) face of a silver single crystal, 3 V/sec, 3×10^{-3} *M* Pb^{2+} + 1.0 *M* KCl; (b) thallium onto electropolished silver, 300 mV/sec, 8.8×10^{-4} *M* Tl^{+} + 1.0 *M* KCl; (c) cadmium onto electropolished lead, 30 mV/sec, 8.3×10^{-3} *M* Cd^{2+} + 1.0 *M* $NaClO^{4}$. Ref. (*93*).

where a_m is the surface activity of the metal on the surface. Therefore from Eqs. (110), (109) and for sweep

$$E = E_a - \beta^* t = E^0 + \frac{RT}{zF} \ln\left(m - \frac{y}{\delta}\right) - \frac{RT}{zF} \ln a_m \tag{111}$$

which gives, on differentiation,

$$i^* = \frac{RT}{z^2 F^2 A \beta^*} i(y) = \left[\frac{1}{(m\delta - y)} + \frac{\partial}{\partial y}(\ln a_m)\right]^{-1} \tag{112}$$

Theoretical curves of i^* can then be plotted as a function of $i^* y/y_s$ where y_s is the saturation monolayer coverage for various isotherms by substituting for a_m. A Langmuir isotherm would predict a curve with two peaks and a linear isotherm one peak.

A second theory (*93*) attributes the prepeak that is less but close in capacity to that required for the formation of a crystalline monolayer of unit cell height on the flat parts of the electrode; the second layer then nucleates on top as a three-dimensional layer. This sequence—monolayer followed by multilayer—is quite analogous to the behavior of some nonmetallic films where it is known that the equilibrium potential for the monolayer is appreciably different from that of the bulk phase. The thermodynamic conditions for the stability of monolayers compared with the bulk phase has been discussed in Sec. III. A more detailed picture of the mechanism can be obtained by pulsing potentiostatically into the monolayer and multilayer regions. To characterize properly the monolayer region as a two-dimensional phase, a current time curve of the form discussed in Sec. III—that is, a peak containing a monolayer charge—should be observed. Unfortunately, a falling transient is seen, possibly because the roughness of the substrate causes a large number of nuclei to be formed initially. The nuclei seem to be restricted to patches that contain a large number of very small nuclei; this restricts the rising portion to very short times and a peak is not observed for a solid substrate system. However, the results for lead on silver and thallium on silver, are consistent with direct deposition from solution, the concentration of ions at the interface being controlled by the Nernst equation and a linearized Langmuir isotherm. Equation (88) of Sec. IIIH4c reproduces this behavior theoretically. The multilayer shows rising transients that are controlled by diffusion in the solution and nucleation. It seems likely that the nucleation is controlled by the formation of defects as the layer grows. Unfortunately, in the cases investigated diffusion in the solution predominates so the detailed mechanism of growth is hidden, but the fact that well-defined mono- and multilayers probably form even on real metal substrates makes a development of this experimental and theoretical approach look very promising. The reason why some systems show a monolayer is probably due to epitaxial factors (*93*).

More complex behavior has been observed when it is suspected that an interaction between substrate and depositing metal occurs, for example, the deposition of lead on gold (*94*). Copper on platinum has been investigated recently by Tindall and Bruckenstein (*95*) where they confirm from linear sweep experiments at a rotating ring and disk that three peaks are observed corresponding to adsorbed copper, thin copper patches and bulk copper (*96*). They also suggest, from potentiostatic evidence, that thin layer copper is an essential precursor to the deposition of bulk copper.

At the moment it is not certain in which systems the monolayer-multilayer growth pattern is dominant and in which the adsorption of the neutral metal atom is formed prior to the thermodynamic potential for bulk metal; this will depend on whether it is energetically favorable for the atoms to aggregate. Very roughly, we might expect substrates that have a large enough number of unsaturated bonds on the surface to favor adsoprtion, and substrates of low valency coupled with a depolarizer of high valency to favor the formation of a crystalline monolayer. It is also uncertain whether a mono- or adsorbed layer preceded multilayer growth and whether this controls the further growth of the multilayer.

Deposits on inert substrates have also been investigated from the nucleation point of view, by comparing the number of nuclei with the Volmer equation. Systems of this type that have been reported on are mercury on to microplatinum single-crystal spheres (*97*), (*97a*); nickel on polished platinum (*81*); silver on platinum (*82*); cobalt on cobalt (*78*). These systems have been compared with the Pangarov theory, which predicts the change in orientation of the deposits with overpotential; no detailed kinetics of growth have been attempted. However, an attempt has been made to compare the growth characteristics of silver and mercury (*59*) on a pyrolitic graphite substrate. A difference is observed when a constant number of nuclei grow; for silver at short times $i \propto t$, and for mercury $i \propto t^{1/2}$. From Fig. 42 and Eqs. (91) and (92) the number of nuclei nucleated instantaneously can be approximately estimated. This is many orders of magnitudes larger than the microscopically observed number, showing that counting methods of estimating A used in the previous references are very unreliable. In view of the ambiguity of Table I it is not certain that $i \propto t$ represents a slow lattice incorporation step of the solid silver metal.

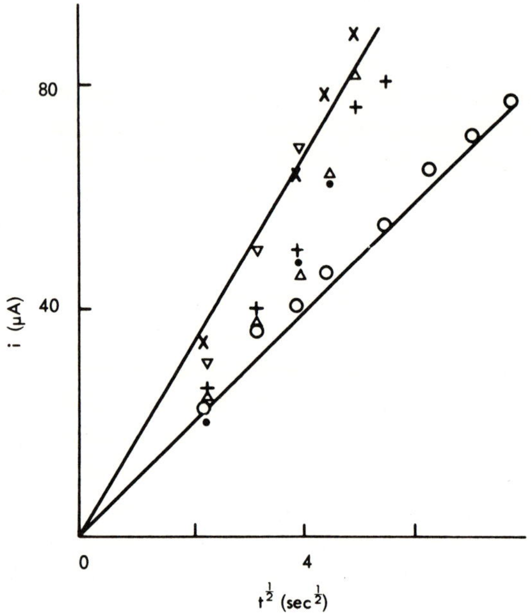

Fig. 42. Plot of i against $t^{1/2}$ for 4.5×10^{-2} M mercurous nitrate (total ionic strength 1 M with KNO_3) deposited on glassy carbon. The gradients are almost independent of η: 0, −40 mV; X, −50 mV; +, −60 mV; •, −70 mV; Δ, −80 mV; ∇, −100 mV; prepulse −180 mV, 20 msecs (potentials wrt 4.5×10^{-2} M Ag^+/Ag). Ref. (*59*).

D. Deposition on Substrates of the Same Metal

1. Silver

The older measurements of metal deposition by straightforward galvanostatic potentiostatic and ac methods have been discussed in Sec. III.

Linear sweep experiments of silver onto silver (*98*) have been compared with the theoretical expression of Berzins and Delahay (*99*) in which the activity a of the surface is assumed equal to one throughout and the concentration of the cation at the surface is given by the Nernst equation,

$$i = \frac{2}{\pi^{1/2}} \frac{n^{3/2}F^{3/2}}{R^{1/2}T^{1/2}} A'C^0D^{1/2}\left(\frac{RT\beta}{\eta F}\right)\Phi[(\beta^* t)^{1/2}]$$

Equation (113) should have a slope between $0.5i_p$ and $0.9i_p$ when $\log(i_p - i)$ is plotted against E. Apart from the difficulties mentioned in Sec. III, three further facts complicate the picture.

(a) The surface of silver does not behave initially as if $a = 1$. Experimental finding is that linearity is observed from $0.7i_p$ to $0.9i_p$. Krebs and Roe (*60*) extended these measurements and showed that in a slow-sweep experiment a peak can be observed on the return anodic sweep before the main dissolution, showing that cathodically nuclei can be formed that are loosely bound to the substrate.

(b) The surface relaxes—that is, becomes smoother—when the current is switched off after deposition for some time (*86–88*).

(c) A fresh surface produced by anodic dissolution, then allowed to stand without the passage of current, changes its activity with time (*100*, *101*). It becomes less active as measured by a current position pulse after standing.

In principle (b) and (c) could be explained by adatom migration, or (c) by the adsorption of impurities. A more likely explanation in view of the expected low value of D_a is local cell action in which the sharp points of pyramids or other surface features become rounded, or large nuclei grow at the expense of smaller ones.

2. Iron, Cobalt, and Nickel

It seems likely that iron, cobalt, and nickel have a similar dissolution and deposition mechanism in noncomplexing solution; the slow step seems to be the electron transfer. The stationary current-potential curves for the dissolution in sulphate or perchlorate depend on pH, the current decreasing at fixed potential with pH decrease. There is controversy about the measured Tafel slope and the value of the reaction order with respect to OH^-.

Heusler (*102*) and associates have suggested in a series of papers the reaction scheme,

$$\text{Ni},K + OH^- \rightarrow NiOH^+ + K + 2e \text{ slow step} \tag{114}$$

$$NiOH^+ + H^+ \rightleftharpoons Ni^{2+} + H_2O \tag{115}$$

The symbol K means a kink site. Partial anodic currents and cathodic currents expected for this scheme are

$$i_a = K_+ a_{K^+} a_{OH^-} \exp\left(\frac{\beta nFE}{RT}\right) \tag{116}$$

$$i_c = K_- a_{K^-} a_{Ni^{2+}} a_{OH^-} \exp\left(\frac{(\beta - 1)nFE}{RT}\right) \tag{117}$$

where β is the anodic transfer coefficient and a_{K+}, and a_{K-} are the activity of kink sites for dissolution or deposition. Assuming that the number of kink sites at a potential is determined by the parallel reactions,

$$\begin{aligned} &Ni + H_2O \rightleftharpoons NiOH_{ad} + H^+ + e \\ &(NiOH)_{ad} \rightleftharpoons K_+ \end{aligned} \tag{118}$$

$$\begin{aligned} &Ni + Ni^{2+} + e \rightleftharpoons Ni(OHNi)_{ad} + H^+ \\ &Ni(OHNi)_{ad} \rightleftharpoons K_- \end{aligned} \tag{119}$$

then

$$a_{K^+} = k a_{OH^-} \exp\left(\frac{FE}{RT}\right) \tag{120}$$

$$a_{K^-} = k' a_{(NiOH)^+} \exp\left(-\frac{FE}{RT}\right) \tag{121}$$

References (*120*), (*121*) in (*116*), (*117*) predict stationary Tafel slopes of ~ 30 mV for anodic and 30 mV cathodic polarizations if $\beta = \frac{1}{2}$. The most recent experimental results for nickel (*102*), cobalt (*105*) in perchlorate buffered with $Ni(OH)_2$, $Co(OH)_2$ agree with these values. The dependence $i_\alpha \propto a^2_{-OH}$, $i_c \propto a^{\,2}_{N2+,i}$ at fixed E is also approximately observed. Earlier measurements of this school suggest similar results for iron (a list of references to previous work on iron, cobalt, and nickel is given in Ref. (*105*)). Galvanostatic pulse measurements agree with Eq. (116). The initial value of η plotted against current shows a 60 mV slope—that is, a_{K^+} is a constant—determined by the original state of the surface.

An alternative scheme has been proposed by Kabanov et al. (*103*) and is discussed in more detail by Bockris (*104*). The difference between it and the previous mechanism lies in the consecutive rather than parallel nature of the electron transfer reaction.

$$\begin{aligned} Fe + H_2O &\rightleftharpoons (FeOH)_{ad} + H^+ + e \\ (FeOH)_{ad} &\rightarrow FeOH^+ + e \qquad \text{slow step} \\ FeOH^+ + H^+ &\rightleftharpoons Fe^{2+} + H_2O \end{aligned} \tag{122}$$

which gives

$$i_a = K_+ a_{OH^-} \exp \frac{(1+\alpha)FE}{RT} \tag{123}$$

$$i_c = K_- a_{FeOH^+} \exp \frac{\alpha FE}{RT} \tag{124}$$

which implies a 40 mV anodic and 120 mV cathodic Tafel slope. It is claimed that experiments on iron (*104*, *107*) in a sulphate medium obey this scheme with $i_a \propto a_{OH^-}$. Other authors have observed a variety of slopes (*105*) and reaction orders and there may be complication from the sulphate anion where this is used (*106*). The details of these systems have not as yet been settled. The form of the galvanostatic and potentiostatic transients suggest nucleation phenomena that have not yet been taken into account.

The pyramids formed during the growth of cobalt and iron onto an initially flat substrate of the same metal have been investigated (*108*); the range is limited between flat pyramids too small to measure at low overpotential and diffusion control at higher overpotentials. A certain thickness of growth must also be chosen to obtain measurable pyramids. Given these limitations, the tangent of the pyramid angle, tan α, is proportional to the overpotential as demanded by the theory (*16*) for the formation of growth pyramids at screw dislocations.

3. Copper

An important system that has been investigated in detail is the deposition of copper from noncomplexing solutions. There now seems little doubt that the mechanism is

$$\begin{aligned} Cu^{2+} + e &\rightarrow Cu^+ \\ Cu^+ + e &\rightleftharpoons Cu \end{aligned} \tag{125}$$

when lattice growth is fast. This has an observed 40 mV anodic and 120 mV cathodic Tafel slopes (*109*, *110*); the crystal growth step is fast and has not been detected or extensively investigated kinetically. A most elegant demonstration of this is the rotating ring and disk experiments of Nekrasov and Berezina (*111*) which allow monitoring of the intermediate Cu^+ at a gold ring electrode after formation at a copper disk. Brown and Thirsk (*112*) have shown that at a rotating copper disk the current can decrease as the rotation speed increases confirming that Cu^+ is an intermediate. At a platinum disk the reduction (*112*) only was observed prior to the equilibrium potential. The principal surface features during deposition at constant current from highly purified 0.25 *M* $CuSO_4$ in 0.1 *M* H_2SO_4 solution have been identified (*113, 114, 115, 116*). The use of constant current could be questioned as it is known that the geometry changes with potential—for example, calomel. Triangular and hexagonal pyramids formed on the {111} planes at a current density of 10 mA cm^{-2}; the {110} planes developed ridges. The {100} planes showed both layer growth and pyramids. It seems that the pyramids could arise from single or groups of dislocations (the number of dislocations was not controlled and was probably large) and the ridges from the piling up of monolayer steps. Exchange currents measured at short times, before the surface had altered from its flat electropolished condition, gave $i^0_{111} < i^0_{100} < i^0_{110}$. The copper deposition is very convenient for electrochemical measurement because of its well-defined Tafel slope. However, because of the uncontrolled dislocation density, the conclusion is not certain.

The change in area during copper dissolution in HCl + HBr solution to form etch pits has been followed by potentiostatic current-time measurement (*117*). Equation (63) of Sec. IIIH 1c predicts the dependence assuming that the rate k of lattice incorporation is constant at a particular potential. Unfortunately the potential dependence was not analyzed, so it is not possible to deduce a mechanism. It was shown, however, that the effective radius of the etch pits increased linearly with time and also with the number of dislocations ($10^5 - 10^7$ cm^{-2}). Optical measurements (*118, 119*) have been made in the Cu + HCl + HBr system for comparison with the theory of Sec. IIIE. The formation of an etch pit at a dislocation is illustrated in Fig. 43. Experimentally single crystals with few dislocations (200–1000 cm^{-2}) were anodically dissolved at constant current. Observation of etch pits allowed graphs of average step velocity $\overline{V}$ as a function of average step spacing $\bar{\varepsilon}$ to be measured. Ledges were ignored and step density was assumed proportional to the measured orientation of the sides of the pit. The graphs ($\overline{V}$ against $\bar{\varepsilon}$) were found to be of two distinct types I and II depending on bromide ion concentration. It was assumed that an electropolished surface had a definite

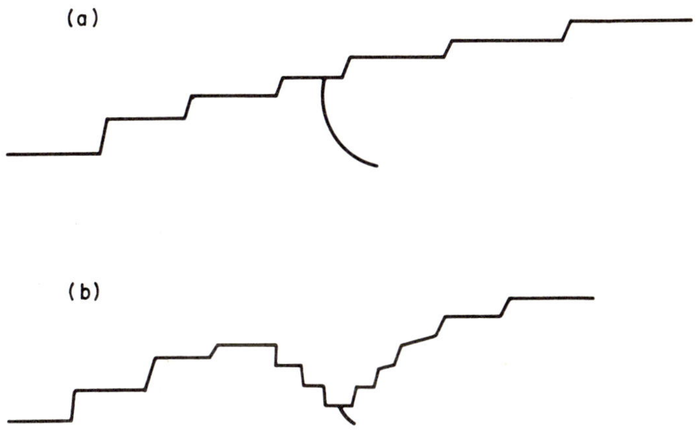

Fig. 43. (a) Terraced surface with a dislocation; (b) formation of an etch pit.

step structure. The f functions $f(\bar{\varepsilon}) = \overline{V}/2$ derived from the experimental curves I and II were then used to predict the change in surface configuration with time for comparison with the experiment. The motion of the nth step Δx_n, for an increment Δt, is given by

$$dX_n = [f(\varepsilon_n) + f(\varepsilon_{n-})]\Delta t \tag{126}$$

The motion was calculated for $n = 100$ by digital computer with cyclic boundary conditions to simulate an infinite surface. Steps were allowed to form multisteps that moved with the velocity of the top step.

Copper morphology in the presence of organic inhibitors has been investigated by Eichkorn and Fischer (*120*). They observe that at a fixed concentration of organic additive, surface structure changes with current density as follows. At low current the copper layer builds up in layers with a preferred orientation; it is assumed that the layers are built up by two-dimensional nucleation and growth. They claim that the organic additive adsorbs initially on the defects on the surface and causes the surface to behave as a dislocation-free surface. Experimental evidence for this is lacking. At intermediate currents the potential oscillates, and the observed structure of the deposit changes from fine grained to coarse grained periodically. Finally, at higher currents three-dimensional nucleation and growth cause a fine-grained deposit with no preferred orientation formed.

Attempts have been made to relate the overpotential-time transients of Fig. 44 to the Volmer equation, that is, the measured number of nuclei (X-rays) were plotted against $1/\eta_{k_3}^2$ for three-dimensional and $1/\eta_{k2}$ for two-

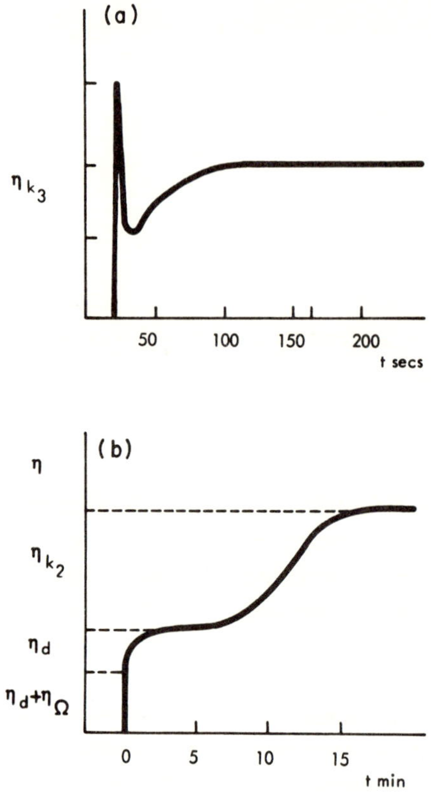

Fig. 44. (a) Initial overvoltage time curve for three-dimensional nucleation on a coarse-grained substrate. (b) Initial overvoltage time curve for two-dimensional nucleation on a fine-grained substrate. Ref. (*120*).

dimensional nucleation. However, the interpretation of the transients is not at all certain unless a mathematical model is set up that can account entirely for the form of the curves and the influence of diffusion can be experimentally determined. We saw in Sec. IIIH that this is very difficult, in principle, for the galvanostatic method.

4. Indium

Studies of both solid and amalgam indium electrodes show the mechanism to be

$$\begin{aligned} \text{In}(0) - 1\text{e} &\rightleftharpoons \text{In}\,(1) \\ \text{In}(1) - 1\text{e} &\rightleftharpoons \text{In}\,(2) \\ \text{In}(2) - 1\text{e} &\rightarrow \text{In}\,(3) \end{aligned} \tag{127}$$

the lattice growth being fast. The first two equations are at equilibrium, the last reaction is slow. The equilibrium potential is such that the second reaction is to the LHS and In(2) is not detected. The mechanism in noncomplexing acid solution has been reviewed by Miller and Visco (*121*) who report ring-disk and split ring-disk experiments. The authors show that homogeneous reaction

$$In(1) + 2H^+ \rightarrow In\,(3) + H_2 \tag{128}$$

or disproportionation of In(1) does not happen in the time that In(1) spends in the diffusion layer.

E. Mercury Substrate, without Electrocrystallization

Advances have been made in the ac method. The theory has been recalculated by Sluyters (*122*), so that it can be directly compared with the measured impedance (as its real and imaginary parts). This has a distinct advantage over the older method of vectorially subtracting the measured double layer capacity in the absence of depolarizer and solution resistance. However, a method in which double layer capacity is selected to give the correct frequency dependence of the faradaic impedance seems equally satisfactory, especially in cases where the depolarizer is specifically adsorbed. Tl^+, Hg^{2+}, K^+ fall into this category and behave to ac as fast reactions with a frequency independent capacity additional to the double layer capacity. Pb^{2+} (*123*), which has a somewhat slower rate, also falls into this group.

It now seems fairly certain that zinc is deposited by a two-electron slow step without the detectable intervention of Zn^+, which has been claimed by some authors. A detailed discussion of the previous literature has been given by Timmer et al. (*124*). Zn^{2+} reduction rate depends strongly on the anion (*125*) associated with it in solution as is well known for other systems (*126*); the transfer remains constant at a value near to 0.7. Teppema et al (*125*) working in solutions containing mixtures of specifically adsorbed anions have suggested that this is due to an associative effect of the discharging Zn^{2+} ions with the adsorbed anion layer effectively lowering the energy of activation rather than the formation of a complex prior to reduction, as observed by Gerischer (*127*), in a number of cases for Zn^{2+} reduction. The investigation of the role of complexes (*128*) is outside the scope of this review.

Cd^{2+} rate of reduction similarly depends on the anion and has been shown by Kooijman et al. (*129*) also to depend somewhat on the pH. However, this is disputed by Hampson et al. (*130*) who do not observe this in very pure solutions.

V. ORGANIC ADDITIVES

The large field of organic additives is of practical importance. Organic compounds affect the morphology of the deposit (leveling and brightening) and also the physical properties (hardness). In general, organic compounds will affect the nucleation and growth characteristics and the charge-transfer reaction. A detailed investigation of an organic compound in an ideal system —that is, the anodic or cathodic process at a dislocation-free metal or mercury—has not yet been attempted. Extensive measurements of the equilibrium amount of organic material adsorbed at mercury have been made by electrocapillary and differential capacity measurements. In some cases also the kinetics of adsorption (*131*) have been investigated. The adsorption on solid metals is more difficult since surface tension measurements cannot be made and differential capacity linear potential sweep and potential pulse measurements are difficult; however, with smooth melted spheres or electropolished single crystals these are possible. Roughness of the surface is likely to introduce an artifact in that the rate of adsorption will appear slow, caused by the outer parts of the surface coming to equilibrium before the inner parts. However, tracer measurements of adsorption on solid metals show similar characteristics to mercury. The amount adsorbed has an approximately bell-shaped dependence on charge or potential. Adsorption falls to zero on either side of the ECM (*132*).

Fischer (*133*) has compared the stationary current-voltage curves obtained during the dissolution and deposition of solid copper in $CuSO_4 + H_2SO_4 +$ organic solution with differential capacity data of the same organic solution, alone and on mercury. The correlation for the V^{2+}/V^{3+} couple (0.01 M) suggests that several compounds are adsorbed in the same way on copper and mercury in the cathodic region, the organic compound then acting as an inhibitor for deposition. Whenever there is evidence for adsorption in the anodic region, the dissolution is also inhibited. The effect of surface-active substances on electron-exchange reactions has been investigated many times in comparison with the few studies of ion reactions.

In the presence of a depositing metal the organic additives are incorporated into the growing metal film and equilibrium coverage with the organic additive can no longer be maintained. Loss of equilibrium means that the concentration of organic additive at the interface is determined by competition between incorporation in the metal and diffusion of organic additive. The most exact measurements of the effect of organic additives have been made under conditions of controlled mass transport using a rotating disk. Only such systems will be discussed here, and the reader is referred to reviews (*134*, *135*, *136*) for a description of earlier work (particularly that of Edwards) mainly carried out at stationary electrodes, or, occasionally, oscillating electrodes. The behavior of organic additive to nickel and copper baths have been those most extensively investigated. At the concentrations used in plating (metal ion 1 M, organic additive 10^{-3} M) the nickel and copper deposition is reaction controlled and it would be expected that the organic deposition would be diffusion controlled. The question to be answered is, what effect does the organic material have on the system. The quantities that can be measured under conditions of controlled mass transfer for a fixed number of coulombs are (a) quantity of metal deposited, (b) quantity of hydrogen evolved, (c) rate of consumption of organic additive, (d) amount of organic additive incorporated into the lattice, and (e) the reaction products on reduction of the organic additive.

If the organic additive is diffusion controlled the flux to a rotating disk is given by

$$f_{\mathrm{org}} = D\left(\frac{C^* - C^0}{\delta}\right) \tag{127}$$

where

$$\delta = 1.62D^{-1/3}\nu^{1/6}\omega^{-1/2} \tag{128}$$

The concentration of the organic additive after time of plating t is given by

$$C_t = C^* \exp\left(-\frac{A'tD}{V\delta}\right) \tag{129}$$

where V is the volume of solution and A' is the area of disk if it is assumed that

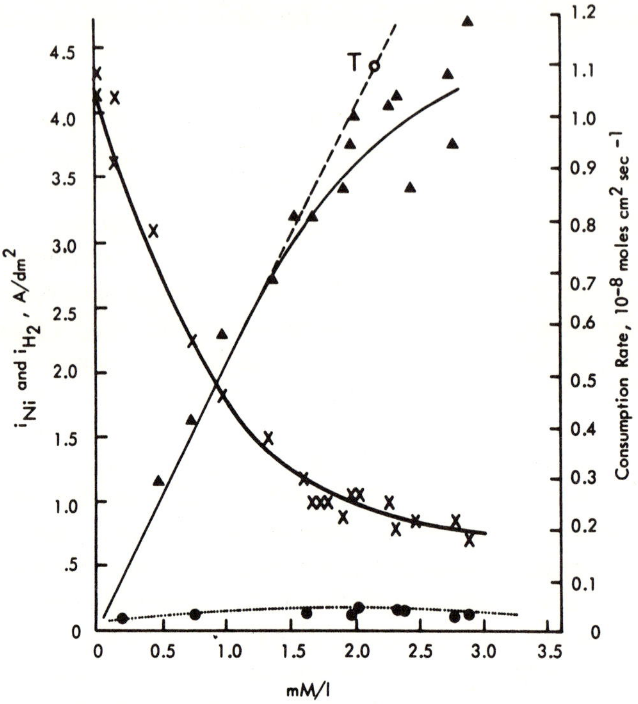

Fig. 45. Rate of coumarin consumption (▲), partial current density for nickel deposition i_{N_i}(X), partial hydrogen evolution current density (●), versus coumarin in Watts nickel bath (−960 mV SCE, 48.5°C, *p*H 4.0, 980 rpm.). Dashed line from Eq. (128). Ref. (*137*), see also (*135*).

$C_0 = 0$. A comparison (*137*) of Eq. (129) with an experimentally determined consumption rate (spectrophotometrically) is shown in Fig. 45. The Fig. 45 also shows the partial currents for nickel H_2 deposition. Up to a concentration of, for example, about 1.5×10^3 *M* coumarin, the organic additive is controlled completely by diffusion with $C_0 = 0$; above this concentration $C_0 = 0$. Under conditions of complete mass transport ($C_0 = 0$) in nickel, deposition is inhibited and the hydrogen reaction stimulated slightly. The reason for this could be (a) secondary effects due to reaction product, or (b) change of the immediate metal layer due to incorporation; (c) is the most likely. In Fig. 45 the nickel deposition rate is approximately 100 times the consumption rate, suggesting specific sites are being blocked by incorporated material.

In the nickel–coumarin system the fate of the coumarin is thought to be

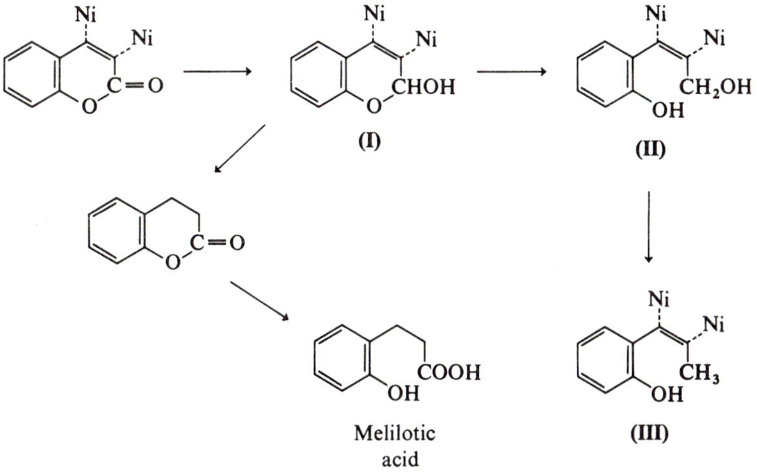

On this model molecules that are reduced further than melilotic acid occupy sites for longer. The ratio of second product to melilotic acid increases with decreasing pH. Similarities of transport control have also been made in the systems nickel-thiourea (*138*), copper-thiourea (*139*), and nickel plus a variety of leveling agents (*140*). The fate of thiourea in nickel is possibly

$$(NH_2)CS_2 + Ni + 2H\bullet \rightarrow NiS + (NH_2)_2CH_2 \qquad (130)$$

the NiS being incorporated into the deposit. This is suggested on the basis of tracer (*141*) experiments to measure the rate of codeposition of sulfur and carbon in the deposit (see also the results of Edwards reviewed in (*135*)). The rate of incorporation of sulphur has also been investigated at a rotating disk (*142*).

Similar results have been obtained for other leveling agents (*135*) such as 2-butyne 1,4 diol, 2-butene 14 diol, quinioline, qumaldine, and chloral hydrate.

Nickel brightening agents on the other hand, saccharin, *p*-toluene sulphonamide, benzene mono- and disulphonic acid, are only diffusion controlled up to 10^{-4} *M*.

VI. LEVELING

Let us consider the model surface shown in Fig. 46. A uniform deposition over the whole surface leads to geometric leveling. The leveling by means of an additive is more efficient. The earlier theories suggested that the preferential adsorption at the peak was responsible—that is, a dependence of adsorption on crystal orientalizes. It now seems accepted on the basis of the experiments in the last section that the mechanism of leveling depends on diffusion control. The peaks are reached by organic additive before the troughs, which consequently grow more quickly.

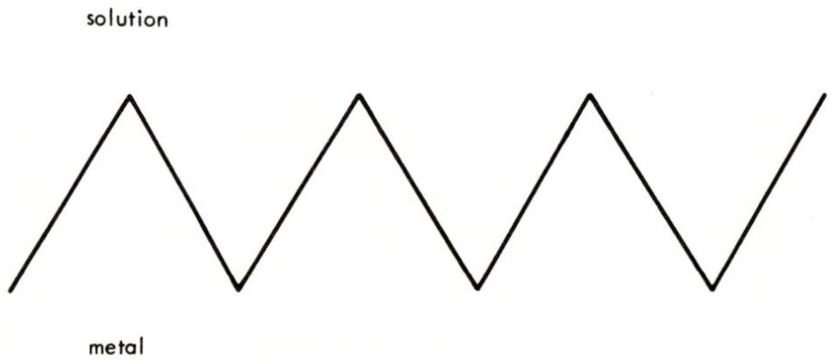

Fig. 46. Model of a rough metal surface.

CONCLUSION

We have tried to show that the threads that make up the problem of metal deposition can be successfully investigated. By choosing conditions so that nucleation and growth (dislocation free substrate) or electron transfer and preceding steps (large number of dislocations or growth sites or deposition into mercury) are rate controlling, a large amount of fundamental information about a metal system can be obtained.

The observation of nonmetal films has lead to an understanding of the theory of phase formation which has, as yet, hardly been applied to metal systems. The time seems ripe for an expansion of knowledge in the general field of metallic and nonmetallic deposition.

ACKNOWLEDGMENT

We would like to thank Dr. R. D. Armstrong for his comments on the manuscript.

SYMBOLS

A	nucleation rate constant
A'	area
C^0, C^*, m	bulk concentrations
C_a^0	equilibrium adatom concentration
C_t	bulk concentration at time t
${}_{\infty}C_a$	adatom concentration
C_o	concentration at $x = 0$
D	solution diffusion coefficient
D_a	adatom diffusion coefficient
E_i	exponential integral
F	the Faraday
i	current per unit area
k_3	forward rate of incorporation at the step line
$k_4 = k_3\, C_a^0$	backward rate of incorporation at the step line
k_1	forward electrochemical rate
k_2	backward electrochemical rate
$2l$	distance apart of step lines
$z = n$	number of electrons
M	molecular weight
N_0	number of nucleation sites at $t = 0$
$\mathcal{N}$	avagadros number
$q_{\text{mon}} = Q_m$	charge in a monolayer $= \dfrac{nF\rho 1}{\text{M}}$
r	radius of a nucleus
S	actual area of nuclei
S_{ex}	area without overlap
t_m	time to reach maximum
V	volume
V_1	molar volume
V_1, V_2, V_3	rate of advance of edge or step
X_0	distance between kinks in steps
Y_0	distance between steps
Z	frequency factor
$\alpha = \alpha_c$	transfer coefficient for cathodic process
βn	$\dfrac{\pi M^2 A k^2}{3\rho^2}$

$\beta^* = \dfrac{\eta F \mu}{RT}$	sweep rate $\sec^{-1}$
$\beta = \alpha_a$	anodic transfer coefficient
δ	Nernst diffusion layer
$\Phi(x)$	$e^{-x^2} \int_0^x e^{-y^2}\, dy$
η_{k2}, η_{k3}	overpotential for two- or three-dimensional nucleation
ρ	density
σ	surface free energy
τ	lifetime of reaction
τ_g	galvanostatic rise time
θ, θ'	see Eq. (79), (89)

REFERENCES

1. J. O'M. Bockris and G. Razumney, *Fundamental Aspects of Electrocrystalization*, Plenum, New York, 1967.

2. J. O'M. Bockris and A. Damjanovic, *Modern Aspects of Electrochemistry*, ed. J. O'M. Bockris and B. E. Conway, Butterworth, London, 1964.

3. M. Fleischmann and H. R. Thirsk, *Advances in Electrochemistry and Electrochemical Engineering*, Vol. 3, ed. P. Delahay, Wiley (Interscience), 1963.

4. K. J. Vetter, *Electrochemical Kinetics*, Academic Press, 1967.

5. K. M. Gorbinova and Yu. M. Polukarov, *Advances in Electrochemistry and Electrochemical Engineering*, Vol. 5, ed. C. W. Tobias, Wiley (Interscience), 1967.

6. K. R. Lawless, *Physics of Thin Films*, Vol. 4, Academic Press, New York, 1967, p. 191.

7. D. J. Arrowsmith, Rep. Prog. Appl. Chem. **51,** 62 (1966).

8. E. Raub and K. Müller, *Fundamentals of Metal Deposition*. Elsevier, Amsterdam, 1967.

9. A. Brenner, *Advances in Electrochemistry and Electrochemical Engineers*, Vol. 5, ed. C. W. Tobias, Wiley (Interscience), 1967.

10. J. P. Hirth and G. M. Pound, *Condensation and Evaporation*, Pergamon, New York, 1963.

11. J. W. Mathews, *Physics of Thin Films*, Vol. 4, Academic Press, New York, 1967. J. P. Hirth, K. L. Moazed, *ibid.* p. 97.

12. H. Heyer, *Angewandte Chem.*, (Internat. ed), **5** (1966).

13. R. D. Gretz and G. M. Pound, *Condensation and Evaporation of Solids*, ed., E. Rutner et al., Gordon and Beach, New York, 1964.

14. T. N. Rhodin and D. Walton, *Single Crystal Films*, ed., M. Francombe, H. Sato, Pergamon, New York, 1964.
J. P. Hirth and S. J. Hruska, and G. M. Pound, *ibid.*

15. R. F. Strickland-Constable, *Kinetics and Mechanism of Crystallisation*, Academic Press, New York, 1968.

16. W. K. Burton, N. Cabreva, and F. C. Frank, *Phil. Trans. Roy. Soc.*, (London) **A243,** 299 (1951).

17. W. Kossel, *Nachr. Gres. Wiss., Gottingen, Jahresber, Math-Physik Kl.*, 135 (1927).
18. I. Stranski, *Z. Physik. Chem.*, **136**, 259 (1928).
19. R. C. Newman, *Phil. Mag.*, **2**, 750, (1951), E. Grunbann, *Proc. Phys. Roy. Soc.*, (London) **72**, 459 (1958).
20. H. M. Montagu-Pollock, T. N. Rhodin, and M. J. Southon, *Surf. Science*, **12**, 1 (1968).
21. J. P. Jones, *Nature*, **211**, 479 (1966).
22. A. J. Melmed, *J. Chem. Phys.*, **43**, 3057 (1965)., J. P. Jones, *Proc. Roy. Soc.* (London), **A 284**, 469, (1965).
23. E. W. Mueller, *Science*, **149**, 591, (1965).
24. G. Ehrlich and F. G. Hudda, *J. Chem. Phys.*, **44**, 1039 (1966).
G. Ehrlich, *J. Chem. Phys.*, **44**, 1050 (1966).
25. M. Fleischmann, S. K. Rangarajan, and H. R. Thirsk, *Trans. Faraday Soc.*, **63**, 1240, 1251, 1256 (1967).
25a. J. A. Harrison, *J. Electroanal. Chem.*, **18**, 377 (1968).
26. W. T. Weeks, *J.A.C.M.*, **13**, 419 (1966).
27. W. Lorenz, Z. *Naturforsch.*, **9a**, 716 (1954).
28. D. A. Vermilyea, *J. Chem. Phys.*, **25**, 1254 (1956).
29. M. Fleischmann, and H. R. Thirsk, *Electrochim. Acta*, **2**, 22 (1960).
30. A. Damjanovic and J. O'M. Bockris, *J. Electrochem. Soc.*, **110**, 1035 (1963).
31. W. Lorenz, *Z. Elektrochem.* **57**, 382 (1953).
32. M. Fleischmann and J. A. Harrison, *J. Electroanal. Chem.*, **12**, 183 (1966).
33. M. Fleischmann and J. A. Harrison, *Electrochim. Acta*, **11**, 749 (1966).
34. K. J. Bachmann and K. J. Vetter, *Electrochim. Acta.*, **11**, 1279 (1966).
35. H. Gerischer and R. P. Tischer, *Z. Elektrochem.*, **61**, 1159 (1957).
35a. M. Fleischmann and J. A. Harrison, *J. Electroanal. Chem.*, **12**, 183 (1966).
36. See Refs. *1*, *2* for summary.
37. S. K. Rangarajan, *J. Electronanal. Chem.*, **16**, 485 (1968).
38. R. de Levie, *Advances in Electrochemistry and Electrochemical Engineering* Vol. 6 ed. P. Delahay, C. W. Tobias, Wiley (*Interscience*), 1967.
39. F. C. Frank, N. Cabrera, and D. A. Vermilyea, *Growth and Perfection of Crystals*, ed. R. H. Doremus et al., Wiley, New York 1958, pp. 411, 393.
A. A. Chirnov, *Dokl. Akad. Nauk SSSR*, **117**, 983 (1957).
N. Cabrera and R. V. Coleman, *The Art and Science of Growing Crystals*, ed. J. J. Gilman, Wiley, New York, 1963 p. 1.
40. M. J. Lighthill and G. B. Whitham, *Proc. Roy. Soc.* (London) **229**, 281, 317 (1955).
41. W. W. Mullins and J. P. Hirth, *J. Phys. and Chem. of Solids*, **24**, 1391 (1963).
42. U. Bertocci, CITCE. Meeting, 1968.
43. S. K. Rangarajan, *J. Electroanal. Chem.*, **17**, 61 (1968).
44. W. Lorenz, *Z. Physik. Chem.* **19**, 377 (1959).
45. J. E. B. Randles, *Discussions Faraday Soc.*, **1**, 11 (1947).
46. H. Gerischer, *Surface*, *66*, ed. N. Ibl. et al., Forster, Zürich, 1967, p. 11.
47. C. Casper, *Trans. Electrochem. Soc.*, **77**, 353 (1940).
48. J. O'M. Bockris and W. Mehl, *J. Chem. Phys.*, **27**, 818 (1957).
49. M. Avrami, *J. Chem. Phys.*, **9**, 177 (1941).
50. R. D Armstrong and J. A. Harrison, *J. Electrochem. Soc.*, **116**, 328 (1969).

51. A. Bewick, M. Fleischmann, and H. R. Thirsk, *Trans. Faraday Soc.*, **58**, 2200 (1962).
52. R. D. Armstrong, M. Fleischmann, and H. R. Thirsk, *J. Electroanal. Chem.*, **11**, 208 (1966).
53. J. Oldfield, *Thesis*, Newcastle upon Tyne, June 1967.
54. D. A. Vermilyea, Advances in Electrochemistry and Electrochemical Engineering ed. P. Delahay, Wiley (Interscience), 1963.
55. M. Fleischmann, J. A. Harrison, and H. R. Thirsk, *Trans. Faraday Soc.*, **61**, 2742 (1965).
56. F. C. Frank, *Proc. Roy. Soc.* (London), **A201**, 586 (1950).
57. P. Delahay and I. Trachtenberg, *J. Amer. Chem. Soc.*, **79**, 2355 (1957).
58. W. H. Reinmuth, *J. Phys. Chem.*, **65**, 473 (1961).
59. D. J. Astley, J. A. Harrison, and H. R. Thirsk, *Trans. Faraday Soc.*, **64**, 192 (1968).
60. W. M. Krebs and D. K. Roe, *J. Electrochem. Soc.*, **114**, 892 (1967).
61. I. N. Stranski and R. Kaishew, *Z. Physik Chem.*, **B126**, 317 (1934).
62. R. D. Armstrong and E. Barr, *J. Electroanal. Chem.*, **20**, 173 (1969).
63. A. Bewick, M. Fleischmann, and H. R. Thirsk, *Trans. Faraday Soc.*, **58**, 2200 (1962).
64. R. D. Armstrong, M. Fleischmann, and H. R. Thirsk, *Trans. Faraday Soc.*, **61**, 2238 (1965).
65. R. D. Armstrong, W. P. Race, and H. R. Thirsk, *J. Electroanal. Chem.*, **19**, 233 (1968).
66. R. D. Armstrong, M. Fleischmann, and H. R. Thirsk, *J. Electroanal. Chem.*, **11**, 208 (1966).
67. R. D. Armstrong, M. Fleischmann, and J. W. Oldfield, *J. Electroanal. Chem.*, **14**, 235 (1967).
68. R. D. Armstrong and M. Fleischmann, *Z. Physik Chem.*, **52**, 131 (1967).
69. R. D. Armstrong, D. F. Porter, and H. R. Thirsk, *J. Phys. Chem.*, **72**, 2300 (1968).
70. M. Fleischmann, J. Pattinson, and H. R. Thirsk, *Trans. Faraday Soc.*, **61**, 1256 (1965).
71. M. Fleischmann, K. S. Rajagopalan, and H. R. Thirsk, *Trans. Faraday Soc.*, **59**, 741, (1963).
72. R. D. Armstrong, J. D. Milewski, and W. P. Race, *J. Electroanal. Chem.*, (in press).
73. R. D. Giles, J. A. Harrison, and H. R. Thirsk, *J. Electroanal. Chem.* (in press).
74. R. D. Armstrong, *J. Electroanal. Chem.* (in press).
75. L. Gurst, P. Herman, *Z. Anal. Chem.*, **216**, 328 (1966).
76. D. J. Astley, J. A. Harrison, and H. R. Thirsk, *J. Electroanal. Chem.*, **19**, 325 (1968).
77. N. A. Pangarov, *J. Electroanal. Chem.*, **9**, 90 (1965).
78. G. Bliznakov and S. Delinechev, *Phys. Stat. Sol.*, **13**, 101 (1966).
79. N. A. Pangarov and S. D. Vitkova, *Electrochim. Acta*, **11**, 1719 (1966).
80. N. A. Pangarov and S. D. Vitkova, *Electrochim. Acta*, **11**, 1733 (1966).
81. N. A. Pangarov and S. D. Vitkova, I. Uzunova, *Electrochim. Acta*, **11**, 1747 (1966).
82. N. A. Pangarov and V. Velinov, *Electrochim. Acta*, **11**, 1753 (1966).
83. N. A. Pangarov, *Phys. Stat. Sol.*, **20**, 365 (1967).
84. N. A. Pangarov and V. Velinov, *Electrochim. Acta*, **13**, 1641 (1968).
85. N. A. Pangarov, *Phys. Stat. Sol.*, **20**, 371 (1967).
86. E. Budewski, V. Bostanov, T. Vitanov, Z. Stoinov, A. Kotseva, and R. Kaischew, *Soviet Electrochem.* **3**, 755 (1967).

87. *Electrochim. Acta*, **11**, 1697 (1966), E. Budewski, T. Vitanov, V. Bostanov, *Phys. Stat. Sol.*, **8**, 369 (1965).

88. E. Budewski, *Electrochim. Metallorum*, **2**, 131 (1966).

88a. *Introduction into the Theory of Crystal Growth*, Bulgarian State Films.

89. D. J. Astley and J. A. Harrison, *Electrochim. Acta*, (to be published).

90. M. Fleischmann, J. Koryta, and H. R. Thirsk, *Trans. Faraday. Soc.*, **63**, 1261 (1967).

91. R. D. Giles, J. A. Harrison, and H. R. Thirsk, *J. Electroanal Chem.*, **20**, 47 (1969).

92. E. Schmidt and H. R. Gygax, *J. Electroanal Chem.*, **12**, 300 (1966).

93. D. J. Astley, J. A. Harrison, and H. R. Thirsk, *J. Electroanal Chem.*, **19** 325 (1968).

94. E. Schmidt and H. R. Gygax, *J. Electroanal Chem.*, **13**, 378 (1967).

95. G. W. Tindall and S. Bruckenstein, *Anal. Chem.*, **40**, 1051 (1968).

96. M. W. Breiter, *J. Electrochem. Soc.*, **114**, 1125 (1967).

97. R. Kaischew and B. Mutaftschiew, *Electrochim. Acta*, **10**, 643 (1965).

97a. S. Toschev and I. Markov, *Ber d. Bunsenges*, **73**, 184 (1969).

98. G. Mammantov, and D. L. Manning, and J. M. Dale, *J. Electroanal. Chem.*, **9**, 253 (1965).

99. T. Berzins and P. Delahay, *J. Am. Chem. Soc.*, **75**, 555 (1953).

100. D. A. Vermilyea, *J. Electrochem. Soc.*, **105**, 286 (1958).

101. J. O'M. Bockris and H. Kita, *J. Electrochem. Soc.*, **109**, 928 (1962).

102. K. E. Heusler and L. Gaiser, *Electrochim. Acta*, **13**, 59 (1968).

103. B. Kabanov, R. Burstein, and A. N. Frumkin, *Discussions Faraday Soc.*, **1**, 259 (1947).

104. J. O'M. Bockris, D. Drasic, and A. R. Despic, *Electrochim. Acta*, **4**, 325 (1961).

105. K. E. Heusler, *Ber. d. Bunsen*, **71**, 620 (1967).

106. W. J. Krawzow, *J. Pren-Tschao Izvest., Leningrad Univ. Ser. Phys. Chem.*, No. 10, 107 (1962).

107. J. O'M Bockris and H. Kita, *J. Electrochem. Soc.* **108**, 676 (1961).

108. K. E. Heusler and R. Knoedler, CITCE Meeting (1968).

109. E. Mattson and J. O'M. Bockris, *Trans. Faraday Soc.*, **55**, 1586 (1959).

110. J. O'M. Bockris and M. Enyo, *Trans. Faraday Soc.*, **58**, 1187 (1962).

111. L. N. Nekrasov and N. P. Berezina, *Dokl. Akad. Nauk SSSR.*, **142**, 855 (1962).

112. O. R. Brown and H. R. Thirsk, *Electrochim. Acta*, **10**, 383 (1965).

113. A. Damjanovic, M. Pannovic, and J. O'M. Bockris, *J. Electroanal Chem.*, **9**, 83 (1965).

114. A. Damjanovic, M. Pannovic, and J. O'M. Bockris, *Electrochim. Acta*, **10**, 111 (1965).

115. A. Damjanovic, T. H. V. Setty, and J. O'M. Bockris, *J. Electrochem. Soc.*, **113**, 429 (1966).

116. U. Bertocci, *J. Electrochem. Soc.*, **113**, 604 (1966).

117. W. Schaarwachter and K. Lücke, *Z. Physik. Chem.*, WF **53** (1967).

118. L. D. Hulett, Jr. and F. W. Young, Jr., *J. Physics Chem. Solids*, **26**, 1287 (1965).

119. L. D. Hulett, Jr. and F. W. Young, Jr., *J. Electrochem. Soc.*, **113**, 410 (1966).

120. G. Eichkorn and H. Fischer, *Z. Physik. Chem.*, **53**, 29 (1967).

121. B. Miller and R. E. Visco, *J. Electrochem. Soc.*, **115**, 251 (1968).

122. J. H. Sluyters, *Rec. Trav. Chim.*, **79**, 1092 (1960).

123. M. Sluyters-Rehbach, B. Timmer, and J. H. Sluyters, *J. Electroanal. Chem.*, **15**, 151 (1967).

124. B. Timmer, M. Sluyters-Rehbach, and J. H. Sluyters, *J. Electroanal. Chem.*, **14**, 181 (1967).

125. P. Teppema, M. Sluyters-Rehbach, and J. H. Sluyters, *J. Electroanal. Chem.*, **16**, 165 (1968).

126. J. E. B. Randles and K. W. Sommerton, *Trans. Faraday Soc.*, **48**, 937, 951 (1952).

127. H. Gerischer, *Z. Physik. Chem.*, **202**, 292, 302 (1953).

128. J. Koryta, *Advances in Electrochemistry and Electrochemical Engineering* Vol. 6, ed. P. Delahay, Wiley (Interscience), 1967.

129. D. J. Kooijmann and J. H. Sluyters, *Electrochim Acta.*, **12**, 693 (1967).

130. N. A. Hampson and D. Larkin, *J. Electroanal. Chem.*, **18**, 401 (1968).

131. A. N. Frumkin and B. B. Damaskin, *Modern Aspects of Electrochemistry*, Vol. 3, ed. J. O'N. Bockris and B. E. Conway, Butterworths, London, 1964.

132. E. Gileadi, L. Duic, and J. O'M. Bockris, *Electrochim. Acta*, **13**, 1915 (1968).

133. H. Fischer and W. Seiler, *Corrosion Science*, **6**, 159 (1966).
M. W. Nippe and H. Fischer, *Electrochim. Acta*, **12**, 369 (1967).

134. O. Kardos and G. Foulke, *Advances in Electrochemistry and Electrochemical Engineering*, Vol. **2**, ed. P. Delahay, C. W. Tobias, Wiley (Interscience), 1962.

135. O. Kardos, *Surface 66*, ed. N. Ibl. et al., Forster, Zurich, 1967.

136. N. Ibl. *Surface 66*, ed. N. Ibl. et al., Forster, Zurich, 1967.

137. G. T. Rogers, K. J. Taylor, *Electrochim. Acta*, **13**, 2189, 109 (1968); **11**, 1685 (1966); **8**, 887, (1963); *Trans. Inst. Met. Fin.*, **43**, 75, (1965).

138. P. L. Javet, N. Ibl, and H. Hintermann, *Galvanotechnik und Oberflächenschutz*, **8**, 231 (1967).

139. P. L. Javet, N. Ibl, and H. Hintermann, *Electrochim. Acta*, **12**, 781 (1967).

140. S. S. Kruglikov, N. T. Kudryavtsev, G. F. Vorobyeva, and A. Ya. Antonov, *Electrochim. Acta*, **10**, 253 (1965).

141. Yu. K. Uyagis, A. I. Bodnevas, and Yu. Yu. Matulis, *Zashchita. Met.*, **1**, 359 (1965); **2**, 201 (1966).

142. S. S. Kruglikov, Yu. I. Sinyakov, and N. T. Kudriavtsev, *Electrokhimii*, **2**, 100 (1966).

Chemical Reactions in Polarography

Rolando Guidelli

INSTITUTE OF ANALYTICAL CHEMISTRY
UNIVERSITY OF FLORENCE
FLORENCE, ITALY

I. INTRODUCTION

A. Entropy Production in a Galvanic Cell

When a finite current flows through a galvanic cell, this latter becomes the site of several irreversible phenomena. Imagine a polarographic cell consisting of a saturated calomel reference electrode and of a dropping cadmium amalgam electrode dipping in a solution of the cadmium complex with nitrilotriacetic acid (Fig. 1). If the sliding contactor of the potentiometer *P* in Fig. 1 is in such a position that the external potential difference applied between the dropping electrode and the SCE is equal to the equilibrium potential difference $\Delta\varphi_{eq}$ between these two electrodes, no current flows through the polarographic circuit and the cell can be considered in mechanical, thermal, and electrochemical equilibrium. If the absolute value of the external

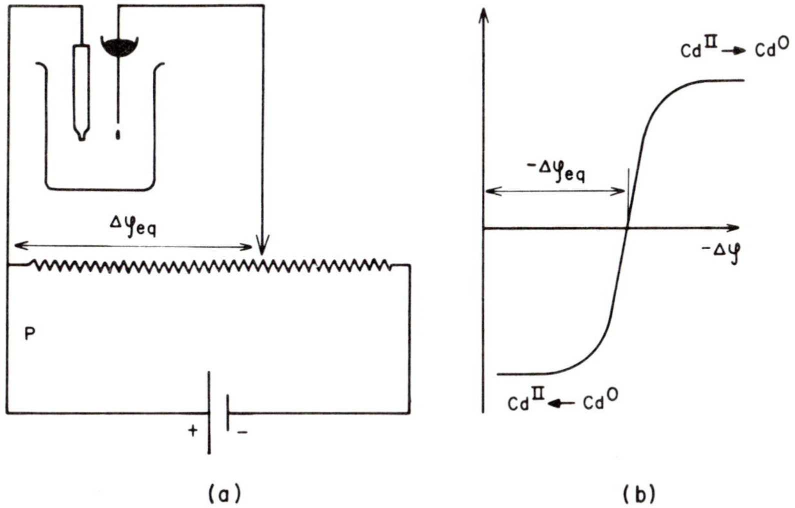

Fig. 1. (a) Scheme of a polarographic circuit. (b) Polarographic wave of Cd^{II} at a cadmium amalgam electrode.

potential difference is made infinitesimally smaller than $|\Delta\varphi_{eq}|$, during the time interval dt there is a transfer of a positive charge $-i\,dt$ through the external circuit from the positive electrode (the SCE) to the negative one (the dropping amalgam electrode) and an infinitesimal amount of metallic cadmium is oxidized to *Cd*(II), passing into solution. The polarographic cell can be considered as a closed thermodynamic system—namely, a system exchanging energy but no matter with its surroundings—insofar as the number of electrons leaving the cell through one lead in a given time interval dt is equal to the number of electrons entering the cell through the other lead in the same time.

During the reversible process previously considered, electrical work is done by the cell on its surroundings. Hence, according to the first law of thermodynamics, the infinitesimal change of internal energy dU taking place within the cell equals the amount of heat dq_R absorbed from the outside minus the mechanical work $P\,dV$ as well as the electrical work $\Delta\varphi_{eq}\,i\,dt$ done on the surroundings (*1*):

$$dU = dq_R - P\,dV - \Delta\varphi_{eq}\,i\,dt \tag{1}$$

It must be noted that in the case under examination $\Delta\varphi_{eq}$ is negative and that the anodic current i is also taken as negative in accordance with the IUPAC convention, so that $\Delta\varphi_{eq}\, i\, dt$ is positive. Equation (1) holds also if the absolute value of the external potential difference is made infinitesimally larger than $\Delta\varphi_{eq}$. In this case electricity is transferred in the opposite direction and cadmium ions are deposited into the amalgam. Thus the cathodic current i flowing through the cell is positive in sign and the contribution $-\Delta\varphi_{eq}\, i\, dt$ to the internal energy change is now positive, in accordance with the fact that electrical work is done on the cell. In view of the second law of thermodynamics, the infinitesimal amount of heat dq_R absorbed by the cell during the reversible processes previously considered equals $T\, dS$, where dS is the entropy change of the system. Thus Eq. (1) can be written in the form

$$dU = T\, dS - P\, dV - \Delta\varphi_{eq}\, i\, dt \tag{2}$$

If the absolute value of the external potential difference $\Delta\varphi$ applied to the cell is now made appreciably larger than $|\Delta\varphi_{eq}|$ by sliding the contactor to the right, a finite cathodic current i flows along the circuit and the following irreversible processes take place in the cell around the dropping electrode: (i) The electrode reaction,

$$Cd^{2+} + 2e \rightleftharpoons Cd(Hg) \tag{3}$$

expressing the reduction of the electroactive free cadmium ions to metallic cadmium, is not in equilibrium, so that the rate of the forward reaction is greater than that of the backward. (ii) The same is true for the homogeneous chemical reaction,

$$CdX^- \rightleftharpoons Cd^{2+} + X^{3-} \tag{4}$$

producing the electroactive species Cd^{2+} from the electroinactive complex CdX^- of cadmium with the nitrilotriacetic ion X^{3-} (*2*). (iii) CdX^- and X^{3-} diffuse, respectively, toward and from the electrode under a finite concentration gradient, whereas Cd atoms formed on the electrode surface diffuse toward the center of the growing drop. (iv) The flow of the charged species, including those constituting the supporting electrolyte, occurs under the partial control of the electric potential gradient (field-assisted diffusion).

Imagine that the same thermodynamic transformation characterized by an infinitesimal energy change dU, a mechanical work $P\, dV$, and a flow of $i\, dt$ charges is carried out both under reversible and irreversible conditions. In the former case Eq. (2) holds, while in the latter, only the first law of thermodynamics can be applied,

$$dU = dq - P\,dV - \Delta\varphi i\,dt \tag{5}$$

where dq is the heat absorbed irreversibly by the cell. Comparing Eqs. (2) and (5) we have

$$T\,dS - dq = (\Delta\varphi_{\text{eq}} - \Delta\varphi)i\,dt \tag{6}$$

It is easy to see that the right-hand side of Eq. (6) is always positive. In fact, if $|\Delta\varphi|$ is larger than $|\Delta\varphi_{\text{eq}}|$, a cathodic current flows through the dropping electrode and i is positive, just as $\Delta\varphi_{\text{eq}} - \Delta\varphi$ is positive. Conversely, if $\Delta\varphi_{\text{eq}} - \Delta\varphi$ is negative, cadmium is oxidized at the dropping electrode and i is negative. From Eq. (6) it can be concluded that the amount of heat dq supplied to the cell by its surroundings during an irreversible process is always less than $T\,dS$, which expresses the heat dq_{R} that would have been absorbed in the same process, had it been carried out reversibly. These conclusions are not surprising if we recall the fundamental differences existing between a "reversible" and an "irreversible" process undergone by a given thermodynamic system in passing from an initial equilibrium state 1 to a final equilibrium state 2.

If the process is reversible, the integral $\int_1^2 dq_{\text{R}}/T$ (where dq_{R} is the infinitesimal amount of heat absorbed by the system in a given infinitesimal portion of the whole process and T is the temperature at which dq_{R} is absorbed) depends exclusively on the initial and final equilibrium states and not on the particular reversible path followed. This circumstance allows an important thermodynamic function, namely, the entropy S, to be defined. Thus the entropy difference, $S_2 - S_1$, between two equilibrium states is simply given by

$$S_2 - S_1 = \int_1^2 \frac{dq_{\text{R}}}{T}$$

or, for an infinitesimal reversible process,

$$T\,dS = dq_{\text{R}} \tag{7}$$

If the process undergone by the system in passing from state 1 to state 2 is "irreversible," we know from thermodynamics that the integral $\int_1^2 dq/T$

along the irreversible path depends on the particular path followed and that the following inequality holds,

$$\int_1^2 \frac{dq}{T} < S_2 - S_1 \tag{8}$$

Analogously for an infinitesimal irreversible process,

$$T\,dS > dq \tag{9}$$

Equations (7) and (9) are two of the several forms in which the second law of thermodynamics can be stated. Clearly, Eq. (6) represents an application of the second law (Eq. (9)) to the particular case of a polarographic cell, although the positive value of the difference $T\,dS - dq$ is expressed by an inequality in Eq. (9) and by an equality in Eq. (6).

1. The Concept of Entropy Production

The temptation to express the second law for irreversible processes under the form of an equation, as is the case for reversible processes, must have been strong among scientists, if already during the last half of the nineteenth century Clausius wrote the following equation for the entropy change of a system undergoing an infinitesimal irreversible process,

$$dS = \frac{dq}{T} + \frac{dq'}{T} \tag{10}$$

where $dq/T \equiv d_e S$ is the entropy "supplied" to the system by its surroundings and $dq'/T \equiv d_i S$ is the entropy "produced" within the system by irreversible processes. The consideration of entropy as a quantity that can flow represents the starting point for the "thermodynamics of irreversible processes."

The concept of entropy "flow" and entropy "production" can be better visualized if we consider a simple thermodynamic system consisting of two heat reservoirs, the former at temperature T_1 and the latter at a lower temperature T_2, and of a metallic wire connecting the two reservoirs and conducting heat from the hotter to the cooler one. Denote by Q the absolute value of the amount of heat transferred through the wire in unit time under steady-state conditions. Since, by definition, the mass of a heat reservoir must be so large that the flow of a finite amount of heat does not appreciably alter its temperature or any other thermodynamic function, it follows that in the unit time the heat Q is rejected reversibly by the hot reservoir and absorbed reversibly by the cool one. Hence, in view of Eq. (7) the entropy decrease suffered by the hot reservoir is $\Delta S_1 = -Q/T_1$, whereas the entropy increase suffered by the cool reservoir is $\Delta S_2 = Q/T_2$. Since the metallic wire does not undergo any entropy change under steady-state conditions, owing to the constancy of its thermodynamic state with time, we can conclude that the entropy change of the whole system per unit time is given by $\Delta S = \Delta S_1 + \Delta S_2 = Q(T_1 - T_2)/(T_1 T_2)$, and therefore is positive. If we set $T_2 = T$ and $T_1 = T + \Delta T$ and we assume that $\Delta T \ll T$, we can write

$$\Delta S = \left(\frac{Q}{T}\right)\left(\frac{\Delta T}{T}\right) \tag{11}$$

The whole system, consisting of the two reservoirs and the wire, is thermally isolated insofar as it does not exchange heat with its surroundings ($\int dq/T = 0$) and, consequently, the positive value of ΔS is in agreement with Eq. (8).

If we consider entropy as a quantity that can flow, we may say that during the flow of an amount Q of heat along the wire the hot reservoir "cedes" an amount $-\Delta S_1 = Q/T_1$ of entropy to the wire while in the same unit time the cool reservoir "gains" an amount $\Delta S_2 = Q/T_2$ of entropy from the wire. Since, however, Q/T_1 is less than Q/T_2, we are led to conclude that the flow of entropy out of the wire is greater than the flow in. Proceeding with this line of reasoning and focusing our attention on the wire, we are forced to assume that entropy is "produced" within the wire at a rate sufficient to compensate for the difference between the rate of outflow and that of inflow. Thus the entropy $\Delta_i S$ produced in unit time within the wire must be given by Eq. (11). This entropy production must be necessarily ascribed to the irreversible flow of heat taking place under the influence of the finite temperature gradient existing in the wire. In fact, if the temperature difference ΔT between the two reservoirs were infinitesimal, the flow of the finite amount Q of heat would require an infinitely long time to take place but the concomitant entropy $\Delta_i S$ produced within the wire would be practically zero (see Eq. (11)).

It is evident in view of the second law (Eq. (9)) and of Eq. (10), that the entropy $d_i S$ produced by irreversible processes is always positive, independent of the positive, zero, or negative value of the entropy $d_e S$ supplied by the surroundings. Thus if we still consider the two reservoirs connected by the wire and we focus our attention on the subsystem consisting of the sole wire, we see that the entropy production

$$\Delta_i S = \frac{d_i S}{dt} = \left(\frac{Q}{T}\right)\left(\frac{\Delta T}{T}\right) \tag{12}$$

is positive whereas the corresponding entropy flow is negative (i.e., there is a net flow of entropy out of the wire), so that the overall entropy change in the wire per unit time is zero. In an analogous way the entropy produced in a polarographic cell during the flow of $-i\,dt$ charges, which is expressed by Eq. (6),

$$d_i S = dS - \frac{dq}{T} = dS - d_e S = \frac{\Delta\varphi_{eq} - \Delta\varphi}{T}(i\,dt) \tag{13}$$

is positive, as previously shown. If we compare Eqs. (12) and (13), we see that in both cases the entropy production $\Delta_i S = d_i S/dt$ is given by the "generalized driving force" giving rise to the irreversible process in question (i.e., the finite temperature gradient divided by T, $\Delta T/T$, in the case of heat transfer, or the finite difference $(\Delta\varphi_{eq} - \Delta\varphi)$ divided by T in the case of charge transfer) multiplied by the corresponding "generalized flow" (i.e., Q/T for heat transfer or i for charge transfer).

In general it can be shown that if more generalized driving forces are simultaneously operative in a given system, the entropy production within the system is given by the sum of the products of each "generalized force" by the corresponding "generalized flow."

2. The Balance Equation for Entropy. The Expression of the Entropy Production Term

The calculation of entropy production during the evolution of any given irreversible process is one of the main goals of the "thermodynamics of irreversible processes" (*3*, *4*). This branch of science treats nonequilibrium processes from a thermodynamic macroscopic point of view without introducing models and assumptions of a molecular or mechanistic nature. In this sense the relationships derivable from thermodynamics of irreversible processes remain valid regardless of how concepts of the molecular structure of a given system may evolve.

The calculation of $d_i S/dt$ rests on one fundamental hypothesis according to which, outside of the thermodynamic equilibrium, the entropy S is assumed to depend on space and time coordinates, not explicitly but rather through the usual variables energy U, volume V, and composition. In other words, the familiar Gibbs' equation

$$T\,dS = dU + P\,dV - \sum_i \mu_i\,dn_i \tag{14}$$

in which P is the pressure, n_i is the number of moles, and μ_i is the molar chemical potential of the ith component of the system, is supposed to remain valid out of equilibrium—that is, out of the domain of classical thermodynamics. The acceptance of Gibbs' relation can be justified by statistical reasonings, but from the standpoint of the thermodynamics of irreversible processes it can be considered as a real postulate.

In this contribution we are particularly concerned with chemical reactions coupled with charge-transfer processes taking place at an electrode; consequently we shall focus our attention on that particular thermodynamic system consisting of the region immediately adjacent to a metal-solution interphase. During the flow of current the intensive state variables defining this system, for example, the concentrations of the various diffusing species, are both time dependent and "nonuniform"—that is, functions of the space coordinates. We then say that the region adjacent to the surface of an electrode is a "continuous system." In order to find the expression for the entropy production within this system, three fundamental relations are required: namely, the law of conservation of mass, the first law of thermodynamics, and the second law in the form of Gibbs' equation. In this connection imagine that the components A_i of the system, n in number, are bounded by m chemical reactions,

$$\sum_{i=1}^{n} \nu_{i,\rho} A_i = 0 \qquad \rho = 1, \ldots, m$$

where $\nu_{i,\rho}$ is the stoichiometric coefficient of the ith component in the ρth reaction. The stoichiometric coefficients are taken as positive for products and negative for reactants. Let us define the rate J_ρ of the ρth reaction by the equation,

$$J_\rho = \frac{1}{\nu_{i,\rho}} \frac{d_\rho C_i}{dt} \qquad i = 1, \ldots n \tag{15}$$

Here $d_\rho C_i/dt$ is the number of moles of the ith component produced per unit of time per unit volume as a consequence of the ρth reaction. Obviously Eq. (15) defines J_ρ independent of the choice of the component A_i participating in the ρth reaction. Let us define by $\mathbf{v}_i$ the mean velocity of A_i. For the sake of simplicity we shall assume that the solution around the electrode is sufficiently dilute and that it is not subject to convective motions. Under these assumptions, provided no external forces act on the solvent molecules, we may reasonably suppose that the weighted average of the velocities of all components, solvent included,

$$\mathbf{v} = \sum_i r_i \mathbf{v}_i \tag{16}$$

is practically zero. In Eq. (16) r_i is the mass of the component A_i per unit mass of the system.

With the previous definitions and simplifying assumptions, the law of conservation of mass for any given component A_i is given by

$$\frac{dC_i}{dt} = -\text{div}\, \mathbf{J}_i + \sum_{\rho=1}^{m} \nu_{i,\rho} J_\rho \tag{17}$$

where

$$\mathbf{J}_i \equiv C_i \mathbf{v}_i \tag{18}$$

expresses the flux of A_i—that is, the number of moles of A_i flowing in unit time through an ideal surface of unit area normal to the direction of motion. We recall that in view of the well-known Gauss' theorem the divergence of the flux of any given substance, or more generally of any given property, equals the total amount of that particular substance or property flowing out of the unit volume per unit time. Hence Eq. (17) expresses the simple fact that the change dC_i/dt in the number of moles of A_i per unit time per unit volume equals the number of moles of A_i entering the unit volume per unit time (i.e., $-\text{div}\, \mathbf{J}_i$) plus the number of moles $\sum_{\rho=1}^{m} \nu_{i,\rho} J_\rho$ of A_i produced per unit time per unit volume as a consequence of the m chemical reactions. The first term on the right-hand side of the "balance equation" (17) is frequently referred to as the "flow term" whereas the second is called the "source term."

The first law of thermodynamics as applied to the unit mass of our continuous system is

$$du = dq - P\,dv + \frac{1}{\rho}\sum_i (\mathbf{F}_i \cdot \mathbf{J}_i)\,dt \tag{19}$$

Here u is the internal energy per unit of mass, v is the specific volume, $\rho = 1/v$ is the density, $\mathbf{F}_i$ is the external force acting on one mole of the substance A_i, and $\mathbf{J}_i$ is the flux of A_i as defined by Eq. (18). Equation (19) expresses the fact that the internal energy change during dt equals the heat dq added per unit of mass during dt, minus the infinitesimal work of expansion $P\,dv$ done by the system and plus the total work done by the external forces $\mathbf{F}_i$ on the various components of the system during the same time interval, dt. This latter work is clearly given by the summation, extended to all components, of the product of the force $C_i\mathbf{F}_i/\rho$ acting on the amount of each component contained in the unit mass by the displacement $\mathbf{v}_i\,dt$ of this component during dt.

The heat dq added per unit of mass is a kind of energy "in flow." Consequently the corresponding heat $\rho\,dq$ added per unit of volume, likewise the number C_i of moles of A_i per unit of volume, can be considered to obey a "balance equation" of the type given in Eq. (17). Since, however, heat is by definition, a kind of energy that can only flow into or out of a given volume element but cannot be produced within it (consider for instance the heat Q transferred through the wire from the hotter to the cooler reservoir in the example of Sec. I.A.1), clearly the balance equation applied to $\rho\,dq$ does not contain the source term,

$$\rho\,\frac{dq}{dt} = -\operatorname{div}(\mathbf{J}_q) \tag{20}$$

Equation (20) defines $\mathbf{J}_q$, which is termed the "flow of heat."

The last equation required in order to obtain the expression of the entropy production is the familiar Gibbs' equation (14), which, for the continuous system under study, takes the form,

$$du = T\,ds - P\,dv + \sum_i \mu_i\,d\,\frac{C_i}{\rho} \tag{21}$$

Here u, v, and s are the specific internal energy, volume, and entropy, respectively, C_i/ρ is the number of moles of A_i per unit of mass, and μ_i is the molar chemical potential. Substituting dq from Eq. (20) into Eq. (19), du from Eq. (19) into Eq. (21), and dC_i from Eq. (17) into Eq. (21), we obtain the following equation for entropy,

$$\rho T \frac{ds}{dt} = -\operatorname{div} \mathbf{J}_q + \sum_i \mathbf{F}_i \cdot \mathbf{J}_i + \sum_i \mu_i \operatorname{div} \mathbf{J}_i - \sum_{\rho=1}^{m} J_\rho \sum_i \nu_{i,\rho} \mu_i \tag{22}$$

Taking into account the well-known vector property,

$$\operatorname{div}(a\mathbf{w}) = a \operatorname{div} \mathbf{w} + \mathbf{w} \cdot \mathbf{grad}\, a$$

Equation (22) can be written in the more compact form (*4*):

$$\rho \frac{ds}{dt} = -\operatorname{div}\left(\frac{\mathbf{J}_q - \sum_i \mu_i \mathbf{J}_i}{T}\right) + \frac{\mathbf{J}_q \cdot \mathbf{X}_q + \sum_i \mathbf{J}_i \cdot \mathbf{X}_i + \sum_\rho A_\rho J_\rho}{T} \tag{23}$$

where

$$\mathbf{X}_q \equiv -\frac{\mathbf{grad}\, T}{T}$$

$$\mathbf{X}_i \equiv \mathbf{F}_i - T\, \mathbf{grad} \frac{\mu_i}{T}$$

$$A_\rho \equiv -\sum_i \nu_{i,\rho} \mu_i$$

Comparing Eq. (23) with the balance equation (17), we immediately see that Eq. (23) also has the form of a balance equation, the first term on the right representing a "flow" term and the second a "source" term. The flow term expresses the external contribution to $\rho\, ds/dt$; $-\operatorname{div}(\mathbf{J}_q/T)$ is the entropy brought into a unit volume of the system by the heat flux whereas $\operatorname{div}(\sum_i \mu_i \mathbf{J}_i)/T$ corresponds to the entropy inflow due to the diffusion fluxes of the various components A_i.

The source term

$$\frac{\mathbf{J}_q \cdot \mathbf{X}_q + \sum_i \mathbf{J}_i \cdot \mathbf{X}_i + \sum_\rho A_\rho J_\rho}{T} \equiv \sigma \tag{24}$$

is more important from our standpoint, expressing the "entropy production" per unit volume of the system per unit time. $T\sigma$ is the sum of the products of the generalized forces $\mathbf{X}_q$, $\mathbf{X}_i$, and A_ρ, representing the direct causes of the various irreversible phenomena by the corresponding "generalized fluxes" $\mathbf{J}_q$, $\mathbf{J}_i$, and J_ρ, which measure the effects of these forces. Thus $(\mathbf{J}_q \cdot \mathbf{X}_q)/T$ is the entropy production due to the heat flow within the system caused by temperature gradients (compare with Eq. (11) for the entropy production within the conducting wire); $\sum_i (\mathbf{J}_i \cdot \mathbf{X}_i)/T$ is the entropy production due to the diffusion flows within the system, caused by concentration gradients and by the external forces $\mathbf{F}_i$; finally $\sum_\rho A_\rho J_\rho/T$ is the entropy production due to the occurrence of the m chemical reactions. In this connection it is worthwhile observing that the driving force $A_\rho \equiv -\sum_i \nu_{i,\rho}\mu_i$ for the ρth chemical reaction is the opposite of the free-energy change ΔG_ρ accompanying one "occurrence" of this reaction at constant temperature and pressure. From classical thermodynamics we know that ΔG_ρ, and also A_ρ, are zero when the ρth reaction is at equilibrium. Now we can see that $-\Delta G_\rho = A_\rho$ not only measures the shift of the ρth chemical reaction from equilibrium, but also expresses the force driving this reaction toward equilibrium. The contribution to the entropy production due to the occurrence of the ρth reaction is therefore given by the product of A_ρ by the rate J_ρ of the reaction in question, divided by T.

3. Entropy Production at Constant Temperature in the Presence of Coulombic Forces

Let us assume for simplicity that temperature is uniform throughout our continuous system and that the only external forces $\mathbf{F}_i$ are coulombic in nature. Hence,

$$\mathbf{F}_i = -z_i F \operatorname{grad} \varphi \tag{25}$$

where z_i is the valence of A_i, F is the faraday, and φ is the inner electric potential. Substitution of $\mathbf{F}_i$ from Eq. (25) into the first law Eq. (19) yields

$$du = dq - P\,dv - \frac{1}{\rho}\,\mathbf{grad}\,\varphi \cdot \mathbf{I}\,dt \tag{26}$$

where

$$\mathbf{I} \equiv \sum_i z_i F \mathbf{J}_i$$

Clearly the vector $\mathbf{I}$, expressing the sum of the products of the fluxes $\mathbf{J}_i$, in moles $cm^{-2} sec^{-1}$, of the various charged species by the corresponding molar charges $z_i F$, is the electric current density. On eliminating du from Eqs. (21) and (26), we have

$$T\,ds + \sum_i \mu_i\, d\,\frac{C_i}{\rho} = dq - \frac{1}{\rho}\,\mathbf{grad}\,\varphi \cdot \mathbf{I}\,dt \tag{27}$$

On the other hand, at constant temperature the entropy balance of Eq. (23) assumes the form

$$T\,ds = dq + \frac{1}{\rho}\,\mathrm{div}\Bigl(\sum_i \mu_i \mathbf{J}_i\Bigr)\,dt + \frac{T\sigma}{\rho}\,dt \tag{28}$$

where account has been taken of Eq. (20). Elimination of $T\,ds$ from Eqs. (27) and (28) results in

$$\sum_i \mu_i\, d\,C_i + \mathbf{grad}\,\varphi \cdot \mathbf{I}\,dt + \mathrm{div}\Bigl(\sum_i \mu_i \mathbf{J}_i\Bigr)\,dt + T\sigma\,dt = 0 \tag{29}$$

Imagine that a volume element of our continuous system is the site of an infinitesimal transformation. Assume that this transformation is characterized by a change dC_i in the concentration of A_i, by the passage of $|\mathbf{J}_i|\,dt$ moles of A_i through a unit surface normal to $\mathbf{J}_i$ as well as by the passage of $|\mathbf{I}|\,dt$ electric charges through a unit surface normal to $\mathbf{I}$. If the above transformation is carried out irreversibly, Eq. (29) applies. Conversely, if the transformation is carried out slowly, under quasistatic conditions and therefore under the equilibrium potential gradient $\mathbf{grad}\,\varphi_{eq}$, no irreversible phenomena take place, and consequently the entropy production σ is zero. Hence Eq. (29) becomes

$$\sum_i \mu_i\, dC_i + \mathbf{grad}\,\varphi_{eq} \cdot \mathbf{I}\,dt + \mathrm{div}\Bigl(\sum_i \mu_i \mathbf{J}_i\Bigr)\,dt = 0 \tag{30}$$

Subtracting Eq. (30) from Eq. (29), we have (*5*)

$$\mathbf{I} \cdot \mathbf{grad}(\varphi_{eq} - \varphi) = T\sigma \tag{31}$$

Noting that the direction of $\mathbf{I}$ is the same as that of the moving positive charges, it is readily seen that the scalar product $\mathbf{I} \cdot \mathbf{grad}(\varphi_{eq} - \varphi)$ is positive in accordance with the positive value of the entropy production. Equation (31), relating the entropy production σ to the local current density $\mathbf{I}$ and to the local electric potential gradient, is perfectly analogous to Eq. (13), relating the entropy d_iS/dt produced per unit time in a polarographic cell to the current intensity i and to the potential difference applied at the terminals of the cell. The main difference between the two above equations lies in the fact that Eq. (31) is in local form and can therefore be applied at any point both of an electrolytic cell and of a galvanic element.

At constant temperature and in the absence of external forces other than those that are coulombic, Eq. (24) yields the following expression for the entropy production,

$$T\sigma = -\sum_i \mathbf{J}_i \cdot \mathbf{grad}(z_i F\varphi + \mu_i) + \sum_\rho A_\rho J_\rho \tag{32}$$

The term $(z_i F\varphi + \mu_i)$ is called the "electrochemical potential," $\tilde{\mu}_i$, of A_i. It consists of the electrical term $z_i F\varphi$ and of the chemical potential $\mu_i = \mu_i^\circ + RT \ln a_i$, where μ_i° and a_i are the standard chemical potential and the activity of A_i, respectively. **Grad** $\tilde{\mu}_i$ is the generalized force driving the species A_i to move at a rate $\mathbf{v}_i = \mathbf{J}_i/C_i$. When $\tilde{\mu}_i$ is uniform (grad $\tilde{\mu}_i = 0$), the influence of the activity gradient **grad** a_i on the flow of the ith species is perfectly counterbalanced by the influence of the electric field strength **grad** φ operating in the opposite direction,

$$RT\,\mathbf{grad}(\ln a_i) = -z_i F\,\mathbf{grad}\,\varphi$$

Consequently the flux $\mathbf{J}_i$ of A_i equals zero.

It is interesting to note that the chemical affinity A_ρ of the ρth chemical reaction, defined in Eq. (23), coincides with the corresponding electrochemical affinity $\tilde{A}_\rho$, defined by the equation,

$$\tilde{A}_\rho \equiv -\sum_i \nu_{i,\rho}\tilde{\mu}_i = A_\rho - \left(\sum_i \nu_{i,\rho} z_i\right) F\varphi \tag{33}$$

In fact the electric charge remains unaltered during a chemical reaction, and consequently $\sum_i \nu_{i,\rho} z_i = 0$.

Combination of Eqs. (31) through (33) yields the equation,

$$T\sigma = \mathbf{I} \cdot \mathbf{grad}(\varphi_{\text{eq}} - \varphi) = -\sum_i \mathbf{J}_i \cdot \mathbf{grad}\,\tilde{\mu}_i + \sum_\rho \tilde{A}_\rho J_\rho \tag{34}$$

which is the starting point for the separation of overpotential into different contributions.

B. Overpotential

1. The Various Types of Overpotential (*5*)

For the sake of simplicity consider a plane electrode and imagine that the various components A_i move only in the direction normal to the electrode surface. Let us focus our attention on a system consisting of an ideal cylinder of unit base normal to the electrode and bounded by the planes $x = 0$ and $x = c$, where x is the distance from the electrode and c a point in the bulk of the solution. During electrolysis, the entropy produced within the above cylinder per unit time is obtained by integrating Eq. (34) over the volume of the cylinder in accordance with the fact that σ is the entropy produced per unit time per unit volume. Noting that at any given instant the magnitude of the current density **I** is uniform throughout the solution up to the electrode surface and that $\mathbf{J}_i$ and grad $\tilde{\mu}_i$ have opposite directions, we obtain

$$-[(\varphi^0 - \varphi^c) - (\varphi^o_{\text{eq}} - \varphi^c_{\text{eq}})]I \equiv -\eta^{0c}I = T\int_0^c \sigma\, dx$$

$$= \int_0^c \sum_i |\mathbf{J}_i| \left|\frac{\partial \tilde{\mu}_i}{\partial x}\right| dx + \int_0^c \sum_\rho \tilde{A}_\rho J_\rho\, dx \quad (35)$$

In this equation I is the scalar current density, φ^0 and φ^c are the inner potentials at $x = 0$ and $x = c$, respectively, whereas η^{0c} is, by definition, the overpotential between $x = 0$ and $x = c$. We shall note that a positive $\mathbf{grad}(\varphi_{\text{eq}} - \varphi)$ causes an anodic process to take place at the electrode surface and consequently **I**, whose direction coincides by definition with that of positive charges, is characterized by a positive component along the x axis. Hence both $\mathbf{grad}(\varphi_{\text{eq}} - \varphi)$ and **I** are directed towards the bulk of the solution and no negative sign should appear in the first two members of Eq. (35). The negative sign has been introduced in order to adhere to the sign convention of Sec.I.A, according to which an anodic current is taken as negative. Analogous conclusions are drawn if $\mathbf{grad}(\varphi_{\text{eq}} - \varphi)$ is negative and the current is consequently cathodic.

Let us assume that the solution contains an inert electrolyte (supporting electrolyte) in strong excess both with respect to electroactive species and to eventual electroinactive species joined to the electroactive ones by chemical reactions. Under these circumstances, usually encountered in polarographic work, the major part of the current flowing in the solution is transported by the supporting electrolyte. This implies that by far the largest contribution to the driving force $\partial\tilde{\mu}_i/\partial x$ of the "field-assisted diffusion" of the species, other than those constituting the supporting electrolyte, is represented by $RT\,\partial \ln a_i/\partial x = \partial\mu_i/\partial x$, whereas the other contribution $z_i F\,\partial\varphi/\partial x$ can be neglected. If we designate the species of the supporting electrolyte by s and if we continue representing the other species by i, Eq. (35) can be written as follows,

$$\eta^{0c} = -\frac{1}{I}\int_0^c \sum_\rho \tilde{A}_\rho J_\rho\,dx - \frac{1}{I}\int_0^c \sum_i |\mathbf{J}_i| \left|\frac{\partial\mu_i}{\partial x}\right| dx - \frac{1}{I}\int_0^c \sum_s |\mathbf{J}_s| \left|\frac{\partial\tilde{\mu}_s}{\partial x}\right| dx = \eta_r^{0c} + \eta_d^{oc} + \eta_\Omega^{oc} \tag{36}$$

In Eq. (36) the term $-1/I\int_0^c \sum_\rho \tilde{A}_\rho J_\rho\,dx$ represents the contribution η_r^{0c} to the total overpotential η^{0c} due to chemical reactions. The term

$$-1/I\int_0^c \sum_i |\mathbf{J}_i||\partial\mu_i/\partial x|\,dx$$

expresses the contribution η_d^{0c} to η^{0c} due to the diffusion of the various i species, which include both the electroactive species and the species bounded to these latter by chemical reactions. η_r^{0c} and η_d^{0c} will be termed, respectively, "overall reaction overpotential" and "diffusion overpotential." The remaining contribution to η^{0c}, attributable to the field-assisted diffusion of the supporting electrolyte, can be identified with the "ohmic overpotential" η_Ω^{0c} between $x = 0$ and $x = c$.

The reaction overpotential η_r^{0c} is the sum of the contributions $-1/I\int_0^c \tilde{A}_\rho J_\rho\,dx$ of all the chemical reactions taking place between $x = 0$ and $x = c$, including the heterogeneous reactions occurring at the metal-solution interphase. In this respect it is worth noting that an overall electrode reaction consists in general of a succession of heterogeneous elementary chemical reactions, some of which necessarily involve the transfer of one or, less frequently, more electrons, whereas the others are purely chemical in nature. Let us denote both the chemical heterogeneous steps and the charge-transfer steps by ρ'. If the various reaction partners are nonspecifically adsorbed, we can define charge-transfer overpotential, η_t, as follows,

$$\eta_t = -\frac{1}{I}\int_0^{a_2} \sum_{\rho'} \tilde{A}_{\rho'} J_{\rho'}\,dx \tag{37}$$

Here $x = a_2$ is to be considered in a somewhat broad sense as the maximum distance from the electrode at which heterogeneous electron-transfer reactions can still occur with an appreciable efficiency. In an approximate manner we can identify the plane $x = a_2$ with the outer Helmholtz plane. Since the order of magnitude of a_2 is not much higher than that of molecular dimensions, we can write, approximately,

$$\eta_t = -\frac{a_2}{I}\sum_{\rho'} \tilde{A}_{\rho'} J_{\rho'} \tag{38}$$

where now $\tilde{A}_{\rho'}$ and $J_{\rho'}$ are to be considered as mean values of the electrochemical affinities and of the reaction rates in the inner layer $(0, a_2)$. If the reacting species are specifically adsorbed and the electron-transfer steps take place only among adsorbed species, the electrochemical affinities retain their statistical significance only in a two-dimensional space and the activities implicitly contained in the expressions for the electrochemical potentials of the various species are surface activities at the inner Helmholtz plane $x = a_1$ rather than volume activities. Under these conditions a more satisfactory definition of the charge-transfer overpotential is given in Eq. (39),

$$\eta_t = -\frac{1}{I}\sum_{\rho'} \tilde{A}_{\rho'} J_{\rho'} \tag{39}$$

where now $J_{\rho'}$ is defined in terms of surface concentrations, Γ_i, rather than in terms of volume concentrations C_i,

$$J_{\rho'} = \frac{1}{v_{i,\rho'}} \frac{d_\rho \Gamma_i}{dt} \tag{40}$$

By the preceding considerations the overall reaction overpotential η_r^{oc} has been separated into a contribution

$$\eta_t = -\frac{1}{I}\int_0^{a_2} \sum_{\rho'} \tilde{A}_{\rho'} J_{\rho'}\, dx$$

due to the heterogeneous reactions taking place between $x = 0$ and $x = a_2$, and a contribution

$$\eta_r^{a_2c} = -\frac{1}{I}\int_{a_2}^{c} \sum_{\rho} \tilde{A}_{\rho} J_{\rho}\, dx$$

due to the homogeneous reactions taking place within the diffuse double layer and outside it up to the plane $x = c$. From now on the latter contribution will simply be referred to as "reaction overpotential." Thus during the electroreduction of the cadmium complex with nitrilotriacetic acid considered in Sec. I.A, the electrode reaction (3) gives rise to a finite charge-transfer overpotential η_t, while the homogeneous chemical reaction (4) originates the reaction overpotential $\eta_r^{a_2c}$.

2. An Approximate Definition of the Various Types of Overpotential

When studying electrode processes through the use of polarography and other electrochemical techniques, some simplifying assumptions are usually made in connection with the concept of overpotential. Thus an attempt is made to locate the various types of overpotential (charge transfer, reaction, diffusion, ohmic overpotential) within regions at different distances from the electrode surface implicitly assuming that at any point of the solution around the electrode only one type of overpotential contributes to the total overpotential in an appreciable way. The formal separation of the solution layer adjacent to the electrode into different regions characterized by the presence of only one type of overpotential must not lead to the erroneous conclusion that actual drops of electrical potential are necessarily located within these regions. What we intend to mean is that these regions are the sites of different types of dissipative phenomena, which on the whole originate the total overpotential actually observed. Thus in the case of the so-called "polarographically reversible" waves, the only dissipative phenomena occurring during electrolysis are due to diffusion and take place within a solution layer much wider than the double layer, where the whole potential drop between the metal and the solution phases is concentrated.

Assume that the electrode process occurring at the electrode surface is expressed by the equation,

$$\sum_{i=1}^{n} \nu_i A_i + ne = 0 \tag{41}$$

Here the stoichiometric coefficients ν_i are positive for products and negative for reactants, so that Eq. (41) represents an anodic process. Imagine for simplicity that all the species A_i participating in the electrode process are soluble in the solution. From the principles of chemical thermodynamics we know that at equilibrium the electrochemical potentials $\tilde{\mu}_i$ of the species A_i are constant throughout the solution phase and that the same is true for the electrochemical potential $\tilde{\mu}_e$ of electrons in the metallic phase. Furthermore, the very existence of the heterogeneous equilibrium of Eq. (41) allows the following relation to be written,

$$\sum_i \nu_i \tilde{\mu}_i + n\tilde{\mu}_e = \sum_i (\nu_i \mu_i^0 + RT \ln {}^cC_i^{\nu_i} + \nu_i z_i F \varphi_{eq}^c) + n\mu_e^0 - nF\varphi_{eq}^o = 0 \quad (42)$$

Here μ_i^o and μ_e^o are the standard chemical potentials of the species A_i and of electrons, respectively, z_i is the ionic charge of A_i, and ${}^cC_i$ is its concentration at the ideal plane $x = c$ placed in the bulk of the solution (for simplicity, activities have been replaced by concentrations). φ_{eq}^o and φ_{eq}^c are the equilibrium potentials at the electrode surface $x = 0$ and at $x = c$. In practice φ_{eq}^c expresses the potential of the solution phase outside the diffuse double layer. Noting that $\sum_i \nu_i z_i = n$, Eq. (42) may be expressed, by simple rearrangement, in the form of Nernst's equation,

$$\varphi_{eq}^0 - \varphi_{eq}^c = E^\circ + \frac{RT}{nF} \ln \prod_i {}^cC_i^{\nu_i} \quad (43)$$

where

$$E^\circ = \frac{\sum_i \nu_i \mu_i^0 + n\mu_e^0}{nF}.$$

Imagine that the difference between the electric potentials at $x = 0$ and at $x = c$ is changed from its equilibrium value to $\varphi^0 - \varphi^c$. Recalling the definition of overpotential η^{0c} given in Eq. (35), we have from Eq. (43),

$$\begin{aligned}\eta^{0c} &= (\varphi^0 - \varphi^c) - (\varphi_{eq}^0 - \varphi_{eq}^c) \\ &= (\varphi^0 - \varphi^c) - \frac{\sum_i \nu_i \mu_i^0 + n\mu_e^0}{nF} - \frac{RT}{nF} \ln \prod_i {}^cC_i^{\nu_i}\end{aligned} \quad (44)$$

According to Eq. (44), $nF\eta^{0c}$ equals the electrochemical affinity $\tilde{A}^{0c}$ which would characterize the electrode reaction (41) if species A_i were at their equilibrium concentrations ${}^cC_i$; however, at the nonequilibrium potential φ^c, at $x = c$ and if electrons were at the nonequilibrium potential φ^0 at the electrode surface, $x = 0$. $\tilde{A}^{0c}$ can be considered approximately equal to the actual electrochemical affinity of reaction (41), with the various species A_i and electrons in the bulk of their respective phases, provided the potential difference across the double layer is remarkably larger than the ohmic drop in the solution phase outside the double layer produced by the flow of current. This latter condition, which is satisfied in the presence of a large excess of supporting electrolyte, guarantees that the potential gradient is practically zero from a relatively small distance from the electrode all the way into the bulk of the solution. It is only under these circumstances that the terms "potential in the bulk of the solution" and "bulk concentrations" may be employed consistently.

From now on the inner potential φ^c in the bulk of the solution will be arbitrarily set equal to zero. The inner potential $\varphi(x)$ at any given distance from the electrode, as referred to φ^c, will be symbolized by $\phi(x) \equiv \varphi(x) - \varphi^c$. According to the Gouy-Chapman model of diffuse layer, ϕ decreases almost exponentially with the distance from the plane $x = a_2$, especially for low values of ϕ,

$$\phi(x) = \phi_2 \exp[-\kappa(x - a_2)] \tag{45}$$

ϕ_2 is the potential at $x = a_2$. The distance $1/\kappa$ from the outer Helmholtz plane at which $\phi(x)/\phi_2 = 1/e$ is called the "thickness of the diffuse layer." Herein the distance $b = a_2 + 9.2/\kappa$ from the electrode at which $\phi(x)/\phi_2 = \exp[-\kappa(b - a_2)] = 0.0001$ will be referred to as the "effective thickness of the double layer." In practice b is about 3×10^{-4} cm for a 10^{-6} M solution of a uniunivalent electrolyte and decreases rapidly with increasing concentration.

The relation $\eta^{0c} = \tilde{A}^{0c}/(nF)$ between overpotential and electrochemical affinity of the electrode process can be conveniently used for formally separating the solution layer around the electrode into regions characterized by the presence of a single type of overpotential. Thus the electrochemical affinity $\tilde{A}^{0c}$ can be written as

$$\begin{aligned}\tilde{A}^{0c} = nF\eta^{0c} &= \Big(-n\tilde{\mu}_e - \sum_i \nu_i{}^{a_2}\tilde{\mu}_i\Big) + \Big(\sum_i \nu_i{}^{a_2}\tilde{\mu}_i - \sum_i \nu_i{}^{b}\tilde{\mu}_i\Big) \\ &+ \Big(\sum_i \nu_i{}^{b}\tilde{\mu}_i - \sum_i \nu_i{}^{c}\tilde{\mu}_i\Big)\end{aligned} \tag{46}$$

where ${}^{a_2}\tilde{\mu}_i$, ${}^{b}\tilde{\mu}_i$, and ${}^{c}\tilde{\mu}_i$ are the electrochemical potentials of the species A_i at the outer Helmholtz plane $x = a_2$, at the boundary $x = b$ of the diffuse layer on the solution side, and at $x = c$, respectively. At this point we might be tempted to identify the three terms on the right-hand side of Eq. (46) with $nF\eta^{0a_2}$, $nF\eta^{a_2b}$, and $nF\eta^{bc}$, respectively, where η^{0a_2}, η^{a_2b}, and η^{bc} denote the overall contributions to the total overpotential in the three regions $(0, a_2)$,

(a_2, b), and (b, c). As a matter of fact the above position involves several simplifying assumptions that can be better understood by comparing the terms in Eq. (46) with the rigorous expressions of η^{0a_2}, η^{a_2b}, and η^{bc} as derived from Eq. (35) and (36).

Starting with η^{0a_2}, we shall note that in the region $(0, a_2)$ the ohmic overpotential η_Ω does not contribute to η^{0a_2}, since in this region the flux $\mathbf{J}_s$ of the supporting electrolyte is zero. Noting that the gradient $\partial\tilde{\mu}_i/\partial x$ of the electrochemical potential of any species A_i is the algebraic sum of the two terms $z_i F\, \partial\phi/\partial x$ and $RT\, \partial \ln C_i/\partial x$, which assume very large values within the double layer, we can make the reasonable assumption that this algebraic sum is negligible with respect to the single terms and, consequently, that $\partial\tilde{\mu}_i/\partial x$ is approximately zero between $x = 0$ and $x = a_2$. Hence, if all the species A_i participating in the electrode process (41) are nonspecifically adsorbed, we have, in view of Eqs. (35), (36), and (38)

$$\eta^{0a_2} = -\frac{1}{I}\int_0^{a_2} \sum_{\rho'} \tilde{A}_{\rho'} J_{\rho'}\, dx - \frac{1}{I}\int_0^{a_2} \sum_i |\mathbf{J}_i| \left|\frac{\partial\tilde{\mu}_i}{\partial x}\right| dx$$

$$-\frac{1}{I}\int_0^{a_2} \sum_s |\mathbf{J}_s| \left|\frac{\partial\tilde{\mu}_s}{\partial x}\right| dx \cong -\frac{a_2}{I} \sum_{\rho'} \tilde{A}_{\rho'} J_{\rho'} = \eta_t \tag{47}$$

Assume that the various heterogeneous steps ρ' composing the overall electrode reaction (41) proceed under steady-state conditions. If we designate by $\nu_{\rho'}$ the number of times the ρ'th step occurs during one occurrence of the overall process of Eq. (41), the rate of this process is given by

$$\frac{I}{nF} = -\int_0^{a_2} \frac{J_{\rho'}}{\nu_{\rho'}}\, dx \cong -a_2 \frac{J_{\rho'}}{\nu_{\rho'}} \qquad \text{for all steps} \tag{48}$$

From Eqs. (47) and (48) it follows that

$$\eta^{0a_2} = -a_2 \frac{J_{\rho'}}{I\nu_{\rho'}} \sum_{\rho'} \nu_{\rho'} \tilde{A}_{\rho'} = \frac{1}{nF} \sum_{\rho'} \nu_{\rho'} \tilde{A}_{\rho'} = \frac{\tilde{A}}{nF}$$

where $\tilde{A}$ is the electrochemical affinity of the overall reaction (41) expressed in terms of the concentrations ${}^{a_2}C_i$ at $x = a_2$. Thus

$$nF\eta^{0a_2} = \tilde{A} = -n\tilde{\mu}_e - \sum_i \nu_i {}^{a_2}\tilde{\mu}_i = nF\eta_t \tag{49}$$

Let us now see which assumptions are involved in the equation,

$$\sum_i \nu_i {}^{a_2}\tilde{\mu}_i - \sum_i \nu_i {}^{b}\tilde{\mu}_i = nF\eta^{a_2 b} \tag{50}$$

In the absence of nonequilibrium homogeneous chemical reactions, the rigorous expression for $\eta^{a_2 b}$ is obtained from Eq. (36),

$$\eta^{a_2 b} = -\frac{1}{I}\int_{a_2}^{b} \sum_i |\mathbf{J}_i| \left|\frac{\partial \tilde{\mu}_i}{\partial x}\right| dx - \frac{1}{I}\int_{a_2}^{b} \sum_s |\mathbf{J}_s| \left|\frac{\partial \tilde{\mu}_s}{\partial x}\right| dx \tag{51}$$

Equation (50) can be identified with Eq. (51) only if the second term on the right-hand side of Eq. (51)—namely, the ohmic overpotential $\eta_\Omega^{a_2 b}$ between $x = a_2$ and $x = b$—is considered negligible. This assumption, which is rather reasonable in view of the fact that $\mathbf{J}_s$ is actually zero at $x = a_2$, is equivalent to the statement that the current flowing between $x = a_2$ and $x = b$ is exclusively transported by the electroactive substances A_i. Thus under steady-state conditions, the absolute value $|I|$ of the current density is given by

$$|I| = nF\frac{|\mathbf{J}_i|}{|\nu_i|} \qquad \text{for all the species } A_i \tag{52}$$

Combining Eqs. (51) and (52) we have

$$\eta^{a_2 b} \cong -\frac{1}{I}\frac{|\mathbf{J}_i|}{|\nu_i|}\int_{a_2}^{b} \sum_i \left|\nu_i \frac{\partial \tilde{\mu}_i}{\partial x}\right| dx = -\frac{|I|}{InF}\sum_i |\nu_i({}^{b}\tilde{\mu}_i - {}^{a_2}\tilde{\mu}_i)| \tag{53}$$

It is easy to see that Eqs. (50) and (53) are also identical in sign. In fact, if the electrode reaction (41) proceeds towards oxidation, I is negative. Furthermore, for a species A_i characterized by a positive value of ν_i, ${}^{a_2}\tilde{\mu}_i$ is larger than ${}^{b}\tilde{\mu}_i$ since the species diffuses toward the solution. Conversely, substances with a negative value of ν_i are such that ${}^{a_2}\tilde{\mu}_i$ is smaller than ${}^{b}\tilde{\mu}_i$. The opposite reasoning holds if the electrode reaction proceeds towards reduction.

Under the previous assumptions $\eta^{a_2 b}$ expresses the contribution to the total overpotential due to the slowness with which electroactive species penetrate into the diffuse double layer. In this sense $\eta^{a_2 b}$ can be termed "penetration overpotential," η_p; η_p cannot be equal to zero during electrolysis because if $\partial\tilde{\mu}_i/\partial x$ were exactly zero for all the species A_i within the diffuse layer the flow of electrical charges across this layer would be nonexisting. However, if we consider, as previously done in connection with

η^{0a_2}, that for $a_2 < x < b$ the two terms $z_i F\,\partial\phi/\partial x$ and $RT\,\partial \ln C_i/\partial x$ are large with respect to their algebraic sum $\partial\tilde{\mu}_i/\partial x$, we may write, approximately,

$$z_i F \frac{\partial \phi}{\partial x} = -RT \frac{\partial \ln C_i}{\partial x}$$

or, also, integrating between $x = a_2$ and $x = b$,

$$^{a_2}C_i = {}^{b}C_i \exp\left[-\frac{z_i F}{RT}(\phi^{a_2} - \phi^{b}) \right] \cong {}^{b}C_i \exp\left[-\frac{z_i F}{RT}\phi^{a_2} \right] \tag{54}$$

where the superscripts a_2 and b denote the locations at which concentrations and potentials are considered. Equation (54) expresses the Frumkin correction (*6*) for the concentrations of electroactive species in the so-called "plane of closest approach." Generally the contribution of $\eta_p \cong \eta^{a_2 b}$ to the total overpotential η^{0c} is assumed negligible and only the other two contributions η^{0a_2} and η^{bc} are taken into account.

The term $(\sum_i \nu_i {}^{b}\tilde{\mu}_i - \sum_i \nu_i {}^{c}\tilde{\mu}_i)$ on the right-hand side of Eq. (46) cannot be considered equal to $nF\eta^{bc}$ without making very rough assumptions. In fact the rigorous expression for η^{bc} as derived from Eqs. (35) and (36) in the absence of reaction overpotential is

$$\begin{aligned}
\eta^{bc} &= -\frac{1}{I}\int_b^c \sum_i |\mathbf{J}_i| \left| \frac{\partial \tilde{\mu}_i}{\partial x} \right| dx - \frac{1}{I}\int_b^c \sum_s |\mathbf{J}_s| \left| \frac{\partial \tilde{\mu}_s}{\partial x} \right| dx \\
&\cong -\frac{1}{I}\int_b^c \sum_i |\mathbf{J}_i| \left| \frac{\partial \mu_i}{\partial x} \right| dx - \frac{1}{I}\int_b^c \sum_s |\mathbf{J}_s| \left| \frac{\partial \tilde{\mu}_s}{\partial x} \right| dx \\
&= \eta_{\mathrm{d}}^{bc} + \eta_{\Omega}^{bc}
\end{aligned}$$

where η_d^{bc} and η_Ω^{bc} are the diffusion and ohmic overpotentials between $x = b$ and $x = c$. It should be noted that the influence of the electric field on the motion of the electroactive substances A_i can be neglected only in the presence of a strong excess of supporting electrolyte, and that under these circumstances the contribution of η_Ω to the total overpotential is far from being negligible within the region (b, c). A formal way of separating the region (b, c) into two zones characterized by the sole presence either of diffusion or ohmic overpotential is the following. Imagine a liquid layer bounded by the fixed plane $x = b$ and by a moving plane $x = \delta$, with $\delta < c$. Let us choose the location of the plane $x = \delta$, at a given instant from the start of electrolysis, in such a way that the current which would flow through the layer (b, δ)

if all and only the diffusion overpotential were concentrated in this layer is equal to the actual current. Within this ideal layer the current should be transported exclusively by the electroactive substances A_i in the absence of an appreciable electric field ($\eta_\Omega^{b\delta} = 0$). Outside this ideal layer (b, δ), which will be called "diffusion layer," the current should be transported exclusively by the supporting electrolyte ($\eta_d^{\delta c} = 0$). Because at a given instant the current must necessarily be uniform throughout the solution, in order that the previous conditions be fulfilled the fluxes $\mathbf{J}_i$ of the electroactive species A_i should also be uniform between $x = b$ and $x = \delta$. Consequently, in view of Fick's first law, which expresses the proportionality between the flux of any given species and the corresponding concentration gradient at zero electric field strength, the concentrations C_i should change linearly with the distance x from the electrode within the diffusion layer.

For the simple electrode process $A_1 \rightarrow A_2 + ne$, the concentrations of A_1 and A_2 as functions of x are represented schematically by the solid curves

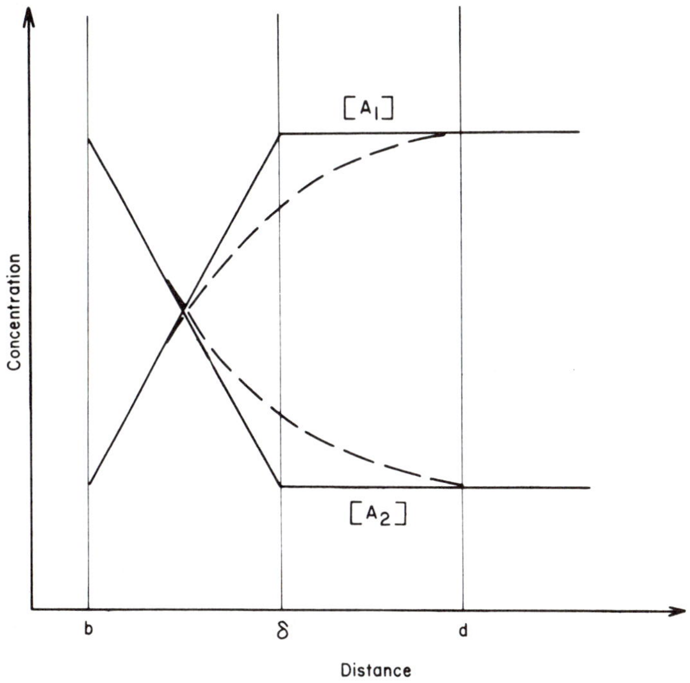

Fig. 2. The solid curves express the idealized concentration-distance profiles for the electrode process $A_1 \xrightarrow{e} A_2$. The dashed curves depict schematically the actual profiles.

in Fig. 2. The dashed curves in the same figure represent the actual concentrations. These are tangent to the corresponding idealized concentrations at $x = b$, where the ohmic overpotential is actually negligible. The moving boundary $x = \delta$ of the diffusion layer is obviously determined by the intersections of the tangents to the actual concentration profiles at $x = b$ with the horizontal straight lines representing the corresponding concentrations in the bulk of the solution. From Fig. 2 it is evident that the actual concentration gradients progressively decrease with increasing x. Consequently, the fraction of the current transported by the supporting electrolyte grows, reaching the unit value at $x = d$. At this distance from the electrode the actual contribution of η_d to the total overpotential becomes completely negligible. According to the ideal behavior this situation is already encountered at $x = \delta$, where the ideal concentration gradients drop abruptly to zero after having remained constant throughout the diffusion layer. On the basis of the above definition of δ we can modify Eq. (46) as follows,

$$\begin{aligned}\tilde{A}^{0\delta} = nF\eta^{0\delta} &= \left(-n\tilde{\mu}_e - \sum_i \nu_i{}^{a_2}\tilde{\mu}_i\right) + \left(\sum_i \nu_i{}^{a_2}\tilde{\mu}_i - \sum_i \nu_i{}^{b}\tilde{\mu}_i\right) \\ &\quad + \left(\sum_i \nu_i{}^{b}\tilde{\mu}_i - \sum_i \nu_i{}^{\delta}\tilde{\mu}_i\right) \\ &\cong nF(\eta^{0a_2} + \eta^{a_2b} + \eta^{b\delta}) \\ &\cong nF(\eta_t + \eta_p + \eta_d). \end{aligned} \tag{55}$$

thus eliminating the contribution of the ohmic overpotential η_Ω from the electrochemical affinity $\tilde{A}^{0\delta}$ of the electrode process.

C. Scope of the Contribution

In the preceding sections we have seen that if the external potential difference $\Delta\varphi$ applied between the dropping electrode and the reference electrode in a polarographic cell is different from the equilibrium potential difference $\Delta\varphi_{eq}$, the polarization of the cell, $\Delta\varphi - \Delta\varphi_{eq}$, causes some irreversible phenomena to take place. In the time scale of polarographic measurements whenever both the charge-transfer overpotential and the reaction overpotential relative to the polarized electrode, namely, the dropping electrode, can be considered negligible, the current-potential curve recorded using the polarographic technique is referred to as a "reversible polarographic wave." It should be noted that this is a polarographic term with no thermodynamic implications, insofar as during the recording of a reversible polarographic wave dissipative phenomena due to migration and diffusion actually take place.

If the solution is the site of one or more chemical reactions, the rate at which they attain their new equilibrium conditions, once the original equilibrium conditions have been disturbed by the flow of matter, may be very high in the time scale of polarographic measurements. The electrochemical affinities $\tilde{A}_\rho$ for these reactions are then practically negligible. Under these conditions the reaction overpotential is practically zero and the chemical reactions are commonly referred to as "perfectly mobile equilibria." This latter denomination is better understood if we consider that once the electrochemical affinity $\tilde{A}_\rho$ for the ρth chemical reaction equals zero, from Eq. (33) it immediately follows that

$$\prod_i C_i^{\nu_{i,\rho}} = \exp\left[-\frac{\sum_i \nu_{i,\rho}\mu_i^0}{RT}\right] \equiv K_\rho \tag{56}$$

where, to simplify matters, activities have been replaced by the corresponding concentrations. Equation (56), which expresses the law of mass action for the ρth reaction, can be applied at any point of the solution. An equilibrium that is perfectly mobile when studied by the polarographic technique may appear sluggish with other electrochemical techniques characterized by a lower diffusion overpotential—for instance, step-function techniques.

Analogous reasoning can be extended to the charge-transfer overpotential. Thus if the rate of the general electrode process of Eq. (41) is very high during electrolysis, then the two terms $-\sum_{\nu_i<0} \nu_i{}^{a_2}\tilde{\mu}_i$ and $\sum_{\nu_i>0} \nu_i{}^{a_2}\tilde{\mu}_i + n\tilde{\mu}_e$ assume values a lot larger than the value of their difference, expressing the electrochemical affinity $\tilde{A}^* = nF\eta^{0a_2}$ of the electrode process (cf. Eq. (49)). In this case we can approximately write

$$-nF\eta^{0a_2} = n\tilde{\mu}_e + \sum_i \nu_i{}^{a_2}\tilde{\mu}_i = 0 \tag{57}$$

Under the reasonable assumption that the penetration overpotential $\eta_p = \eta^{a_2b}$ is practically zero, we have

$$-nF(\eta^{0a_2} + \eta^{a_2b}) = -nF\eta^{0b} = n\tilde{\mu}_e + \sum_i \nu_i{}^b\tilde{\mu}_i = 0 \tag{58}$$

or also, after simple rearrangement

$$\varphi^0 - \varphi^b \cong \phi^0 = \frac{(\sum_i \nu_i\mu_i^0 + n\mu_e^0)}{nF} + \frac{RT}{nF}\ln\prod_i {}^bC_i^{\nu_i} \tag{59}$$

Equation (59) expresses Nernst's equation as applied to the concentrations of the electroactive species at $x = b$ and is the equivalent of Eq. (56) for a charge-transfer reaction. An electrode process to which Eq. (59) may be applied, when using the polarographic technique, is said to be "polarographically reversible." Here, too, we must note that the charge-transfer overpotential η_t for a polarographically reversible process can possibly appear different from zero when the process is studied through the use of a step-function technique.

If all the chemical equilibria coupled with a charge-transfer process are perfectly mobile, the polarographic investigation of such a process obviously cannot elucidate the kinetics and mechanism of the chemical reactions taking place in the solution. Nevertheless, it can provide valuable information about the number and stoichiometry of the chemical reactions at the same time allowing the determination of their equilibrium constants. The kinetics of a given chemical reaction ρ can possibly be clarified with the polarographic technique if the electrochemical affinity $\tilde{A}_\rho$ of this reaction is appreciably different from zero during the flow of current—that is, if the reaction overpotential η_r is at least comparable with the omnipresent diffusion overpotential η_d. Under these circumstances the polarographic study of a chemical reaction is more easily performed by focusing attention on the polarographic limiting current, which is independent of the potential applied to the dropping electrode. The advantage of such a procedure is that usually the charge-transfer overpotential η_t is zero along the limiting current, so that the study of the chemical overpotential appears simplified. In order to better understand why η_t equals zero when the limiting current is reached, consider the simple charge-transfer process $\mathrm{Red} \rightleftharpoons \mathrm{Ox} + ne$ and assume that the charge-transfer overpotential is different from zero along the rising portion of the polarographic wave. In this case Nernst's equation,

$$\phi^0 = \frac{\mu^0_{\mathrm{Ox}} + n\mu^0_e - \mu^0_{\mathrm{Red}}}{nF} + \frac{RT}{nF} \ln \frac{{}^bC_{\mathrm{Ox}}}{{}^bC_{\mathrm{Red}}} \tag{60}$$

does not apply. However, the more the potential applied to the dropping electrode is made positive with respect to the equilibrium potential, the higher becomes the rate for the electrooxidation of the reduced form and the lower its concentration, ${}^bC_{\mathrm{Red}}$, at $x = b$. At the same time the concentration, ${}^bC_{\mathrm{Ox}}$, of the oxidized form increases at its expense. This simply means that when the limiting current is approached, the actual concentrations of Ox and Red at $x = b$ as well as the corresponding electrochemical potentials tend to the values that they would assume were Eq. (60) applicable—that is, where the charge-transfer process reversible.

The above considerations are not always applicable. In this respect imagine an overall anodic process including some purely chemical heterogeneous steps. If the potential E applied to the electrode is made sufficiently positive, the rates of the charge-transfer elementary steps (which are those directly affected by E) become very high, so that some purely chemical heterogeneous steps may become rate determining. In this case the charge-transfer overpotential η_t as defined in Eq. (37) is different from zero, not only along the rising portion of the wave but also in the portion corresponding to the limiting current. The heterogeneous chemical reactions responsible for the finite value of η_t under limiting current conditions are strongly influenced by the structure of the double layer, within which they take place. Since the double layer structure changes when the potential applied to the electrode is varied, the rates of these heterogeneous chemical reactions and, consequently, the value of the limiting current may also change with potential, giving rise to ill-defined limiting currents.

In the following sections we shall examine in some detail the applications of the polarographic technique to the study of homogeneous chemical reactions when only diffusion and ohmic overpotentials contribute to the total overpotential. The complications due to a finite contribution of charge-transfer overpotential will subsequently be examined. Then the more involved case corresponding to the simultaneous presence of diffusion, charge-transfer, and chemical overpotentials will be considered. Last, but not least, heterogeneous chemical reactions and the effect of the double layer structure upon very fast homogeneous chemical reactions will be examined.

II. PURE DIFFUSION OVERPOTENTIAL. PERFECTLY MOBILE HOMOGENEOUS EQUILIBRIA

A. General Formulation of the Diffusional Problem

The Diffusion Equations

Let us assume that the charge-transfer process taking place at the surface of a dropping electrode is coupled with m homogeneous perfectly mobile equilibria, and that the solution contains a supporting electrolyte in strong excess with respect to all the species A_i, n in number, involved in the m equilibria. Under these circumstances the flow of any species A_i outside the double layer is exclusively determined by the driving force of diffusion, $\partial\mu_i/\partial x$. The component $J_{i,x}$ of the flux $\mathbf{J}_i$ of the ith species, relative to the x axis, is then proportional to the corresponding concentration gradient, $\partial C_i/\partial x$,

$$J_{i,x} = C_i v_{i,x} = -D_i \frac{\partial C_i}{\partial x} \qquad i = 1, \ldots, n \tag{61}$$

where $v_{i,x}$ is the mean velocity of A_i in the direction normal to the electrode. Equation (61) is a phenomenological relation due to Fick (Fick's first law). The proportionality constant D_i is the diffusion coefficient of the ith species. The minus sign in Eq. (61) accounts for the fact that the direction of $\mathbf{v}_i$ is opposite to that of the corresponding concentration gradient, **grad** C_i.

The expression for the instantaneous polarographic current at constant potential as a function of time and that for the mean polarographic current as a function of the applied potential can be readily derived, once the dependence of the concentrations C_i of all the species A_i on the distance x from the electrode surface and on the time t measured from the beginning of the drop life is exactly known. Therefore, the main problem to be tackled in the polarographic treatment of chemical equilibria consists in determining $C_i = C_i(x, t)$ from the n equations of conservation of mass relative to the various species A_i. In the case of diffusion towards a dropping electrode, these equations can be written

$$\frac{\partial C_i}{\partial t} = -\frac{\partial}{\partial x} J_{i,x} + \frac{2x}{3t}\frac{\partial C_i}{\partial x} + \frac{d_i C_i}{dt} \qquad i = 1, \ldots, n \tag{62}$$

Equation (62) expresses the fact that the increase in the number of moles of the ith species per unit of time per unit volume, $\partial C_i/\partial t$, equals the net number of moles of this species entering the unit volume in the same unit time, $-\partial J_{i,x}/\partial x$, plus the number of moles, $d_i C_i/dt$, formed within the unit volume as a consequence of the m chemical reactions. The term $\frac{2}{3}x/t\ \partial C_i/\partial x$ takes into account the convective motions due to the compression of the liquid layer around the electrode by the expanding drop. Combination of Eqs. (61) and (62) results in

$$\frac{\partial C_i}{\partial t} = D_i \frac{\partial^2 C_i}{\partial x^2} + \frac{2x}{3t}\frac{\partial C_i}{\partial x} + \frac{d_i C_i}{dt} \qquad i = 1, \ldots, n \tag{63}$$

If the ith species is not involved in any of the m chemical reactions, the "source term," $d_i C_i/dt$, is zero and Eq. (63) reduces to the Ilkovic diffusion equation (*7*). Having assumed that the m homogeneous reactions remain at equilibrium during electrolysis, we can make use of the corresponding expressions for the law of mass action (see Eq. (56)),

$$\prod_i C_i^{\nu_{i,\rho}}(x, t) = K_\rho \qquad \rho = 1, \ldots, m \tag{64}$$

which hold for any value of t and x. K_ρ is the constant of the ρth equilibrium. The expressions of the law of chemical equilibrium represent m relations among the concentrations C_i. Hence for the mathematical problem under investigation to be physically sound, only $n - m$ of the n differential equations (63) must be linearly independent. A convenient way of determining these $n - m$ linearly independent equations is based on the concept of "atomic clusters" (*8*).

We shall designate by the term "atomic cluster" any grouping of the greatest number of atoms contained in the various diffusing species A_i which remains unaltered in all the chemical reactions taking place in the solution. The meaning of "unaltered" is purely stoichiometric and has no structural implications. The oxidation number of the atoms constituting a given cluster does not represent a criterion of differentiation among clusters, thus the same cluster may bear different charges. As an example consider a strongly alkaline solution of TlY^{3-}, where the symbol Y^{4-} denotes the EDTA anion. The only equilibrium that is established among the species Tl^+, TlY^{3-}, and Y^{4-} in the present case is the dissociation equilibrium,

$$TlY^{3-} \rightleftharpoons Tl^+ + Y^{4-} \tag{65}$$

and the only clusters that are directly or indirectly involved in the reduction of Tl^I to Tl(Hg) on a dropping electrode are Y and Tl. The latter cluster includes both the thallium atoms with oxidation number +1 and 0. If the medium is neutral or acid, the four protonation equilibria of Y^{4-} must also be taken into account, as they cooperate in subtracting the ligand Y^{4-} from the complex TlY^{3-}. In the present case the clusters to be considered are Y, Tl, and H.

If we apply the law of conservation of mass to clusters rather than to molecular species, we easily realize that the source terms in Eq. (63) equal zero, since no cluster can be produced or consumed in a volume element of the solution by definition of cluster. On the other hand the concentration of a cluster can be immediately expressed in terms of the concentrations of the molecular species that contain it. Thus if the clusters are h in number and $a_{i,j}$ designates the number of times the jth cluster is contained in a molecule of the species A_i, the concentration of this latter cluster is given by

$$\sum_{i=1}^{n} a_{i,j} C_i \tag{66}$$

On applying the law of conservation of mass as expressed in Eq. (63) to the jth cluster, we have

$$\sum_{i=1}^{n} a_{i,j}\, \delta_i C_i = 0 \qquad j = 1, \ldots, h \tag{67}$$

where the symbol δ_i denotes the operator,

$$\delta_i = D_i \frac{\partial^2}{\partial x^2} + \frac{2x}{3t} \frac{\partial}{\partial x} - \frac{\partial}{\partial t} \tag{68}$$

The h differential equations (67), together with the m equations (64), constitute a set of $m + h$ relations among the concentrations of the various diffusing species.

In the case of the thallium complex with EDTA previously considered we see that in a strongly alkaline medium, the species of interest are Tl^0, Tl^+, Y^{4-}, TlY^{3-}, that is, four in number, whereas the number m of the chemical equilibria is one and the number h of clusters is two. In neutral or acid media the species to be taken into account are Tl^0, Tl^+, Y^{4-}, TlY^{3-}, H^+, HY^{3-}, H_2Y^{2-}, H_3Y^-, H_4Y, that is, nine in number, while we have $h = 3$ and $m = 5$. In both cases the sum, $m + h$, of the number of chemical equilibria and of the number of clusters is less than the number n of the diffusing species by one unit,

$$m + h = n - 1 \tag{69}$$

Equation (69) is valid in general, as could be demonstrated by simple arguments (*8*). It follows that for the complete formulation of the diffusional problem, one further relation among the concentrations C_i must be found. In this connection consider that in general a charge-transfer process involves the change in the charge of a particular cluster, which we shall denote by A. If A^O and A^R designate the oxidized and reduced forms of A, respectively, the species being directly oxidized and reduced at the dropping electrode can be considered for convenience as multiples, A_a^R and A_b^O, of A^R and A^O, provided suitable numbers of formal dissociation equilibria are introduced. The charge-transfer reaction can therefore be written in the form,

$$bA_a^R \rightleftharpoons aA_b^O + ne \tag{70}$$

As concerns eventual interactions between A^O and A^R in the solution phase, we shall distinguish three cases.

Case 1: No interactions between A^O and A^R take place in the solution phase.

Case 2: A^O is converted into A^R and vice versa in the solution phase due to a redox equilibrium.

Case 3: A_b^O and A_a^R interact in the solution phase through a disproportionation equilibrium.

In Case 1 the mutual conversion of A^O into A^R occurs only on the electrode surface. Hence the differential equation (67) relative to the cluster A,

$$\sum_{i=1}^{n} a_{i,\,A}\, \delta_i C_i = 0 \tag{71}$$

can be replaced by the two linearly independent differential equations,

$$\sum_i a_{i,\,O}\, \delta_i C_i = 0; \qquad \sum_i a_{i,\,R}\, \delta_i C_i = 0 \tag{72}$$

In Eqs. (71) and (72) $a_{i,A}$, $a_{i,O}$, and $a_{i,R}$ denote, respectively, the number of times the cluster A, its oxidized, and its reduced forms are contained in one molecule of the ith species. One of the two equations of (72), chosen at will, furnishes the lacking relation among the concentrations C_i of the n diffusing species.

Let us now assume that the solution contains a redox couple, Ox/Red, causing the mutual conversion of A^O into A^R (Case 2). In the present case Eq. (72) is no longer valid and the material balance implicit in the derivation of this equation must now be replaced by a suitable electron balance. Denote by B^{Ox} and B^{Red} the oxidized and reduced forms of the cluster B being contained in the redox couple Ox/Red and exchanging electrons with the couple A^O/A^R. Let z_O, z_R, z_{Ox}, and z_{Red} designate the charges of A^O, A^R, B^{Ox} and B^{Red}, respectively. If we focus our attention on the totality of electrons that the reduced forms of the two couples A^O/A^R and B^{Ox}/B^{Red} have in excess with respect to the corresponding oxidized forms, it is manifest that the number of these electrons is not altered by the occurrence of the redox reaction between A^O/A^R and B^{Ox}/B^{Red}. It follows that the law of conservation of mass as applied to these electrons does not contain the source term. Noting that the concentration of the above electrons is given by $(z_O - z_R)\sum_i a_{i,R} C_i + (z_{Ox} - z_{Red})\sum_i a_{i,Red} C_i$, where $a_{i,Red}$ designates the number of times the reduced form of the cluster B is contained in one molecule of the ith species, the application of Eq. (63) yields

$$(z_O - z_R)\sum_i a_{i,R}\,\delta_i C_i + (z_{Ox} - z_{Red})\sum_i a_{i,Red}\,\delta_i C_i = 0 \tag{73}$$

Equation (73) is the missing equation that, added to the h differential equations (67) and to the m equations (64), furnishes the whole set of relations among C_i for the present case.

Case 3 is encountered when the molecular species A_b^O and A_a^R interact mutually in the solution through a homogeneous disproportionation equilibrium, giving rise to a product A_c^J containing the cluster A in an oxidation state A^J intermediate between those of A^O and A^R,

$$a_1 A_a^R + a_2 A_b^O \rightleftharpoons a_3 A_c^J \tag{74}$$

Here $c = (a_1 a + a_2 b)/a_3$. In the present case an electron balance can be effected with the electrons that the two forms A^R and A^J of the cluster A contain in excess with respect to the oxidized form A^O. The concentration of these electrons is equal to $(z_O - z_R)\sum_i a_{i,R} C_i + (z_O - z_J)\sum_i a_{i,J} C_i$, where z_J designates the charge of A^J and $a_{i,J}$ the number of times A^J is contained in one molecule of the ith species. On applying Eq. (63) without the source term to these electrons, we obtain

$$(z_O - z_R)\sum_i a_{i,R}\,\delta_i C_i + (z_O - z_J)\sum_i a_{i,J}\,\delta_i C_i = 0 \tag{75}$$

which is the equivalent of Eq. (73) for the case of a disproportionation reaction.

2. The Boundary and Initial Conditions

In order to obtain the concentrations C_i of the n diffusing species as functions of x and t and consequently the expression for the polarographic current as a function of the applied potential, it is now necessary to solve the set of the $h+1$ second-order differential equations previously derived. If A^O and A^R do not interact mutually in the solution (Case 1), this set consists of the equations (67) together with one of the two equations of (72). In the case of a reciprocal interaction between A^O and A^R in the solution phase, the set of the $h+1$ differential equations to be solved is composed of the equations (67) together with Eq. (73) (Case 2) or alternatively with Eq. (75) (Case 3). In order to proceed to the mathematical solution of the above $h+1$ differential equations, we must first determine the $2(h+1)$ formulas that will express the initial and boundary conditions satisfied by the concentrations C_i. The boundary conditions represent the mathematical statement of hypotheses concerning the behavior of the various C_i at the boundary between the electrode and the solution phases at any instant t from the start of electrolysis. On the contrary the initial conditions describe the behavior of the concentrations C_i at the beginning of the electrolysis ($t=0$), for any value of the distance x from the electrode surface.

As concerns the initial conditions we shall assume that the concentrations C_i are uniform throughout the solution up to the electrode surface at $t=0$—that is, at the beginning of the drop life. Thus we have

$$C_i = C_i^* \qquad \text{for} \begin{cases} t=0, x \geqslant 0 \\ t>0, x \to \infty \end{cases} \qquad i=1,\ldots,n \tag{76}$$

where the asterisks denote bulk concentrations. The conditions of Eq. (76) are quite simple and apparently obvious, but it should be noted that they are not fully satisfied when using a conventional vertical capillary on account of the local depletion of the solution produced near the capillary tip by electrolysis at each drop and not completely removed after the fall of the drop. In order to apply the conditions of Eq. (76) to Cases 1, 2, and 3, it is convenient to rewrite them as follows:

$$\sum_i a_{i,j} C_i = \sum_i a_{i,j} C_i^* \qquad \text{for} \begin{cases} t=0, x \geqslant 0 \\ t>0, x \to \infty \end{cases}$$
$$j=1,\ldots,h \text{ (Cases 1, 2, and 3)} \tag{77a}$$

$$\sum_i a_{i,\mathrm{O}} C_i = \sum_i a_{i,\mathrm{O}} C_i^*$$

or alternatively: (Case 1) (77b)

$$\sum_i a_{i,\mathrm{R}} C_i = \sum_i a_{i,\mathrm{R}} C_i^*$$

$$\left.\begin{aligned}&(z_{\mathrm{O}} - z_{\mathrm{R}}) \sum_i a_{i,\mathrm{R}} C_i + (z_{\mathrm{Ox}} - z_{\mathrm{Red}}) \sum_i a_{i,\mathrm{Red}} C_i \\ &\quad = (z_{\mathrm{O}} - z_{\mathrm{R}}) \sum_i a_{i,\mathrm{R}} C_i^* + (z_{\mathrm{Ox}} - z_{\mathrm{Red}}) \sum_i a_{i,\mathrm{Red}} C_i^*\end{aligned}\right\} \text{(Case 2)} \qquad (77\text{c})$$

$$\left.\begin{aligned}&(z_{\mathrm{O}} - z_{\mathrm{R}}) \sum_i a_{i,\mathrm{R}} C_i + (z_{\mathrm{O}} - z_{\mathrm{J}}) \sum_i a_{i,\mathrm{J}} C_i \\ &\quad = (z_{\mathrm{O}} - z_{\mathrm{R}}) \sum_i a_{i,\mathrm{R}} C_i^* + (z_{\mathrm{O}} - z_{\mathrm{J}}) \sum_i a_{i,\mathrm{J}} C_i^*\end{aligned}\right\} \text{(Case 3)} \qquad (77\text{d})$$

In considering the boundary conditions we shall assume that none of the diffusing species accumulates at the electrode surface during electrolysis. This amounts to saying that the flux of any cluster at the electrode surface equals zero. In view of Eq. (61) the previous statement can be expressed as

$$\sum_i a_{i,j} D_i \frac{\partial C_i}{\partial x} = 0 \qquad \text{for } x = 0,\ t > 0 \qquad j = 1, \ldots, h \qquad (78)$$

More properly the conditions of Eq. (78) should be applied at the boundary $x = b$ of the double layer. In fact it is only outside this layer that the fluxes of the species A_i are exclusively determined by the corresponding concentration gradients. Within the double layer concentration gradients are very high, but their effect on fluxes is almost completely counterbalanced by that of the electric potential gradient (cf. Sec. I.B.2). It must be noted that in the time scale of polarographic measurements, the thickness of the whole double layer is considerably smaller than that of the diffusion layer. Consequently when the solution of diffusional problems is involved, the plane $x = b$ can be confused with the plane $x = 0$ with full confidence. The $(h + 1)$th condition which completes the set of boundary conditions relative to the $h + 1$ diffusional equations is obtained by assuming that the charge-transfer overpotential η_t equals zero. Thus applying Nernst's equation in the form expressed by Eq. (59) to the charge-transfer process (70), we have

$$E = E^\circ + \frac{RT}{nF} \ln \frac{\overline{[A_b^{\mathrm{O}}]^a}}{\overline{[A_a^{\mathrm{R}}]^b}} \qquad (79)$$

or also

$$\theta = \exp\left[\frac{nF}{RT}(E - E^\circ)\right] = \frac{\overline{[A_b^{\mathrm{O}}]^a}}{\overline{[A_a^{\mathrm{R}}]^b}} \qquad (80)$$

where E° is the standard potential of the A_b^O/A_a^R couple and $\overline{[A_b^O]}$, $\overline{[A_a^R]}$ denote volume concentrations at $x = 0$. Here, too, the plane $x = b$ is confused with plane $x = 0$ for mathematical convenience. In the following part of this contribution volume concentrations at $x = b$ will be denoted indifferently by ${}^bC$ and $\bar{C}$. Analogously, bulk concentrations will be designated either by C^* or by ${}^\delta C$.

The mathematical problem relative to the diffusion of the n species A_i is now completely stated.

B. The Mathematical Solution of the Diffusional Problem

1. Case of Equal Diffusion Coefficients

In order to determine the expression for the polarographic current as a function of the electrolysis time t and of the applied potential E, we note that the current density is given by the sum of the fluxes at $x = 0$ of all the species being reduced at the electrode surface, each flux being multiplied by the electric charge involved in the reduction of one mole of the corresponding substance. Thus, designating the Faraday by F and the area of the electrode surface by A, the polarographic current i is given by the equation,

$$i = FA(z_O - z_R)\sum_i a_{i,O} D_i\left(\frac{\partial C_i}{\partial x}\right)_{x=0} \quad \text{for Case 1} \tag{81a}$$

$$i = FA\sum_i [(z_O - z_R)a_{i,O} + (z_{Ox} - z_{Red})a_{i,Ox}]D_i\left(\frac{\partial C_i}{\partial x}\right)_{x=0} \quad \text{for Case 2} \tag{81b}$$

$$i = FA\sum_i [(z_O - z_R)a_{i,O} + (z_J - z_R)a_{i,J}]D_i\left(\frac{\partial C_i}{\partial x}\right)_{x=0} \quad \text{for Case 3} \tag{81c}$$

The solution of the differential equations (67), (72), (73), and (75) with the initial and boundary conditions of Eqs. (77), (78), and (80) is quite complicated and requires recourse to numerical procedures. It is therefore a pleasant surprise to realize how simple the mathematical solution of the diffusional problem becomes if we assume that the diffusion coefficients D_i of all the n species are equal,

$$D_1 = D_2 = \cdots = D_i = \cdots = D_n \equiv D \tag{82}$$

The position in Eq. (82) constitutes a reasonable hypothesis if we consider that the diffusion coefficients of the majority of solutes in aqueous solutions at room temperature are of the order of magnitude of 10^{-5} cm^2/sec and that species bounded by chemical equilibria are often characterized by having approximately the same size. A remarkable exception is represented by complexation equilibria of metal ions with organic ligands. Here the coordination compounds are usually much more bulky than the free metal ion and consequently the simplifying assumption of Eq. (82) is less satisfactory. If the position in Eq. (82) is assumed valid, all the differential operators, δ_i, in Eq. (67) become equal to

$$\delta = D\frac{\partial^2}{\partial x^2} + \frac{2x}{3t}\frac{\partial}{\partial x} - \frac{\partial}{\partial t} \tag{83}$$

so that Eq. (67) can be written as

$$\delta \sum_i a_{i,j} C_i = 0 \tag{84}$$

Analogously, Eq. (78) takes the form,

$$D\frac{\partial}{\partial x}_{x=0} \sum_i a_{i,j} C_i = 0 \tag{85}$$

In view of Eqs. (84), (85), and (77a), the linear combination $\sum_i a_{i,j} C_i$ behaves like the concentration of a simple species S obeying the diffusion Eq. (63) uncomplicated by the source term (i.e., $\delta[S] = 0$) and characterized by a uniform concentration before electrolysis and by a zero flux at the electrode surface during electrolysis. Clearly the concentration $[S]$ of the species S is not altered by the flow of current, and the same is true with the linear combination $\sum_i a_{i,j} C_i$, for which we can write

$$\sum_i a_{i,j} C_i = \sum_i a_{i,j} C_i^* \qquad j = 1, \ldots, h \tag{86}$$

The h relations of Eq. (86), together with the expressions of Eq. (64) for the m equilibrium constants K_ρ, constitute a set of $h + m = n - 1$ relations that hold for any value of t and x. Applying these $n - 1$ equations at $x = 0$ and taking into account the potential dependent boundary condition of Eq. (80), we obtain a set of n algebraic relations among the n surface concentrations $\bar{C}_i$.

Hence through simple passages, the concentrations $\bar{C}_i$ can be derived as functions of the applied potential E, the m equilibrium constants K_ρ, and the bulk concentrations C_i^*. Clearly, in the presence of the sole diffusion overpotential the surface concentrations $\bar{C}_i$ are time independent, provided the potential is kept constant. In conventional polarography the latter circumstance is encountered in practice during the growth of a single drop.

Let us now see how the surface concentrations $\bar{C}_i$ can be used in order to derive the expression for the polarographic current.

a. Case 1. Considering Case 1 first, we write Eq. (81a) in the form,

$$i = (z_{\text{O}} - z_{\text{R}})FAD\left[\frac{\partial(\sum_i a_{i,\text{O}} C_i)}{\partial x}\right]_{x=0} \tag{87}$$

The linear combination of concentrations $\sum_i a_{i,\text{O}} C_i$ satisfies the following requirements:

(1) It obeys the diffusion equation uncomplicated by the source term, as appears from Eq. (72) in which the various coefficients D_i are set equal to D.

(2) From Eq. (87) the ratio $i/[\text{FAD}(z_{\text{O}} - z_{\text{R}})]$ equals the gradient of $\sum_i a_{i,\text{O}} C_i$ at the electrode surface.

(3) Provided the charge-transfer overpotential is zero, the linear combination under study is constant at the electrode surface during the drop life. The preceding conditions are fulfilled by the concentration of any of the two forms Ox and Red of a simple redox couple not complicated by chemical reactions in solution, provided Ox and Red exchange electrons reversibly with the electrode, according to the equation,

$$\text{Ox} + (z_{\text{O}} - z_{\text{R}})\text{e} \rightleftharpoons \text{Red} \tag{88}$$

The differential equation;

$$\delta C_{\text{Ox}} = 0 \tag{89}$$

with the boundary and initial conditions,

$$C_{\text{Ox}} = \bar{C}_{\text{Ox}}\begin{cases} x = 0 \\ t > 0; \end{cases} \qquad C_{\text{Ox}} = C_{\text{Ox}}^*\begin{cases} x \geqslant 0 \\ t = 0 \end{cases} \tag{90}$$

was solved by Ilkovic (7) who obtained the following expression for the concentration gradient $\partial C_{\text{Ox}}/\partial x$ at $x = 0$,

$$\left(\frac{\partial C_{\text{Ox}}}{\partial x}\right)_{x=0} = \frac{(C_{\text{Ox}}^* - \bar{C}_{\text{Ox}})}{\sqrt{\frac{3}{7}\pi Dt}} \tag{91}$$

(see the mathematical appendix for a derivation of Eq. (91)). Generalizing Eq. (91) to Case 1 and taking Eq. (87) into account, the polarographic current is given by

$$i = \sqrt{\frac{7D}{3\pi t}}(z_O - z_R)FA\sum_i a_{i,O}(C_i^* - \bar{C}_i) \tag{92}$$

b. Case 2. The application of Eq. (84) to the clusters A and B results in

$$\sum_i (a_{i,O} + a_{i,R})\,\delta C_i = 0$$
$$\sum_i (a_{i,Ox} + a_{i,Red})\,\delta C_i = 0$$

If the former equation is multiplied by $(z_O - z_R)$ and the latter by $(z_{Ox} - z_{Red})$, the comparison of the resulting equations with Eq. (73) yields the differential equation,

$$\delta \sum_i [(z_O - z_R)a_{i,O} + (z_{Ox} - z_{Red})a_{i,Ox}]C_i = 0 \tag{93}$$

In view of Eqs. (93), (81b), and (82) the linear combination $\sum_i [(z_O - z_R)a_{i,O} + (z_{Ox} - z_{Red})a_{i,Ox}]C_i$ fulfills the three requirements already listed for Case 1. By an entirely analogous procedure the polarographic current for the present case is expressed by the equation,

$$i = \sqrt{\frac{7D}{3\pi t}}FA\sum_i [(z_O - z_R)a_{i,O} + (z_{Ox} - z_{Red})a_{i,Ox}](C_i^* - \bar{C}_i) \tag{94}$$

c. Case 3. Let us apply Eq. (84) to the cluster A in its three forms A^O, A^J, and A^R,

$$\delta \sum_i (a_{i,O} + a_{i,J} + a_{i,R})C_i = 0 \tag{95}$$

Comparing Eq. (95) with Eq. (75) results in the differential equation,

$$\delta \sum_i [(z_O - z_R)a_{i,O} + (z_J - z_R)a_{i,J}]C_i = 0 \tag{96}$$

From Eqs. (96), (81c), and (82) it is manifest that the combination $\sum_i [(z_O - z_R)a_{i,O} + (z_J - z_R)a_{i,J}]C_i$ satisfies the three conditions considered in the previous cases. Hence, proceeding as usual, the current for Case 3 is given by

$$i = \sqrt{\frac{7D}{3\pi t}}\, FA \sum_i [(z_O - z_R)a_{i,O} + (z_J - z_R)a_{i,J}](C_i^* - \bar{C}_i) \tag{97}$$

It should be born in mind that the application of the general procedure previously outlined to the polarographic study of homogeneous chemical equilibria does not require the solution of differential equations, but rather the recourse to simple algebraic manipulations. In fact, once the $n = h + m + 1$ volume concentrations of the various diffusing species at $x = 0$ have been derived from the h linear combinations of Eq. (86), the expressions of the m equilibrium constants in Eq. (64), and the boundary condition in Eq. (80), then the polarographic currents for Cases 1, 2, and 3 are immediately obtained from Eqs. (92), (94), and (97), respectively.

d. One Example. As an example of Case 1, consider the reversible reduction of the complexes of the metal ion M_O, leading to the formation of the amalgam $M_R(Hg)$ within the mercury drop. If we symbolize a rapid chemical equilibrium by $\rightleftharpoons$ and a rapid charge transfer by $\Leftrightarrow$, the problem at hand can be represented through the scheme,

$$\begin{array}{c} \underbrace{M_O \rightleftharpoons M_OX \rightleftharpoons M_OX_2 \rightleftharpoons \cdots \rightleftharpoons M_OX_n} \\ {\scriptstyle +ne}\Big\Updownarrow \\ M_R\ (Hg) \end{array} \tag{98}$$

where X denotes the ligand of the complex compound M_OX_n. Assume that the ligand gives rise, on account of its basicity, to m protonation equilibria,

$$X \rightleftharpoons HX \rightleftharpoons H_2X \rightleftharpoons \cdots \rightleftharpoons H_mX \tag{99}$$

If X is not present in strong excess with respect to the metal ion M_O and the solution is unbuffered, both the n complexation equilibria of Eq. (98) and the m protonation equilibria of Eq. (99) must be taken into account in the solution of the diffusional problem. In fact, during the electro-reduction of M_O to M_R, the concentration of the free ligand increases in the neighborhood of the electrode, and consequently the hydrogen ion concentration decreases. Hence the species X, HX, ..., H_mX, which diffuse away from the electrode, and the hydrogen ions H^+, diffusing towards the electrode, contribute, although electroinactive, to the diffusion overpotential η_d (cf. Eq. (36)). Application of Eq. (77a) to the three clusters X, H, and M (this latter in its two forms M_O and M_R) leads to the three material balances,

$$[M_O] + [M_R] + \sum_{i=1}^{n} [M_OX_i] = [M_O]^* + \sum_{i=1}^{n} [M_OX_i]^* \tag{100}$$

$$[H^+] + \sum_{j=1}^{m} j[H_jX] = [H^+]^* + \sum_{j=1}^{m} j[H_jX]^* \tag{101}$$

$$[X] + \sum_{i=1}^{n} i[M_OX_i] + \sum_{j=1}^{m} [H_jX] = [X]^* + \sum_{i=1}^{n} i[M_OX_i]^* + \sum_{j=1}^{m} [H_jX]^* \tag{102}$$

It should be noted that during electrolysis M_R diffuses within the metal phase, whereas all the other species diffuse within the solution phase. The general treatment described in Sec.II.B.1.a can be applied to this case, provided the distance x from the interphase is taken as positive both in the solution and the metal phase. With this convention Eq. (100) expresses the constancy of the combination $[M_O] + [M_R] + \sum_i [M_OX_i]$ at any given distance x in both the phases. The boundary condition in Eq. (80) as applied to the concentrations $\overline{[M_O]}$ and $\overline{[M_R]}$ at $x = 0$ yields

$$\exp\left[\frac{nF}{RT}(E - E^\circ)\right] = \frac{\overline{[M_O]}}{\overline{[M_R]}} \tag{103}$$

Equations (100) through (103), together with the expressions of the n stability constants and of the m protonation constants, form a set of $m + n + 4$ algebraic relations among the $m + n + 4$ volume concentrations of the diffusing species at $x = 0$. On deriving these $m + n + 4$ concentrations, the expression for the polarographic current is obtained by applying Eq. (92) to the present problem,

$$\begin{aligned} i &= \sqrt{\frac{7D}{3\pi t}}\, nFA\left[[M_O]^* - \overline{[M_O]} + \sum_i ([M_OX_i]^* - \overline{[M_OX_i]})\right] \\ &= \sqrt{\frac{7D}{3\pi t}}\, nFA\overline{[M_R]} \end{aligned} \tag{104}$$

2. Case of Different Diffusion Coefficients

a. Electroactive and Electroinactive Substances. In the general treatment of chemical equilibria of Sec.II.A.1 we have written the charge-transfer reaction (41) in the form of Eq. (70). This way of considering the charge-transfer reaction is purely formal and does not express the actual state of affairs in the pre-electrode layer—namely, the layer at which the charge-transfer actually takes place. In fact any of the m homogeneous reactions (64) can be combined with the electrode reaction (70) giving rise to a different electrode reaction, which a priori is equally acceptable. This situation is due to the fact that η_t is completely negligible with respect to η_d in the time scale of polarographic measurements. Under these circumstances there is no way of determining, at least with the polarographic technique, which of the various diffusing species containing the cluster A^O (or A^R) is electroreduced (or electro-oxidized) at the highest rate. In practice it is this species that is directly reduced (or oxidized) in the preelectrode layer, while the other species containing A^O (or A^R) are reduced (or oxidized) only indirectly, through the perfectly mobile equilibria of Eq. (64).

It must be realized that if the equilibria established in the solution are sufficiently rapid, both the electroactive substances and the electroinactive substances joined to the electroactive ones by these equilibria are characterized by nonzero fluxes at the boundary $x = b$ of the diffuse layer during electrolysis. Consider as an example the complexation equilibrium,

$$M_O + X \rightleftharpoons M_OX \tag{105}$$

in the presence of a strong excess of the free ligand X, and assume that the complex M_OX is very stable. If the rate at which the complex M_OX is reduced directly to the metal M_R at the electrode surface is sufficiently high, owing to the strong excess of M_OX with respect to the free metal ion M_O, the electrode process actually responsible for the electroreduction to M_R is

$$M_OX + ne \rightleftharpoons M_R + X \tag{106}$$

Conversely, if the rate at which M_OX is directly discharged is very low, then the actual state of affairs in the pre-electrode layer is expressed by the cathodic process,

$$M_O + ne \rightleftharpoons M_R \tag{107}$$

and it is the free ion M_O that is actually discharged at the electrode surface. If equilibrium (105) is perfectly mobile, the law of mass action,

$$[M_O] = \frac{K}{[X]}[M_OX]$$

can be applied at any point of the solution, at least outside the diffuse double layer. Assuming that [X] remains constant in the solution up to the double layer during electrolysis, then we have

$$\left(\frac{\partial[M_O]}{\partial x}\right)_{x=0} = \frac{K}{[X]}\left(\frac{\partial[M_OX]}{\partial x}\right)_{x=0} \tag{108}$$

where the notation $x = 0$ is equivalent, as usual, to the other, $x = b$. Equation (108) expresses the proportionality between the fluxes $D(\partial[M_O]/\partial x)$ and $D(\partial[M_OX]/\partial x)$ at the boundary $x = b$ of the diffuse layer, regardless of whether M_O, or M_OX, or both are actually electroactive. We easily see that Eq. (108) cannot be perfectly valid up to the pre-electrode layer. In this connection let us assume that only M_O is electroactive and that the ligand X is a neutral substance, so that M_O and M_OX are equally charged and consequently equally influenced by the electric field in the double layer. The validity of Eq. (108) up to the plane of closest approach, $x = a_2$, would imply that the ratio of the amount of M_O reaching this plane in a given time to the corresponding amount of M_OX is constant. Since, however, only M_O is consumed in the pre-electrode layer whereas the electroinactive species M_OX is left unaltered, M_OX must necessarily be transformed into M_O at $x = a_2$ in order to avoid an irreversible accumulation of M_OX. This implies that the chemical reaction (105) must necessarily be out of equilibrium in the pre-electrode layer. It should probably be mentioned that the situation is further complicated by the influence of the high electric field, existing in the inner part of the double layer, upon equilibrium constants (*9*). In conclusion, for a chemical equilibrium to be perfectly mobile in a polarographic sense it is sufficient that its electrochemical affinity be practically zero up to the diffuse layer, at least in the time scale of polarographic measurements.

The analogy between the behaviors of M_OX and M_O with respect to diffusion outside the double layer does not imply that it is always impossible to determine on experimental grounds which of the two species M_O and M_OX is directly discharged at the electrode surface. In fact, if the rate of the charge-transfer step (namely, the step (106) or (107) according to whether M_OX is directly discharged or not) is so low that η_t is comparable with η_d, then the experimental current is partially controlled by this rate. On the other hand the rate for the forward electrode reaction will presumably depend on the concentration at $x = a_2$ of that of the two species M_OX and M_O which is directly discharged in the pre-electrode layer. Thus if the ligand is present in strong excess with respect to the metal ion and the complex M_OX is stable, an increase in [X] will not appreciably change the rate for the forward reaction at a given potential if M_OX is electroactive, while it will strongly decrease it if this is not the case. Consequently the behavior of the polarographic current will be different in the two cases. If, however, the electrochemical affinity of the charge-transfer step is practically zero, no definite conclusions can be drawn as to the nature of the species directly discharged, since Nernst's equation (59) as applied to Eq. (107),

$$\theta = \frac{\overline{[M_O]}}{\overline{[M_R]}}$$

differs from Nernst's equation as applied to Eq. (106):

$$\theta' = \frac{\overline{[M_OX]}}{\overline{[M_R]}\,\overline{[X]}}$$

by the factor $\overline{[M_OX]}/(\overline{[M_O]}[X])$, which simply expresses the equilibrium constant for the homogeneous equilibrium (105).

b. The Diffusion Layer. In polarographic literature many diffusional problems are solved by having recourse to the approximate concept of diffusion layer (*10*). In the presence of diffusion overpotential only, the procedure based on the use of this concept leads to results that are analogous to those derivable by the rigorous treatment previously described. When, however, charge-transfer and reaction overpotentials contribute to the total overpotential, the procedure based on the diffusion layer concept represents a valuable method for rapidly obtaining an approximate expression for the polarographic current-potential characteristic.

According to the definition of diffusion layer thickness δ given in Sec.I.B.2, the concentration gradient of a diffusing species A_i at the plane $x = b$ is given by

$$\left(\frac{\partial C_i}{\partial x}\right)_{x=0} = \frac{C_i^* - \bar{C}_i}{\delta} \qquad (109)$$

Comparing Eq. (109) with the expression for the concentration gradient at $x = 0$ derived by Ilkovic (Eq. (91)), it immediately follows that in the case of diffusion toward a dropping electrode the thickness δ of the diffusion layer is given by $\sqrt{3\pi Dt/7}$ and, consequently, increases with the square root of electrolysis time. The use of the relation

$$\left(\frac{\partial C_i}{\partial x}\right)_{x=0} = \frac{C_i^* - \bar{C}_i}{\delta} \qquad \text{with} \qquad \delta = \sqrt{\frac{3\pi Dt}{7}} \tag{110}$$

is perfectly legitimate only in the field of applicability of the Ilkovic equation (91)—that is, in the case of a diffusing substance A_i obeying the diffusion equation $\delta C_i = 0$ (and, consequently, not involved in homogeneous chemical reactions) and characterized by a constant concentration $\bar{C}_i$ at $x = 0$. The partial derivative $(\partial C_i/\partial x)_{x=0}$ can be approximately replaced by the right-hand side of Eq. (110) in the boundary conditions defining the particular diffusional problem under study even when the previous conditions are not fulfilled ($\bar{C}_i$ changing during the drop life as a consequence of a slow charge-transfer process, presence of homogeneous chemical reactions, and so on). This is the essence of the approximate treatment based on the concept of diffusion layer thickness. It must be emphasized that the use of Eq. (110) reduces boundary conditions to simple algebraic relations among the applied potential and the various concentrations $\bar{C}_i$ at $x = 0$. Thus once the proper number of boundary conditions has been written, the solution of the set of algebraic relations so obtained allows the polarographic current-potential characteristic to be derived by simple passages without having recourse to the solution of the diffusional differential equations (63). In the presence of reaction overpotential such an approximate procedure requires the introduction of the further concept of reaction layer thickness (cf. Sec. IV.C.1).

As an application of the concept of diffusion layer thickness, let us re-examine the general case of m perfectly mobile equilibria interconnecting n diffusing species A_i. On considering directly the h boundary conditions in Eq. (85) and making use of Eq. (110), we immediately obtain

$$\sum_i a_{i,j}\bar{C}_i = \sum_i a_{i,j}C_i^* \qquad j = 1, \ldots, h \tag{111}$$

which represents the application of Eq. (86) to the electrode surface. Equation (111) together with Eq. (64) and the boundary condition in Eq. (80) allow the volume concentrations $\bar{C}_i$ at $x = 0$ to be obtained as functions of the applied potential. In order to derive the current-potential characteristic, consider the expression (81a) of the polarographic current for Case 1 (Sec.II.B.1). By replacing the concentration gradients at $x = 0$ from Eq. (110) into Eq. (81a) we directly obtain Eq. (92). We can conclude that in the presence of perfectly mobile equilibria the theoretical current-voltage curves obtained through the use of the diffusion layer thickness δ coincide with those obtained by the rigorous procedure outlined in Sec. II.B.1, provided the diffusion coefficients of the various substances A_i are assumed to be equal.

It is useful to consider the form assumed by the diffusion overpotential between $x = b$ and $x = \delta$:

$$\eta_{\mathrm{d}}^{b\delta} = -\frac{1}{I}\int_b^\delta \sum_i |\mathbf{J}_i| \left|\frac{\partial \mu_i}{\partial x}\right| dx \tag{112}$$

in the diffusion layer thickness approximation. In view of Eq. (61), Eq. (112) can be written in the form,

$$\eta_{\mathrm{d}}^{b\delta} = -\frac{DRT}{I}\int_b^\delta \sum_i \left(\frac{\partial C_i}{\partial x}\frac{\partial \ln C_i}{\partial x}\right) dx \tag{113}$$

Noting that the dependence of C_i on the distance x from the electrode is expressed by

$$C_i = \bar{C}_i + \frac{C_i^* - \bar{C}_i}{\delta} x$$

Eq. (113) becomes

$$\eta_{\mathrm{d}}^{b\delta} = -\frac{DRT}{I}\sum_i \frac{C_i^* - \bar{C}_i}{\delta} \ln \frac{C_i^*}{\bar{C}_i} \tag{114}$$

If we choose to express $\eta_{\mathrm{d}}^{b\delta}$ in terms of the concentrations $[\mathrm{A^O}]$ and $[\mathrm{A^R}]$ of the two forms of the electroactive cluster, it is convenient to add and subtract the two terms $(DRT/I) \cdot [\sum_i a_{i,\mathrm{O}}(C_i^* - \bar{C}_i)/\delta] \cdot \ln([\mathrm{A^O}]^*/\overline{[\mathrm{A^O}]})$ and $(DRT/I) \cdot [\sum_i a_{i,\mathrm{R}}(C_i^* - \bar{C}_i)/\delta] \cdot \ln([\mathrm{A^R}]^*/\overline{[\mathrm{A^R}]})$ on the right-hand side of Eq. (114),

$$\eta_{\mathrm{d}}^{b\delta} = \frac{DRT}{I}\left\{\sum_i \left(a_{i,\mathrm{O}}\frac{C_i^* - \bar{C}_i}{\delta}\right)\ln\frac{\overline{[\mathrm{A^O}]}}{[\mathrm{A^O}]^*} + \sum_i \left(a_{i,\mathrm{R}}\frac{C_i^* - \bar{C}_i}{\delta}\right)\right.$$

$$\left.\ln\frac{\overline{[A^{\mathrm{R}}]}}{[A^{\mathrm{R}}]^*} - \sum_i \left(\frac{C_i^* - \bar{C}_i}{\delta}\ln\frac{C_i^*\,\overline{[A^{\mathrm{O}}]}^{a_{i,\mathrm{O}}}\,\overline{[A^{\mathrm{R}}]}^{a_{i,\mathrm{R}}}}{\bar{C}_i\,[A^{\mathrm{O}}]^{*a_{i,\mathrm{O}}}[A^{\mathrm{R}}]^{*a_{i,\mathrm{R}}}}\right)\right\} \tag{115}$$

On applying Eq. (85) to the cluster A in its two forms A^O and A^R, from Eqs. (87) and (110) it follows that

$$D\sum_i \left(a_{i,O}\frac{C_i^* - \bar{C}_i}{\delta}\right) = -D\sum_i \left(a_{i,R}\frac{C_i^* - \bar{C}_i}{\delta}\right) = \frac{I}{(z_O - z_R)F} \tag{116}$$

Recalling Eq. (69), we can now express the n concentrations C_i as functions of the m equilibrium constants (Eq. (64)), of the $h-1$ concentrations of all the clusters, other than A, as well as of the concentrations $[A^R]$ and $[A^O]$ of the two forms of A,

$$C_i = K_i'[A^O]^{a_{i,O}}[A^R]^{a_{i,R}}\prod_{j\neq A}\left(\sum_i a_{i,j}C_i\right)^{a_{i,j}} \qquad i = 1, \ldots, n \tag{117}$$

Here $\sum_i a_{i,j}C_i$ is the concentration of the jth cluster. Equation (117) formally expresses the dissociation of the ith species into the clusters composing it. Replacing Eqs. (116) and (117) into Eq. (115), we obtain

$$\eta_d^{b\delta} = \frac{RT}{(z_O - z_R)F}\ln\frac{\overline{[A^O]}\,[A^R]^*}{\overline{[A^R]}\,[A^O]^*} - \frac{DRT}{I}\sum_i\left(\frac{C_i^* - \bar{C}_i}{\delta}\ln\frac{\prod_{j\neq A}(\sum_i a_{i,j}C_i^*)^{a_{i,j}}}{\prod_{j\neq A}(\sum_i a_{i,j}\bar{C}_i)^{a_{i,j}}}\right)$$

In view of Eq. (111),

$$\eta_d^{b\delta} = \frac{RT}{(z_O - z_R)F}\ln\frac{\overline{[A^O]}\,[A^R]^*}{\overline{[A^R]}\,[A^O]^*} \tag{118}$$

An expression for $\eta_d^{b\delta}$ analogous to Eq. (118) would have also been obtained if we had chosen a redox couple consisting of any two species containing the oxidized and the reduced forms of the electroactive cluster A respectively, as for instance the A_b^O/A_a^R couple of Eq. (80). Equation (118) is identical with the expression of the diffusion overpotential derived by Vetter (*11*) for a simple redox coule A^O/A^R by subtracting the equilibrium potential,

$$E = E^0 + \frac{RT}{(z_O - z_R)F}\ln\frac{[A^O]^*}{[A^R]^*}$$

from the nonequilibrium potential as expressed by Nernst's equation applied to the concentrations at $x = 0$,

$$E = E^0 + \frac{RT}{(z_O - z_R)F}\ln\frac{\overline{[A^O]}}{\overline{[A^R]}}$$

c. The Diffusion Layer Approximation and Its Limitations. When the differences among the diffusion coefficients D_i of the n diffusing species cannot be neglected, the approximate treatment based on the diffusion layer concept can still be applied quite easily by replacing the partial derivatives $(\partial C_i/\partial x)_{x=0}$ in the boundary conditions of Eq. (78) and in the expressions (81a), (81b), and (81c) with the relations,

$$\left(\frac{\partial C_i}{\partial x}\right)_{x=0} = \frac{C_i^* - \bar{C}_i}{\delta_i}$$

where

$$\delta_i = \sqrt{\frac{3\pi D_i t}{7}} \qquad i = 1, \ldots, n \tag{119}$$

Thus for Case 1 of Sec. II.B.1 we obtain the equations,

$$\sum_i a_{i,j} D_i^{1/2} C_i^* = \sum_i a_{i,j} D_i^{1/2} \bar{C}_i \qquad j = 1, \ldots, h \tag{120}$$

and

$$i = (z_O - z_R) FA \sum_i a_{i,O} D_i^{1/2} \frac{C_i^* - \bar{C}_i}{(3\pi t/7)^{1/2}} \tag{121}$$

Combining Eq. (120) with Eqs. (64) and (80) yields the concentrations $\overline{C_i}$ at $x = 0$; their replacement into Eq. (121) leads directly to an approximate expression for the current-potential characteristic. The above procedure amounts to assuming that each species diffuses across its own diffusion layer, the thickness of which is the higher the larger the diffusion coefficient of the species being considered.

Relations identical with Eqs. (120) and (121) are obtained if we follow the "apparently rigorous" procedure based on the assumption that each substance diffusing in the solution obeys the diffusion equation uncomplicated by the source term,

$$\frac{\partial C_i}{\partial t} = D_i \frac{\partial^2 C_i}{\partial x^2} + \frac{2x}{3t}\frac{\partial C_i}{\partial x} \qquad i = 1, \ldots, n \tag{122}$$

As a matter of fact Eq. (122), which has been applied by several investigators (*12*, *13*) to the diffusion of substances involved in perfectly mobile equilibria, is not wholly correct because it does not account for the mutual conversion of the substances bounded by the homogeneous reactions. It is interesting to note that whereas the validity of Eq. (122) implies the validity of the differential equations (62), (72), (73), and (75), the contrary is not necessarily true. On temporarily assuming the validity of Eq. (122), let us take into account the boundary conditions in Eq. (78). Making the substitution (*14*),

$$z_i = \frac{x}{D_i^{1/2}}$$

equations (122) and (78) become

$$\frac{\partial C_i}{\partial t} = \frac{\partial^2 C_i}{\partial z_i^2} + \frac{2z_i}{3t}\frac{\partial C_i}{\partial z_i}; \qquad \sum_i a_{i,j} D_i^{1/2} \left(\frac{\partial C_i}{\partial z_i}\right)_{z_i=0} = 0 \qquad j = 1, \ldots, h \quad (123)$$

Let us now define the following function of z,

$$F_j(z) = \sum_i a_{i,j} D_i^{1/2} C_i(z_i = z) \qquad j = 1, \ldots, h \quad (124)$$

The value of $F_j(z)$ for a given value of the independent variable z is obtained by making the n variables z_i equal to z. Combining Eqs. (123) and (124) we have

$$\frac{\partial F_j}{\partial t} - \frac{\partial^2 F_j}{\partial z^2} - \frac{2z}{3t}\frac{\partial F_j}{\partial z} = \sum_i a_{i,j} D_i^{1/2} \left[\frac{\partial C_i}{\partial t} - \frac{\partial^2 C_i}{\partial z_i^2} - \frac{2z_i}{3t}\frac{\partial C_i}{\partial z_i}\right]_{z_i=z} = 0 \qquad j = 1, \ldots, h \quad (125)$$

and

$$\left(\frac{\partial F_j}{\partial z}\right)_{z=0} = 0 \qquad j = 1, \ldots, h \quad (126)$$

Thus the linear combination F_j formally behaves like the concentration of a simple substance diffusing towards the dropping electrode along the generalized coordinate z, with a "formal" dimensionless diffusion coefficient equal to one and being characterized by a zero flux at the electrode surface. Since F_j is uniform throughout the solution before electrolysis, this amounts to saying that $F_j\,(z)$ remains constant during the drop life. Hence

$$\sum_i a_{i,j} D_i^{1/2} C_i(z_i = z) = \sum_i a_{i,j} D_i^{1/2} C_i(z_i = \infty) \equiv \sum_i a_{i,j} D_i^{1/2} C_i^* \qquad \text{for any value of } z \quad (127)$$

On applying Eq. (127) at the electrode surface ($x = 0$) and noting that for $x = 0$ all values of z_i are zero, independent of the different values for D_i, we immediately obtain Eq. (120). Confining ourselves to Case 1 of Sec. II.B.1, the expression (81a) for the polarographic current can be written in the form,

$$i = (z_O - z_R)FA\left(\frac{\partial F_O}{\partial z}\right)_{z=0} \tag{128}$$

where $F_O \equiv \sum_i a_{i,O} D_i^{1/2} C_i(z_i = z)$ is the function $F_j(z)$ as applied to the oxidized form A^O of the electroactive cluster. In view of Eqs. (125), (127), and (128), the linear combination $F_O(z)$ satisfies the three conditions listed in Sec. II.B.1.a. Following the same line of reasoning employed in that section we then have, in view of the Ilkovic equation (Eq. (91)),

$$i = (z_O - z_R)FA\frac{F_O^* - \bar{F}_O}{(3\pi t/7)^{1/2}} \tag{129}$$

with

$$F_O^* = \sum_i a_{i,O} D_i^{1/2} C_i^* \qquad \text{and} \qquad \bar{F}_O = \sum_i a_{i,O} D_i^{1/2} \bar{C}_i$$

Equation (129) is identical to Eq. (121), showing that both the procedure based on the diffusion layer concept and that based on the improper use of Eq. (122) lead to the same results when Case 1 is concerned. Analogous conclusions could be drawn for Cases 2 and 3.

It must be emphasized that the two above procedures are approximate. Thus they fail to explain, even from a qualitative point of view, the non-additivity of some polarographic waves. An example is offered by the aqueous system $I^- - I_2 - ICl$ on platinum (*15–17*). Here, if only iodide is present in the bulk of the solution, two anodic waves are observed, the former due to the oxidation of I^- to I_2 and the latter to the successive oxidation of I_2 to ICl. Taking into account that each step involves the transfer of one electron from an iodine atom to the electrode, the two steps must be of equal heights, provided the diffusion coefficients of the various species involved in the electrode process, namely, I^-, I_2, I_2Cl^-, ICl, ICl_2^-, are equal. As a matter of fact the $I^- \rightarrow I_2$ step is about 116/100 of the successive $I_2 \rightarrow ICl$ step. Analogously, if only ICl is present in the bulk of the solution, the first cathodic step, due to the electroreduction of ICl to I_2, is about 106/100 of the following $I_2 \rightarrow I^-$ step. On the contrary the anodic step $I_2 \rightarrow ICl$ obtained by polarographing an iodine solution has the same height as the corresponding cathodic step $I_2 \rightarrow I^-$. This particular behavior can be qualitatively understood by considering the concentration-distance profiles for the various diffusing species (Fig. 3). These profiles were obtained (*17*) by following the

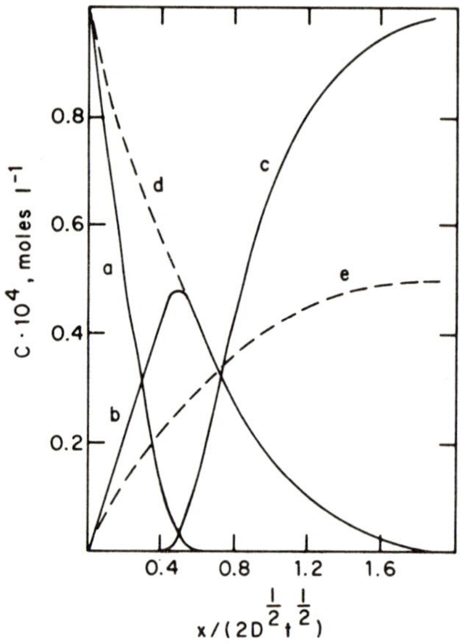

Fig. 3. Concentrations as functions of $x/(2D^{1/2}t^{1/2})$, where x is the distance from the electrode and t the electrolysis time. Anodic (cathodic) limiting current of 10^{-4} M I^- (10^{-4} M ICl) in 10^{-1} M chloride: curve a, C_{ICl} ($[I^-]$); curve b, C_{I_2}; curve c, $[I^-]$ (C_{ICl}). Anodic (cathodic) limiting current of $5 \cdot 10^{-5}$ M I_2 in 10^{-1} M chloride: curve d, C_{ICl} ($[I^-]$); curve e, C_{I_2}. (From *17*)

procedure of Sec. II.B.1.c, on assuming equal diffusion coefficients. The dashed curve e expresses the total concentration of iodine, $C_{I_2} \equiv [I_2] + [I_2Cl^-]$, corresponding to the anodic as well as the cathodic limiting current of $5 \times 10^{-5} M$ I_2 in $10^{-1} M$ chloride. The dashed curve d expresses the iodide concentration $[I^-]$, or alternatively the total concentration, $C_{ICl} \equiv [ICl] + [ICl_2^-]$, of iodine chloride in the same solution according to whether the cathodic or the anodic limiting current is considered. It is readily seen from the figure that both the anodic and cathodic limiting currents of iodine are exclusively determined by the flux of I_2 towards the electrode. Consequently these currents have equal heights regardless of whether the diffusion coefficients of the various species are equal or not. Curves c, b, and a in Fig. 3 represent, respectively, the concentrations $[I^-]$, C_{I_2}, and C_{ICl} corresponding to the limiting current of the second step, $I_2 \rightarrow ICl$, of the electrooxidation of $10^{-4}M$ I^- in $10^{-1}M$ chloride. It is evident from the figure that when ICl and ICl_2^- diffuse from the electrode, they react with incoming iodide to give I_2 and I_2Cl^-. In practice, passing from the limiting current of the first anodic step, $I^- \rightarrow I_2$, to the total limiting current of I^-, we have a gradual replacement of iodide ions with I_2 molecules and I_2Cl^- ions as charge-transfer carriers towards the electrode and, consequently, the total limiting current is determined both by the flux of I^- and by those of I_2 and I_2Cl^- at the electrode surface. A different situation is encountered within the range of potentials corresponding to the limiting current of the first anodic step. Here the tendency of I_2 molecules and I_2Cl^- ions diffusing away from the electrode to react with incoming I^-, resulting in I_3^- formation, is negligible and the current is almost exclusively determined by the flux of I^- towards the electrode. The experimental value 116/110 for the ratio of the first anodic wave of I^- to the second can therefore be justified by considering that the diffusion coefficient of I^- is greater than the weighted average of the diffusion coefficients of I_2 and I_2Cl^-. Analogous considerations can be drawn for the two cathodic steps of iodine chloride.

Let us calculate the total limiting current of I^- by the approximate procedure outlined in the present section, which takes into account the differences among the various diffusion coefficients. The problem under study is an example of Case 3. Hence, on substituting $(\partial C_i/\partial x)_{x=0}$ from Eq. (119) into Eq. (81c), which expresses the polarographic current for Case 3, we have

$$i = FA \sum_i [(z_O - z_R)a_{i,O} + (z_J - z_R)a_{i,J}]D_i^{1/2} f(t)(C_i^* - \bar{C}_i) \qquad (130)$$

where $f(t) = 1/(3\pi t/7)^{1/2}$. As a matter of fact the electrooxidation of I^- is carried out on platinum, since the potentials at which it takes place are prohibitively high for mercury electrodes. It follows that owing to the somewhat different conditions under which the renewal of the diffusion layer is performed at solid electrodes with respect to dropping electrodes, the time dependence of the potentiostatic current is expressed by a function different from $f(t)$ (*17*). Nevertheless, this difference is of no importance in the present discussion. In an iodide solution both $C_{I_2}^*$ and C_{ICl}^* are zero. Noting that the electroactive cluster is represented by the iodine atom I, in Eq. (130) we have $a_{I^-,O} = 0$, $a_{I_2,O} = 0$, $a_{ICl,O} = 1$, $a_{I^-,J} = 0$, $a_{I_2,J} = 2$, $a_{ICl,J} = 0$, $a_{I^-,R} = 1$, $a_{I_2,R} = 0$, $a_{ICl,R} = 0$. When the total anodic limiting current i_d is reached, the volume concentrations $\overline{[I^-]}$ and $\bar{C}_{I_2}$ at $x = 0$ are zero, whereas $\bar{C}_{ICl}$ is equal to $(D_{I^-}^{1/2}/D_{ICl}^{1/2})[I^-]^*$, in view of Eq. (120) as applied to the cluster I. Making the proper substitutions in Eq. (130), i_d is given by

$$i_d = -2FAf(t)D_{I^-}^{1/2}[I^-]^*$$

and depends exclusively on the value for the diffusion coefficient of iodide ion. According to the approximate procedure, the two anodic waves of I^- should therefore be of equal heights, independent of the values for D_{ICl} and D_{I_2}. These theoretical predictions are in contrast with experiment. The only way of interpreting the experimental data from a quantitative point of view is to solve the rigorous differential equations of Sec. II.A by a numerical procedure, as done by Beilby and Crittenden (*16*) under somewhat simplifying assumptions.

d. Treatment of Complexation Equilibria by a Rigorous Procedure. An important case where the differences among diffusion coefficients can be taken into account very easily by the rigorous procedure, without having recourse to numerical methods, is represented by a reversible electrode process coupled with complexation equilibria of the type,

$$M_OX_k + X \rightleftharpoons M_OX_{k+1}$$

in the presence of a strong excess of the complex forming agent (*8*). Under these conditions the concentration of the free ligand is not appreciably altered by electrolysis and, consequently, the consideration of the cluster X is superfluous.

It must be realized that the flux $D_x(\partial[X]/\partial x)_{x=0}$ of X at the electrode surface, or also the increment $[X]^* - \overline{[X]}$, is comparable with those of the other diffusing species. The main distinctive feature of a diffusing species X present in strong excess is the very small value of the ratio $([X]^* - \overline{[X]})/[X]^*$. From a thermodynamic point of view this situation can be expressed by saying that the contribution of X to diffusion overpotential is negligible. In fact, Eq. (36) shows that this contribution is given by

$$\int_0^c |\mathbf{J}_x| \left| \frac{\partial \mu_x}{\partial x} \right| dx = D_x RT \int_0^c \frac{1}{[X]} \left(\frac{\partial [X]}{\partial x} \right)^2 dx \tag{131}$$

Since the concentration gradients of all the diffusing species are comparable, the concentration [X] in the denominator in Eq. (131) renders the contribution of X to η_d much smaller than those of the other diffusing species.

Considering Eq. (98), the number of molecular species whose concentrations vary appreciably in the neighborhood of the electrode during electrolysis is $n + 2$, the number of equilibria is n, whereas the number of clusters to be taken into account is 1 (the electroactive cluster M). If we designate the diffusion coefficient of a given species $M_O X_i$ by D_i^O and that of M_R by D^R, application of Eq. (72) to the present case yields the differential equations,

$$\sum_i D_i^O \, \delta'[M_O X_i] - \sum_i \delta''[M_O X_i] = 0 \tag{132}$$

$$D^R \, \delta'[M_R] - \delta''[M_R] = 0 \tag{133}$$

where

$$\delta' \equiv \frac{\partial^2}{\partial x^2} \qquad \text{and} \qquad \delta'' \equiv \frac{\partial}{\partial t} - \frac{2x}{3t}\frac{\partial}{\partial x}$$

In order to find a linear combination S of the various concentrations $[M_O X_i]$ which satisfies the diffusion equation $D^O \, \delta' S - \delta'' S = 0$, uncomplicated by the source term, Eq. (132) will be written as

$$D^O \sum_i \frac{D_i^O}{D^O} \delta'[M_O X_i] = \sum_i \frac{D_i^O}{D^O} \delta''[M_O X_i] + \delta'' \sum_i \left(1 - \frac{D_i^O}{D^O}\right)[M_O X_i] \tag{134}$$

where D^O is a factor to be determined, having the dimensions of a diffusion coefficient. The linear combination $\sum_i (D_i^O/D^O)[M_O X_i]$ satisfies the diffusion equation without the source term only if D^O is given a value that makes $\delta'' \sum_i [1 - (D_i^O/D^O)][M_O X_i]$ equal to zero. Designate by $\beta_1, \beta_2, \ldots, \beta_i, \ldots, \beta_n$ the equilibrium constants for the complexation equilibria of scheme (98),

$$[M_O X_i] = \beta_i [X]^i [M_O] \qquad i = 1, \ldots, n \tag{135}$$

The second term on the right-hand side of Eq. (134) may then be written in the form,

$$\delta'' \frac{[M_O]}{D^O} \sum_i (D^O - D_i^O)\beta_i [X]^i \tag{136}$$

If the expression (136) is made equal to zero, D^O is given by

$$D^O = \frac{\sum_i D_i^O \beta_i [X]^i}{\sum_i \beta_i [X]^i} \tag{137}$$

whereas Eq. (134) becomes

$$D^{\mathrm{O}}\delta' \sum_i \frac{D_i^{\mathrm{O}}}{D^{\mathrm{O}}}[\mathrm{M_OX}_i] = \delta'' \sum_i \frac{D_i^{\mathrm{O}}}{D^{\mathrm{O}}}[\mathrm{M_OX}_i] \tag{138}$$

In view of Eqs. (135) and (137) we shall note that

$$\sum_i \frac{D_i^{\mathrm{O}}}{D^{\mathrm{O}}}[\mathrm{M_OX}_i] = [\mathrm{M_O}]\sum_i \frac{D_i^{\mathrm{O}}}{D^{\mathrm{O}}}\beta_i[\mathrm{X}]^i = [\mathrm{M_O}]\sum_i \beta_i[\mathrm{X}]^i = \sum_i [\mathrm{M_OX}_i] \tag{139}$$

so that Eq. (138) takes the more simplified form,

$$D^{\mathrm{O}}\delta' \sum_i [\mathrm{M_OX}_i] = \delta'' \sum_i [\mathrm{M_OX}_i] \tag{140}$$

Taking Eq. (135) into account, Nernst's equation (80) can be written as follows,

$$\theta' = \theta \sum_i \beta_i[X]^i = \frac{\overline{\sum_i [\mathrm{M_OX}_i]}}{\overline{[\mathrm{M_R}]}} \tag{141}$$

where

$$\theta = \frac{\overline{[\mathrm{M_O}]}}{\overline{[\mathrm{M_R}]}} = \exp\left[\frac{nF}{RT}(E - E^0)\right] \tag{142}$$

The application of Eq. (78) to the cluster M leads to the boundary condition,

$$\sum_i D_i^{\mathrm{O}}\left(\frac{\partial[\mathrm{M_OX}_i]}{\partial x}\right)_{x=0} + D^{\mathrm{R}}\left(\frac{\partial[\mathrm{M_R}]}{\partial x}\right)_{x=0} = 0$$

or also, taking Eq. (139) into account,

$$D^{\mathrm{O}}\frac{\partial}{\partial x_{x=0}}\sum_i [\mathrm{M_OX}_i] + D^{\mathrm{R}}\frac{\partial}{\partial x_{x=0}}[\mathrm{M_R}] = 0 \tag{143}$$

In view of Eqs. (81a), (139), and (143), the expression for the current is given by

$$i = nFAD^{\mathrm{O}}\frac{\partial}{\partial x_{x=0}}\sum_i [\mathrm{M_OX}_i] = -nFAD^{\mathrm{R}}\frac{\partial}{\partial x_{x=0}}[\mathrm{M_R}] \tag{144}$$

Equations (133) and (140) are in the form of the diffusion equation (122) uncomplicated by the source term. Consequently, on dealing with the linear combination $\sum_i [M_O X_i]$ as if it were a simple concentration, the application of the procedure followed in Eqs. (122) through (127) to the concentration $[M_R]$ as well as to $\sum_i [M_O X_i]$ is perfectly legitimate. Thus from Eq. (127) we obtain

$$D^{O^{1/2}} \sum_i [M_O X_i]^* + D^{R^{1/2}}[M_R]^* = D^{O^{1/2}} \sum_i \overline{[M_O X_i]} + D^{R^{1/2}}\overline{[M_R]} \quad (145)$$

The linear combination $\sum_i \overline{[M_O X_i]}$ at the electrode surface and the concentration $\overline{[M_R]}$ at $x = 0$ are bounded by the two algebraic relations (141) and (145). Hence, provided the applied potential and consequently θ' and θ are kept constant, then $\sum_i \overline{[M_O X_i]}$ and $\overline{[M_R]}$ also remain constant. The boundary and initial conditions relative to the diffusional equations (133) and (140) are identical with conditions (90), employed by Ilkovic (7) for the solution of Eq. (89), so that from Eq. (91) we have

$$\left.\begin{aligned} \frac{\partial \sum_i [M_O X_i]}{\partial x} &= \frac{\sum_i [M_O X_i]^* - \sum_i \overline{[M_O X_i]}}{\sqrt{\frac{3}{7}\pi D^O t}} \\ \frac{\partial [M_R]}{\partial x} &= \frac{[M_R]^* - \overline{[M_R]}}{\sqrt{\frac{3}{7}\pi D^R t}} \end{aligned}\right\} \quad (146)$$

The combination of Eqs. (146) and (144) yields

$$i = K^O\left(\sum_i [M_O X_i]^* - \sum_i \overline{[M_O X_i]}\right) = -K^R([M_R]^* - \overline{[M_R]}) \quad (147)$$

where

$$K^O = \frac{nFAD^{O^{1/2}}}{(3\pi t/7)^{1/2}}, \qquad K^R = \frac{nFAD^{R^{1/2}}}{(3\pi t/7)^{1/2}} \quad (148)$$

When the cathodic diffusion limiting current $i_{d,c}$ is reached, $\sum_i \overline{[M_O X_i]} = 0$ so that from Eq. (147) we have

$$i_{d,c} = K^O \sum_i [M_O X_i]^* \quad (149)$$

Analogously, the anodic diffusion limiting current $i_{d,a}$, characterized by $\overline{[M_R]} = 0$, is given by

$$i_{d,a} = -K^R[M_R]^* \tag{150}$$

From Eqs. (141), (142), and (147) through (150) we have, after simple passages,

$$E = E^O - \frac{RT}{nF} \ln \sum_i \beta_i[X]^i + \frac{RT}{2nF} \ln \frac{D^R}{D^O} + \frac{RT}{nF} \ln \frac{i_{d,c} - i}{i - i_{d,a}} \tag{151}$$

Equation (151) expresses the current-potential characteristic for the present case.

C. Diagnostic Criteria for the Distinction of Polarographic Currents Characterized by the Exclusive Contribution of Diffusion Overpotential to Total Overpotential (Reversible Waves)

Usually the experimental data gathered both from the measurement of mean polarographic currents as functions of the applied potential and from the measurement of instantaneous currents as functions of electrolysis time are necessary and sufficient for distinguishing diffusion-controlled currents from other types of currents, partially controlled either by a slow charge-transfer step or by a slow purely chemical reaction. In this respect the shapes of current-potential characteristics must be in accord with those predicted theoretically upon application of Nernst's equation to volume concentrations at the electrode surface (cf. Sec. II.A.2). It must be emphasized that such an accord is by no means sufficient for characterizing diffusion-controlled currents. In fact many polarographic waves partially limited by the rate of a chemical reaction and characterized by a very fast charge-transfer step exhibit the same shapes (apart from an eventual shift of the whole wave along the potential axis) as the polarographic curves that would be observed were the chemical reaction infinitely rapid. If a change in the profile of the current-potential characteristic with a variation of pH or of the concentration of a complex-forming agent is theoretically expected, the experimental verification of such a change is a further proof of the validity of the starting hypotheses. No general rules can be given as to the behavior of the half-wave potential

$E_{1/2}$ of a "reversible" wave. In fact, if it is true that for a simple electrode process of the type Ox + $ne \rightleftharpoons$ Red, $E_{1/2}$ remains constant with changing the concentration of the depolarizer, this situation is no longer encountered if the electrode process is asymmetrical as in the reduction of I_2 to I^- (*8*). An important distinctive feature of diffusion-controlled currents is represented by the fact that the ratio $\bar{\imath}(E)/\bar{\imath}_d$ of the mean current $\bar{\imath}(E)$ at any given potential E along the rising portion of the wave to the mean diffusion limiting current $\bar{\imath}_d$ is independent of the drop time t_d. This follows immediately from Eq. (92), according to which the functional dependence of the instantaneous current i on the electrolysis time t, and consequently of the mean current $\bar{\imath}$ on the drop time t_d, is unaffected by a change in the applied potential.

Additional information on diffusion-controlled currents may be obtained by observing how instantaneous currents vary with electrolysis time during the life of a single drop. From an experimental point of view this can be done by using either a good oscillograph or a string galvanometer with a rapid response. Equation (92) shows that along the whole rising portion of a diffusion-controlled wave the instantaneous current at constant potential is proportional to $A/t^{1/2}$, where A is the instantaneous area of the dropping electrode. We recall that the flow rate m of mercury is directly proportional and the drop time t_d inversely proportional to the height h of the mercury head, corrected for the back pressure of the solution,

$$m = k_1 h; \qquad t_d = \frac{k_2}{h} \tag{152}$$

Hence for a given value of h, m is constant with time and the volume V of the drop at a given instant t is given by

$$V = \frac{mt}{d} = \frac{4\pi r^3}{3}$$

where d is the density of mercury and r the instantaneous radius of the drop. It immediately follows that the instantaneous area $A = 4\pi r^2$ is proportional to $(mt)^{2/3}$,

$$A = 0.85\, m^{2/3} t^{2/3} \tag{153}$$

where A is expressed in cm^2, m in g sec^{-1}, and t in sec. From Eq. (92) we may conclude that whereas the instantaneous current density $I = i/A$ decreases with time as $t^{-1/2}$, the current intensity i increases with time as $t^{1/6}$,

$$i \propto m^{2/3} t^{1/6} \tag{154}$$

owing to the progressive growth of the electrode surface. For this reason a plot of log i against log t should yield a straight line with a slope of $0.167 = \frac{1}{6}$ at any potential along the polarographic wave. In reality a slightly greater slope is observed, due to the sphericity of the drop which was disregarded in the derivation of Eq. (92). Thus under normal experimental conditions ($t = 1$ to 6 sec, $D = 10^{-5}$ cm^2 sec^{-1}, $m = 1$ mg sec^{-1}), the slope of the log i to the log t plot is of about 0.190. More significant deviations from the slope of 0.167 are observed if the instantaneous current is not recorded on the first drop, counted from the start of electrolysis. Measurement of current on the second, third, or subsequent drops introduces an error due to the concentration polarization transferred from the earlier drops. A number of authors have described methods for excluding this "depletion effect" from the measurement of current-time curves (*18–21*).

A more rapid, although less precise, method for verifying Eq. (154) consists in measuring the mean current intensity,

$$\bar{i} = \frac{1}{t_d} \int_0^{t_d} i \, dt \propto m^{2/3} t_d^{1/6} \tag{155}$$

as a function of m and t_d. The constancy of the ratio $\bar{i}/m^{2/3}$ at constant t_d and of the ratio $\bar{i}/t_d^{1/6}$ at constant m can be verified through the use of an electromagnetic tapper, which knocks off the drop by hitting the capillary at regular intervals of time. With this device m and t_d can be varied independently. More commonly, t_d and m are varied simultaneously by changing the height of the mercury head. From Eqs. (152) and (155) it is readily seen that under these circumstances $\bar{i}$ is proportional to $h^{1/2}$. Thus a plot of $\bar{i}$ against $h^{1/2}$ must yield a straight line passing through the origin of the plot (obviously $\bar{i} = 0$ for $h = 0$).

Diffusion-limiting currents increase linearly with the concentration of the depolarizer, but this is not sufficient to characterize diffusion-controlled currents, since all the currents controlled by the rate of a charge-transfer step and the majority of the currents controlled by the rate of a chemical reaction exhibit the same behavior.

III. DIFFUSION AND CHARGE-TRANSFER OVERPOTENTIALS. PERFECTLY MOBILE EQUILIBRIA COUPLED WITH A SLOW CHARGE-TRANSFER STEP

A. General Formulation of the Diffusional Problem

When a slow charge-transfer step partially controls a polarographic current, we have seen in Sec. II.B.2.a that it is possible, at least in principle, to determine from experimental data the nature of the electroactive substances—namely, the substances actually produced or consumed at the plane of closest approach as a consequence of the charge transfer. Assume that the stoichiometric equation of the electrode process proper, involving exclusively the electroactive species, is expressed by the general Eq. (41). When the electrochemical affinity $\tilde{A}_{\rho'}$ of one or more of the elementary steps ρ' composing reaction (41) cannot be neglected in the time scale of polarographic measurements (cf. Eqs. (47) and (48)), Nernst's equation cannot be applied to the volume concentrations of electroactive species at $x = b$. The inapplicability of the boundary condition in Eq. (79) constitutes the main effect of a nonzero charge-transfer overpotential on the mathematical formulation of the diffusional problem relative to n diffusing species A_i bounded by m perfectly mobile equilibria. In fact the diffusional equations (71) through (73) and (75), together with the initial conditions of Eq. (77) and the boundary conditions in Eq. (78), remain unchanged. Let the rate v_t of the electrode process denote the number of $|\nu_i|$ moles of each substance A_i which are either produced ($\nu_i > 0$) or consumed ($\nu_i < 0$) per unit of time per unit surface of the electrode. Clearly, the current intensity i is given by

$$i = -nFAv_t \qquad (156)$$

The minus sign in Eq. (156) is due to the fact that i is negative when the electrode process (41) proceeds toward the right—that is, toward oxidation. If the stoichiometric equation (41) of the electrode process is multiplied by an integer f, the rate v_t of the electrode process, as previously defined, is divided by the same factor, whereas the number of electrons involved in the electrode process is now given by fn. From Eq. (156) it is therefore manifest that the current density remains unaltered. In general v_t is a function of the concentrations $\bar{C}_i$ of the diffusing species A_i at $x = b$, of the applied potential E, and of some kinetic parameters k_h,

$$v_t = v_t(\{C_i\}, E, \{k_h\}) \tag{157}$$

In Eq. (157) $\{\bar{C}_i\}$ designates the set of concentrations at $x = b$ and $\{k_h\}$ the set of kinetic parameters. On combining Eqs. (156) and (157) with Eq. (81), which expresses the current in terms of the fluxes of the diffusing species at the electrode surface, we have

$$nv_t(\{\bar{C}_i\}, E, \{k_h\}) = -(z_O - z_R) \sum_i a_{i,O} D_i \left(\frac{\partial C_i}{\partial x}\right)_{x=0} \quad \text{for Case 1} \tag{158a}$$

$$nv_t(\{\bar{C}_i\}, E, \{k_h\}) = -\sum_i [(z_O - z_R)a_{i,O} + (z_{Ox} - z_{Red})a_{i,Ox}] \times D_i\left(\frac{\partial C_i}{\partial x}\right)_{x=0} \quad \text{for Case 2} \tag{158b}$$

$$nv_t(\{\bar{C}_i\}, E, \{k_h\}) = -\sum_i [(z_O - z_R)a_{i,O} + (z_J - z_R)a_{i,J}] \times D_i\left(\frac{\partial C_i}{\partial x}\right)_{x=0} \quad \text{for Case 3} \tag{158c}$$

Equations (158a,b,c) are, respectively, the potential dependent boundary conditions for Cases 1, 2, and 3 of Sec. II.B.1. These relations replace Nernst's equation (80) when the charge-transfer overpotential is different from zero. Noting that the summations in Eq. (158) are extended to the fluxes of all diffusing species, we might be tempted to equate to zero the fluxes of the species characterized by a zero stoichiometric coefficient in the equation for the electrode process (41). As a matter of fact, in Sec. II.B.2.a we have noted that during electrolysis not only the fluxes of electroactive species but also those of electroinactive species are different from zero at the boundary $x = b$ of the diffuse layer, provided the homogeneous equilibria are perfectly mobile. It is important to realize that the nature of the electroactive substances can be inferred from the phenomenological dependence of the rate v_t on the concentrations $\bar{C}_i$ at $x = b$ and not on the basis of the erroneous assumption that the fluxes of electroinactive species at $x = b$ are zero.

The diffusional problem expressed by the differential Eqs. (71) through (73) and Eq. (75), with the initial conditions of Eq. (77) and the boundary conditions in Eqs. (78) and (158), cannot be solved in general other than by following a numerical procedure. The problem is strongly simplified if we assume that the diffusion coefficients D_i of the various substances are equal (cf. Eq. (82)). Under this assumption, Eq. (86), which was derived from Eqs. (71), (77), (78), and (82), still applies. Equations (86) and the expressions of Eq. (64) for the m equilibrium constants constitute $h + m = n - 1$ algebraic relations among the n concentrations C_i, allowing all these concentrations but one—let us say C_1—to be easily expressed as functions of this latter. Consequently the rate v_t of the electrode process can be expressed in terms of the single concentration C_1,

$$v_t = v_t'(\bar{C}_1, E, \{k_h\}) \tag{159}$$

Limiting ourselves to Case 1, the linear combination $\psi = \sum_i a_{i,O} C_i$ can also be determined as a function of the concentration C_1 and, vice versa, C_1 can be obtained in terms of ψ,

$$C_1 = f(\psi) \tag{160}$$

On substituting C_1 from Eq. (160) into Eq. (159) and taking into account Eqs. (158a) and (82), we obtain the boundary condition,

$$nv_t'(f(\psi), E, \{k_h\}) = -(z_O - z_R)D\left(\frac{\partial\psi}{\partial x}\right)_{x=0} \tag{161}$$

Thus the diffusional problem is reduced to the solution of the single differential equation uncomplicated by the source term,

$$\delta\psi = 0 \tag{162}$$

together with the initial condition of Eq. (77b) and the boundary condition in Eq. (161). Equation (162) is immediately derived from Eq. (72) by placing the various D_i equal to D. Once ψ is obtained as a function of x and t, the instantaneous current is given by Eq. (87),

$$i = (z_O - z_R)FAD\left(\frac{\partial\psi}{\partial x}\right)_{x=0}$$

A perfectly analogous procedure can be followed for Cases 2 and 3. The boundary value problem in Eqs. (161) and (162) can be tentatively solved by integration in series (see the appendix for an outline of the mathematical procedure).

B. Approximate Solution of the Diffusional Problem on the Basis of the Diffusion Layer Concept

An approximate solution of the problem formulated in Sec. III.A can be obtained quite easily by using the treatment based on the diffusion layer concept. Thus in the case that the diffusion coefficients for the various species A_i are different, the replacement of $(\partial C_i/\partial x)_{x=0}$ from Eq. (119) into the boundary conditions in Eq. (78) leads to the h algebraic relations of Eq. (120). Confining ourselves to the consideration of Case 1, an analogous replacement in the boundary condition (158a) yields

$$nv_t(\{\bar{C}_i\}, E, \{k_h\}) = -\frac{(z_O - z_R)}{\sqrt{\frac{3}{7}\pi t}} \sum_i a_{i,O} D_i^{1/2}(C_i^* - \bar{C}_i) = -\frac{i}{FA} \qquad (163)$$

Equations (120) and (64) and the first two members of Eq. (163) constitute a set of $h + m + 1 = n$ algebraic relations among the n concentrations $\bar{C}_i$ at $x = 0$, which can be used in order to obtain these concentrations as functions of the electrolysis time t and of the applied potential E. Once the expressions for the n concentrations $\bar{C}_i$ have been derived, their replacement in the last two members of Eq. (163) leads directly to an approximate expression of i as a function of t and E. It should be noted that the use of Eq. (163) implies an intrinsic inconsistency. In fact Eq. (163) was derived by making direct use of Eq. (119), and one of the conditions on which the validity of Eq. (119) relies is the constancy of $\bar{C}_i$ during the drop life. On the contrary, the various concentrations $\bar{C}_i$ at $x = 0$ as obtained from Eqs. (64), (120), and (163) are in general dependent on the electrolysis time t. In spite of this inconsistency, the agreement between current-potential characteristics obtained by this approximate procedure and those obtained by rigorous procedures is often surprisingly satisfactory.

An alternative approximate procedure more commonly employed by researchers in the field consists in integrating the rigorous equation (156) over the drop time, t_d, under the assumption that v_t, as given by Eq. (157), is time independent,

$$\frac{1}{t_d}\int_0^{t_d} i\,dt = -\frac{nF}{t_d}\int_0^{t_d} Av_t\,dt \cong -\frac{nFv_t}{t_d}\int_0^{t_d} A\,dt \qquad (164)$$

This procedure amounts to forcing the concentrations $\bar{C}_i$, which appear in the expression (157) for v_t, to remain constant during the drop life. Noting that the area A of the drop surface is given by $A_O t^{2/3}$, where $A_O = 0.85\ m^{2/3}$, from Eqs. (163) and (164) we obtain

$$nv_t(\{\bar{C}_i\}, E, \{k_h\}) \frac{1}{t_d} \int_0^{t_d} t^{2/3}\, dt = -\frac{z_O - z_R}{\sqrt{\frac{3}{7}\pi}} \sum_i a_{i,O} D_i^{1/2}(C_i^* - \bar{C}_i) \frac{1}{t_d} \int_0^{t_d} t^{1/6}\, dt$$

and consequently

$$nv_t = -10 \frac{(z_O - z_R)}{\sqrt{21\pi t_d}} \sum_i a_{i,O} D_i^{1/2}(C_i^* - \bar{C}_i) \tag{165}$$

The concentrations $\bar{C}_i$ as derived from Eqs. (64), (120), and (165) are apparently constant during the drop life, but in reality their values depend on the drop time. This inconsistency reveals the approximate nature of this further procedure. Once the expressions for the n concentrations $\bar{C}_i$ are obtained, in view of Eqs. (164) and (165) the mean polarographic current $\bar{i}$ is given by

$$\bar{i} = \frac{1}{t_d} \int_0^{t_d} i\, dt = \frac{6}{\sqrt{21\pi}} (z_O - z_R) F A_O t_d^{1/6} \sum_i a_{i,O} D_i^{1/2} (C_i^* - \bar{C}_i) \tag{166}$$

As an application of the diffusion layer thickness approximation, we shall consider the following electrode process:

$$\begin{array}{c} M_O \rightleftharpoons M_OX \rightleftharpoons M_OX_2 \rightleftharpoons \cdots\cdots \rightleftharpoons M_OX_k \rightleftharpoons \cdots\cdots \rightleftharpoons M_OX \\ \Downarrow + ne \\ M_R \end{array} \tag{167}$$

where the ligand X is in strong excess with respect to the metal ion M_O. According to the foregoing mechanism the species M_OX_k is electroreduced much more rapidly than the other metal complexes, thus constituting in practice the only electroactive oxidized species. The unidirectional arrow symbolizes a slow electron transfer. We shall assume that the overall electrode reaction,

$$M_OX_k + ne \underset{k_b}{\overset{k_f}{\rightleftharpoons}} M_R + kX \tag{168}$$

coincides with the rate-determining elementary charge-transfer step. This assumption, which is reasonable in the case of the simple electrode reaction (168), is far from being obvious with more complicated overall electrode processes. We then have

$$-v_t = k_f[\overline{M_O X_k}] - k_b[\overline{M_R}][X]^k \tag{169}$$

where k_f and k_b are the potential dependent rate constants for forward and backward processes. Upon noting that $z_O - z_R = n$ and substituting v_t from Eq. (169) into Eq. (165), we obtain

$$k_f\overline{[M_OX_k]} - k_b\overline{[M_R]}[X]^k = \frac{10}{\sqrt{21\pi t_d}} \sum_i D_i^{1/2}([M_OX_i]^* - \overline{[M_OX_i]}) \quad (170)$$

The application of Eq. (120) to the cluster M yields

$$\sum_i D_i^{1/2}[M_OX_i]^* = \sum_i D_i^{1/2}\overline{[M_OX_i]} + D^{R^{1/2}}\overline{[M_R]} \quad (171)$$

under the assumption that $[M_R]^* = 0$. Taking into account the equilibrium constants of Eq. (135) for the n complexation equilibria, the expression of $\overline{[M_OX_k]}$ as a function of the linear combination $\sum_i D_i^{1/2}\overline{[M_OX_i]}$ is given by

$$\overline{[M_OX_k]} = \frac{\beta_k[X]^k}{\sum_i D_i^{1/2}\beta_i[X]^i} \sum_i D_i^{1/2}\overline{[M_OX_i]} \quad (172)$$

The replacement of $\overline{[M_OX_k]}$ from Eq. (172) and of $\overline{[M_R]}$ from Eq. (171) into Eq. (170) allows the linear combination $\sum_i D_i^{1/2}\overline{[M_OX_i]}$ to be readily obtained as a function of the potential dependent parameters k_f and k_b,

$$\sum_i D_i^{1/2}\overline{[M_OX_i]} = \sum_i D_i^{1/2}[M_OX_i]^* \frac{\dfrac{10}{\sqrt{21\pi t_d}} + \dfrac{k_b[X]^k}{D^{R^{1/2}}}}{\dfrac{10}{\sqrt{21\pi t_d}} + \dfrac{k_f\beta_k[X]^k}{\sum_i D_i^{1/2}\beta_i[X]^i} + \dfrac{k_b[X]^k}{D^{R^{1/2}}}} \quad (173)$$

From Eqs. (166) and (173) the mean current $\bar{\imath}$ is given by

$$\bar{\imath} = \frac{6}{\sqrt{21\pi}}(z_O - z_R)FA_O t_d^{1/6}\left(\sum_i D_i^{1/2}[M_OX_i]^* - \sum_i D_i^{1/2}\overline{[M_OX_i]}\right)$$

$$= \frac{6}{\sqrt{21\pi}}(z_O - z_R)FA_O t_d^{1/6} \sum_i D_i^{1/2}[M_OX_i]^*$$

$$\times \frac{0.81 \dfrac{k_f\beta_k[X]^k}{\sum_i D_i^{1/2}\beta_i[X]^i}\sqrt{t_d}}{1 + 0.81\left(\dfrac{k_f\beta_k[X]^k}{\sum_i D_i^{1/2}\beta_i[X]^i} + \dfrac{k_b[X]^k}{D^{R^{1/2}}}\right)\sqrt{t_d}} \quad (174)$$

On assuming that the complex with the highest coordination number is in strong excess with respect to the other complexes, the following inequality holds,

$$\frac{[M_O X_n]}{[M_O]} = \beta_n [X]^n \gg \frac{[M_O X_i]}{[M_O]} = \beta_i [X]^i \qquad \text{for } i \neq n \tag{175}$$

If the charge-transfer process of Eq. (168) is so slow that the backward process can be neglected, Eq. (174) takes the simplified form,

$$\bar{\imath} = \frac{6}{\sqrt{21\pi}} (z_O - z_R) F A_O t_d^{1/6} \sum_i D_i^{1/2} [M_O X_i]^* \frac{0.81 \dfrac{k_f \beta_k [X]^{k-n}}{D_n^{1/2} \beta_n} \sqrt{t_d}}{1 + 0.81 \dfrac{k_f \beta_k [X]^{k-n}}{D_n^{1/2} \beta_n} \sqrt{t_d}} \tag{176}$$

The mean diffusion limiting current $\bar{\imath}_d$ is obtained from Eq. (174) by letting the rate constant k_f for the forward process tend to ∞,

$$\bar{\imath}_d = \frac{6}{\sqrt{21\pi}} (z_O - z_R) F A_O t_d^{1/6} \sum_i D_i^{1/2} [M_O X_i]^* \tag{177}$$

On combining Eqs. (176) and (177) we have, after simple passages

$$\frac{\bar{\imath}}{\bar{\imath}_d - \bar{\imath}} = 0.81 \frac{k_f \beta_k [X]^{k-n} \sqrt{t_d}}{D_n^{1/2} \beta_n} \tag{178}$$

It immediately follows that

$$-\frac{\partial}{\partial \ln[X]} \ln \frac{\bar{\imath}}{\bar{\imath}_d - \bar{\imath}} = n - k \tag{179}$$

Equation (179) can be used conveniently for determining k and consequently the nature of the electroactive coordination compound.

C. Treatment of Complexation Equilibria by a Rigorous Procedure

The diffusional problem relative to the depolarization scheme of Eq. (167) can also be solved by following the rigorous method (*8*). The particular confidence with which complexation equilibria of the type in Eq. (135) in the presence of a strong excess of the free ligand can be handled is due to the fact that within a group of complexes bounded by perfectly mobile complexation equilibria, the concentration of any given coordination compound is proportional to the concentration of any other complex of the same group. The proportionality constant depends in general both on the stability constants β_i and on the free ligand concentration.

In view of Eq. (169) the application of the general boundary condition in Eq. (158a) to the depolarization scheme of Eq. (167) results in

$$\sum_i D_i\left(\frac{\partial[M_OX_i]}{\partial x}\right)_{x=0} = k_f\overline{[M_OX_k]} - k_b\overline{[M_R]}[X]^k \tag{180}$$

Recalling from Eq. (140) that the linear combination $\sum_i [M_OX_i]$ satisfies the diffusion equation uncomplicated by the source term, $[M_OX_k]$ is conveniently expressed in terms of the above combination. Thus taking equations (135) into account, we have

$$[M_OX_k] = \frac{\beta_k[X]^k}{\sum_i \beta_i[X]^i}\sum_i [M_OX_i] \tag{181}$$

On combining Eqs. (180) and (181) and noting from Eq. (139) that the linear combination $\sum_i (D_i/D^O)[M_OX_i]$, where D^O is defined by Eq. (137), equals $\sum_i [M_OX_i]$, we obtain,

$$D^O\left(\frac{\partial \sum_i [M_OX_i]}{\partial x}\right)_{x=0} = k_f\frac{\beta_k[X]^k}{\sum_i \beta_i[X]^i}\sum_i \overline{[M_OX_i]} - k_b\overline{[M_R]}[X]^k \tag{182}$$

The linear combination $\sum_i [M_OX_i]$ and the concentration $[M_R]$ satisfy the two differential equations (140) and (133) and the boundary conditions of Eqs. (143) and (182). These conditions are identical with those fulfilled by a simple redox couple Ox/Red exchanging charges with the electrode according to the reaction,

$$\text{Ox} + ne \underset{k'_b}{\overset{k'_f}{\rightleftharpoons}} \text{Red}$$

provided the diffusion coefficients of Ox and Red are D^O and D^R, respectively, and the rate constants for the forward and backward electrode processes are given by

$$k'_f = k_f\frac{\beta_k[X]^k}{\sum_i \beta_i[X]^i} \qquad \text{and} \qquad k'_b = k_b[X]^k$$

Koutecký (*22*) solved the above boundary value problem by using the method of dimensionless parameters. The application of his results to the present problem yields, for the instantaneous current i,

$$i = -(z_O - z_R)FA\sqrt{\frac{7}{3}\frac{D^O}{t}}\,\beta_1 \sum_{i=1}^{\infty} \gamma_i \xi^i$$

where

$$\beta_1 = -\frac{1}{\sqrt{D^O}}\frac{k_f' \sum_i [M_O X_i]^*}{k_f'/\sqrt{D^O} + k_b'/\sqrt{D^R}}; \qquad \xi = \left(\frac{k_f'}{\sqrt{D^O}} + \frac{k_b'}{\sqrt{D^R}}\right)\sqrt{\frac{3}{7}t}$$

and $\sum_{i=1}^{\infty} \gamma_i \xi^i$ is a power series for which the reader is referred to the mathematical appendix (Sec. VII.D. Example III). If the charge-transfer process is totally irreversible, the application of Koutecký's formula (*22*) for the mean current to the problem at hand leads to the equation,

$$\frac{\bar{i}}{\bar{i}_d - \bar{i}} = 0.886\sqrt{\frac{t_d}{D^O}}\frac{k_f \beta_k [X]^k}{\sum_i \beta_i [X]^i} \tag{183}$$

If the inequality (175) holds, recalling that D^O is defined by Eq. (137) Eq. (183) becomes

$$\frac{\bar{i}}{\bar{i}_d - \bar{i}} = 0.886\sqrt{\frac{t_d}{D_n}}\frac{k_f \beta_k [X]^{k-n}}{\beta_n}$$

which is identical with the approximate Eq. (178), apart from the numerical factor of 0.886.

D. Diagnostic Criteria for the Distinction of Polarographic Currents Characterized by Comparable Contributions of Diffusion and Charge-Transfer Overpotential to Total Overpotential (Irreversible Waves)

As already shown for diffusion-controlled currents, likewise for currents simultaneously controlled by the rate of diffusion and by that of a charge-transfer step, measurements of current-potential characteristics and those of current-time curves are complementary in ascertaining the nature of polarographic currents. In general we can say that with a shift of the applied potential towards negative values for a cathodic wave or towards positive values for an anodic wave, the contribution of η_d to total overpotential increases with respect to that of η_t. Thus when the limiting current is reached, this latter is exclusively controlled by diffusion. A plot of the logarithm of the instantaneous limiting current against log t must, therefore, have a slope of about 0.190 and the mean limiting current must be proportional to the square root of the height h of the mercury head (cf. Sec. II.C). Furthermore, the limiting current must be proportional to the concentration of the depolarizer.

In order to distinguish between diffusion-controlled currents and currents characterized by comparable contributions of η_d and η_t, it is necessary to examine the rising portion of the particular wave under study. The shapes of current-potential characteristics must be in accord with theoretical expectations based on the assumption of a slow charge-transfer step. No general rules can be given as to the theoretical behavior of current-potential curves, owing to the great variety of cases. Thus the half-wave potential of a wave due to the irreversible electroreduction of a depolarizer Ox is independent of the bulk concentration [Ox]* of Ox if the charge-transfer step is first order with respect to Ox, but varies with [Ox]* if this condition is not fulfilled.

In the proximity of the foot of an irreversible wave, η_d is practically negligible with respect to η_t, so that no appreciable depletion of the depolarizer occurs around the electrode and the concentration of the depolarizer is uniform from the bulk of the solution up to the diffuse double layer. Hence the current density is practically constant during the drop life. This is shown schematically by curve *a* in Fig. 4. Curve *b* expresses the current density ($\propto 1/t^{1/2}$) that would be observed at the same potential if the charge-transfer step were infinitely fast. With increasing t, the diffusion overpotential η_d progressively increases with respect to η_t, so that for $t \to \infty$ it would become comparable and ultimately much greater than η_t. At this point the current would become practically diffusion controlled, so that curves *a* and *b* in Fig. 4 would coincide. In practice the current density i/A is time independent in the time scale of polarographic measurements, so that the variation of the current intensity i with electrolysis time is exclusively determined by the variation of A. From Eq. (153) it immediately follows that

$$i \propto m^{2/3} t^{2/3} \tag{184}$$

Hence, at the foot of the wave the plot of log i against log t yields a straight line with a slope of $\frac{2}{3}$. Proceeding along the rising portion of the wave, η_d tends to increase with respect to η_t and the slope of the log i against log t plot

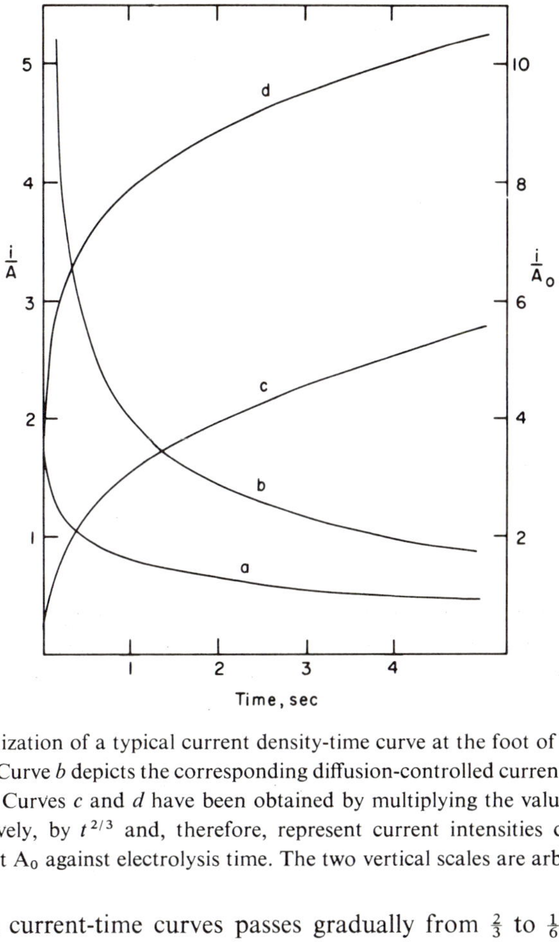

Fig. 4. Schematization of a typical current density-time curve at the foot of an irreversible wave (curve *a*). Curve *b* depicts the corresponding diffusion-controlled current density at the same potential. Curves *c* and *d* have been obtained by multiplying the values on curves *a* and *b*, respectively, by $t^{2/3}$ and, therefore, represent current intensities divided by the constant A_0 against electrolysis time. The two vertical scales are arbitrary.

derived from current-time curves passes gradually from $\frac{2}{3}$ to $\frac{1}{6}$. This latter value is reached when the limiting current is attained. Since the functional dependence of i on t, and consequently that of $\bar{i}$ on t_d, vary along the rising portion of the wave, the ratio $\bar{i}(E)/\bar{i}_d$ of the mean current $\bar{i}$ at a given potential E to the mean limiting current $\bar{i}_d$ must depend on the drop time t_d. Noting from Eqs. (154) and (184) that the rate of growth of the instantaneous current i with t decreases as we proceed from the foot of the wave to the limiting current, it is apparent that with increasing t_d the ratio $\bar{i}(E)/\bar{i}_d$ at a given potential increases and consequently the whole wave shifts towards decreasing overvoltages (Fig. 5). In principle, for infinitely large values of t_d the wave would become perfectly reversible and its shape would no longer change with a further increase in t_d.

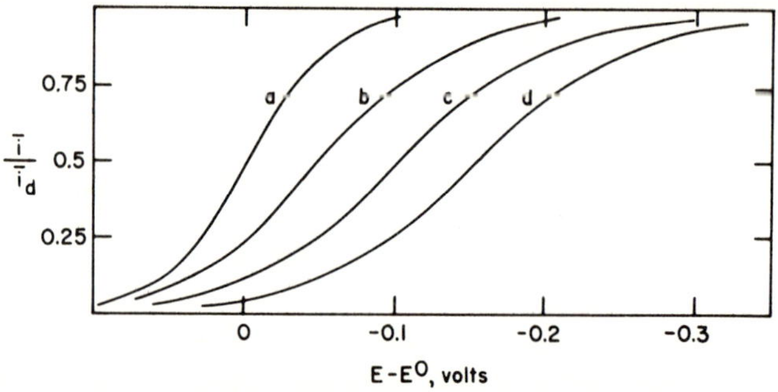

Fig. 5. Ratio $\bar{i}/\bar{i}_d$ of the mean polarographic current $\bar{i}$ to the corresponding diffusion limiting current $\bar{i}_d$ against $E - E^\circ$, for the irreversible electroreduction $0 \rightleftharpoons R$. The various polarographic waves were computed for a standard rate constant $k_s = 3.16 \cdot 10^{-4}$ cm sec^{-1}, a diffusion coefficient $D = 10^{-5}$ cm^2 sec^{-1}, a transfer coefficient $\alpha = 0.5$ and for a drop time t_d equal to : 0.3 sec (curve d), 3 sec. (curve c), 30 sec (curve b) and ∞ (curve a). Curve a depicts the reversible behavior ($\eta_t = 0$).

IV. DIFFUSION, CHARGE-TRANSFER, AND REACTION OVERPOTENTIALS. SLOW HOMOGENEOUS CHEMICAL REACTIONS NOT INFLUENCED BY THE DIFFUSE LAYER STRUCTURE

A. General Formulation of the Diffusional Problem

In the present section we shall consider nonequilibrium homogeneous chemical reactions coupled with a charge-transfer reaction. For simplicity we shall confine ourselves to the consideration of the general overall electrode process of Eq. (41) coupled with the general homogeneous chemical reaction,

$$\sum_{i=1}^{n} \nu_i' A_i + \sum_{k=n+1}^{n'} \nu_k' B_k = 0 \tag{185}$$

Here B_k are the species taking part in the chemical reaction but not in the electrode reaction. The electroactive substances A_i not involved in reaction (185) are characterized by a zero stoichiometric coefficient ν_i'. We shall assume that the stoichiometric equation (185) of the homogeneous reaction is written in such a way that under steady-state conditions, one occurrence of this reaction is simultaneous to one occurrence of the charge-transfer reaction, written as in Eq. (41). The presence of more nonequilibrium homogeneous reactions is difficult to disclose on experimental grounds since usually one reaction is sufficiently slow (case of consecutive reactions) or sufficiently fast (case of parallel reactions) with respect to the others to obscure the effects due to these latter. Suffice here to say that the case of more nonequilibrium homogeneous reactions coupled with an electrode process (*14*) although more involved from a mathematical point of view, relies on the same considerations that hold for the more particular case of a single homogeneous chemical reaction. The combination of the stoichiometric equations (41) and (185) gives the equation of a new overall electrode reaction, characterized by an electrochemical affinity different from that of the original electrode process Eq. (41). In fact, in the present case, as opposed to the case of an electrode process coupled with perfectly mobile equilibria, the electrochemical affinity of the chemical reaction is different from zero. The combination of the stoichiometric equations (41) and (185) is by no means convenient since it replaces an electrode process with an electrochemical affinity "located" in the pre-electrode layer by a new electrode process whose electrochemical affinity is located both in the pre-electrode layer and in a much thicker liquid layer around the electrode. The above considerations are necessary in order to understand that the choice of the stoichiometric equation for the electrode process is not arbitrary, as in the case of the formal charge-transfer process of Eq. (70) relative to diffusion-controlled polarographic currents. Consequently, the stoichiometric coefficients ν_i in Eq. (41) are definite, apart from a common multiplicative factor.

The chemical reaction of Eq. (185) has a nonzero electrochemical affinity during electrolysis, and consequently the law of mass action (Eq. (64)) cannot be applied to it other than at a sufficient distance from the electrode. The algebraic relation of Eq. (64) among the concentrations of the reacting species, which was so profitably used with perfectly mobile equilibria, must therefore be replaced by an equation containing the kinetic parameters of reaction (185). Unfortunately this latter equation is of the differential type, as will be shown later. Designate by ρ the rate of the chemical reaction (185), expressed by the number of $|\nu_i'|$ moles of A_i or $|\nu_k'|$ moles of B_k which are produced or consumed per unit of time per unit volume,

$$\rho = \frac{1}{\nu_i'}\frac{d_i C_i}{dt} = \frac{1}{\nu_k'}\frac{d_i C_k}{dt} \tag{186}$$

In general ρ is a function of the concentrations of A_i and B_k,

$$\rho = \rho(\{C_i\}, \{C_k\})$$

Here $\{C_i\}$ and $\{C_k\}$ denote the sets of concentrations of A_i and B_k, respectively. On substituting $d_i C_i/dt$ and $d_i C_k/dt$ from Eq. (186) into Eq. (63), we obtain n' differential equations expressing the law of conservation of mass,

$$\frac{\partial C_\alpha}{\partial t} = D_\alpha \frac{\partial^2 C_\alpha}{\partial x^2} + \frac{2}{3}\frac{x}{t}\frac{\partial C_\alpha}{\partial x} + \nu'_\alpha \rho \qquad \alpha = i, k \tag{187}$$

one for each diffusing species. The n' initial conditions for the above differential equations are

$$\begin{matrix} C_i = C_i^* \\ C_k = C_k^* \end{matrix} \quad \text{for} \begin{cases} x \geqslant 0, & t = 0 \\ x \to \infty, & t > 0 \end{cases} \tag{188}$$

where C_i^* and C_k^* denote bulk concentrations.

The n' boundary conditions required for the solution of Eq. (187) vary according to whether the electrode process (41) has a finite or a practically zero electrochemical affinity. In the first case, in order to exclude the possibility of accumulation or desorption of the various electroactive species A_i at the electrode, we must assume that the amount of any substance A_i which reaches the electrode by diffusion per unit of time equals the amount of A_i which is consumed in the same unit of time as a consequence of the slow electrode reaction. Thus recalling the definition of the rate v_t of the electrode process given in Sec. III.A and taking Eqs. (61) and (157) into account, we have

$$-v_t(\overline{\{C_i\}}, E, \{k_h\}) = \frac{1}{\nu_i} D_i \left(\frac{\partial C_i}{\partial x}\right)_{x=0} \qquad i = 1, \ldots, n \tag{189}$$

The minus sign in Eq. (189) is due to the fact that when the electrode reaction (41) proceeds towards the right ($v_t > 0$), an electrode product ($\nu_i > 0$) is characterized by a negative concentration gradient $(\partial C_i/\partial x)_{x=0}$. As concerns the substances B_k, not participating in the electrode process, their fluxes at the electrode surface equal zero,

$$D_k \left(\frac{\partial C_k}{\partial x}\right)_{x=0} \qquad k = n + 1, \ldots, n' \tag{190}$$

The boundary conditions in Eq. (190) are valid only if the chemical reaction (185) is sufficiently slow. In fact we have seen in Sec. II.B.2.a that the fluxes at $x = b$ of electroinactive substances joined to electroactive ones by perfectly mobile equilibria are actually different from zero during electrolysis. Equations (189) and (190) are the n' boundary conditions for the case of a slow charge transfer.

If the electrochemical affinity of the electrode process (41) is negligible, Nernst's equation can be applied to the concentrations $\bar{C}_i$ at $x = 0$,

$$\prod_i \bar{C}_i^{\nu_i} = \exp\left[\frac{nF}{RT}(E - E^0)\right] \tag{191}$$

where E° is the standard potential. In the present case the expression v_t for the rate of the electrode process, contained in Eq. (189), cannot be used because it has been replaced by the equilibrium condition (191). On equating the term $D_1/\nu_1(\partial C_1/\partial x)_{x=0}$ relative to the electroactive substance A_1 to the remaining right-hand sides of the conditions of Eq. (189), we obtain the following $n - 1$ boundary conditions,

$$\frac{D_1}{\nu_1}\left(\frac{\partial C_1}{\partial x}\right)_{x=0} = \frac{D_i}{\nu_i}\left(\frac{\partial C_i}{\partial x}\right)_{x=0} \qquad \text{for } i \neq 1 \tag{192}$$

Equation (192) expresses the proportionality relations among the fluxes of the various electroactive species A_i, as deduced from the stoichiometry of the electrode reaction (41). Equations (190), (191), and (192) are the n' boundary conditions for the case of a fast electrode process.

The solution of the system of differential equations (187) is strongly simplified if we have recourse to the concept of atomic clusters (cf. Sec. II.A.1), at the same time assuming that the diffusion coefficients of all the species A_i and B_k are equal. The law of conservation of mass as applied to clusters does not contain the source term, so that Eqs. (84) and (85) are still valid. The resulting h algebraic relations (86) among the n' concentrations, one for each cluster, allow h concentrations to be expressed as functions of the remaining $n' - h$. In view of the general equation (69), which we now write as,

$$m + h = n' - 1$$

we have

$$n' - h = m + 1 = 2$$

because in the present case the number m of homogeneous reactions is 1. If the two concentrations in terms of which we express the remaining concentrations are those C_1, C_2 of the species A_1 and A_2, the rates v_t and ρ become functions of C_1 and C_2,

$$\rho = \rho'(C_1, C_2); \qquad v_t = v_t'(\bar{C}_1, \bar{C}_2)$$

Thus the diffusional problem is reduced to the solution of the two differential equations,

$$\delta C_1 + \nu_1' \rho'(C_1, C_2) = 0 \tag{193a}$$

$$\delta C_2 + \nu_2' \rho'(C_1, C_2) = 0 \tag{193b}$$

where δ is defined by Eq. (83), together with the boundary conditions;

$$\left.\begin{aligned} -v_t'(\bar{C}_1, \bar{C}_2) &= \frac{D}{\nu_1}\left(\frac{\partial C_1}{\partial x}\right)_{x=0} \\ -v_t'(\bar{C}_1, \bar{C}_2) &= \frac{D}{\nu_2}\left(\frac{\partial C_2}{\partial x}\right)_{x=0} \end{aligned}\right\} \tag{194}$$

The problem can be further simplified by replacing the concentration C_2 by the linear combination $\varphi = C_1/\nu_1' - C_2/\nu_2'$. It is readily seen from Eqs. (193) and (194) that φ obeys the diffusion equation uncomplicated by the source term,

$$\delta\varphi = 0 \tag{195}$$

with the boundary condition,

$$D\left(\frac{\partial \varphi}{\partial x}\right)_{x=0} = -\left(\frac{\nu_1}{\nu_1'} - \frac{\nu_2}{\nu_2'}\right) v_t'\left(\bar{C}_1, \frac{\nu_2'}{\nu_1'}\bar{C}_1 - \nu_2'\bar{\varphi}\right) \tag{196}$$

The linear combination φ is not mathematically significant if one of the two species A_1 and A_2 is not involved in the chemical reaction (185) (ν_1' or $\nu_2' = 0$). However, in this case one of the two differential equations (193) is already characterized by a zero source term.

Comparing the formulation of the diffusional problem for the case of perfectly mobile equilibria with the present formulation for a slow chemical reaction, we see that the diffusion equation (196), obeyed by the combination

φ, is analogous to Eq. (162). The algebraic relation expressing the law of mass action, which is applicable to an equilibrium reaction, is here replaced by one of the two differential equations (193), which contains the kinetic parameters for the particular nonequilibrium reaction under study.

If the chemical reaction is first order with respect to one of the various species A_i or B_k,

$$\rho = kC_i \qquad \text{or} \qquad \rho = k'C_k$$

where k and k' are rate constants, it is frequently possible to find a linear combination ϕ of the various concentrations, which obeys the law of conservation of mass in the form,

$$\frac{\partial \phi}{\partial t} = D\frac{\partial^2 \phi}{\partial x^2} + \frac{2x}{3t}\frac{\partial \phi}{\partial x} + \lambda\phi \tag{197}$$

Here, the source term is proportional to ϕ. Equation (197) is conveniently used in place of Eq. (193a) or (193b) and solved together with the differential equation (195), uncomplicated by the kinetic term. In fact through the transformation

$$\psi = \phi \exp(-\lambda t)$$

the source term in Eq. (197) is eliminated,

$$\delta\psi = 0 \tag{198}$$

while the kinetic parameter k or k' is embodied in the new variable ψ. In this way the boundary value problem reduces to the solution of two diffusion equations uncomplicated by source terms.

B. Rigorous Solution of the Diffusional Problem. Examples

Only a very few cases are known in which the diffusional problem previously formulated has been solved rigorously (*23–25*). Before considering some of these cases it will be convenient to make a classification of electrode processes coupled with a nonequilibrium homogeneous reaction according to the position of this latter with respect to the electron-transfer reaction. If the chemical reaction is "preceding"—that is, supplies an electroactive substance from an electroinactive one—the resulting current is termed "prekinetic." If the chemical reaction is "consecutive"—that is, deactivates a product of the electrode reaction—the current is termed "postkinetic." Finally, if the chemical reaction is "parallel"— that is, regenerates the initial electroactive substance from a product of the electrode reaction—the current is called "catalytic."

1. Total Regeneration of the Depolarizer by a First-Order Reaction

Imagine that the product A_2 of the reversible electrooxidation of a depolarizer A_1 reacts in the solution phase with a substance B_3 regenerating A_1 as follows:

$$\left.\begin{array}{l}\text{(a) } A_2 \underset{kK}{\overset{k}{\rightleftharpoons}} A_1 \\ \text{(b) } A_1 \rightleftharpoons A_2 + ne\end{array}\right\} \tag{199}$$

B_3 is electroinactive and is present in strong excess with respect to A_1 and A_2. Hence the depletion of B_3 in the neighborhood of the electrode during electrolysis is negligible and the concentration C_3 of this species is practically constant throughout the solution up to the electrode. The rate for the pseudo-first-order forward reaction is kC_2, where k embodies the concentration C_3. The rate for the backward reaction is kKC_1, where K is the equilibrium constant for the chemical reaction. Since the number of diffusing species is two, there is no need for considering atomic clusters. In the present case the diffusion equations (187), together with the initial conditions (188) and the boundary conditions (191) and (192), are written as follows:

$$\left.\begin{array}{l}\delta C_1 + k(C_2 - KC_1) = 0 \\ \delta C_2 - k(C_2 - KC_1) = 0\end{array}\right\} \tag{200}$$

$$\left.\begin{array}{l}C_1 = C_1^* \\ C_2 = C_2^*\end{array}\right\} \quad \text{for}\begin{cases} t = 0 & x \geqslant 0 \\ t > 0 & x \to \infty\end{cases} \tag{201}$$

$$\left.\begin{array}{r}D(\partial C_1/\partial x) + D(\partial C_2/\partial x) = 0 \\ C_2/C_1 = \theta = \exp[nF(E - E^\circ)/RT]\end{array}\right\} \quad \text{for } t > 0 \quad x = 0 \tag{202}$$

The linear combination φ satisfying the diffusion equation (195) uncomplicated by the source term is $\varphi = C_1 + C_2$,

$$\delta\varphi = 0 \tag{203}$$

whereas the combination ϕ satisfying the diffusion equation (197) is $\phi = C_2 - KC_1$, with $\lambda = -k(1 + K)$. Setting

$$\psi = \phi \exp(-\lambda t) = (C_2 - KC_1) \exp[k(1 + K)t] \tag{204}$$

we have

$$\delta\psi = 0 \tag{205}$$

The boundary and initial conditions of Eqs. (201) and (202) are now written in terms of φ and ψ as

$$\left.\begin{aligned} &\text{(a)} \quad \varphi = C_1^* + C_2^* = \varphi^* \\ &\text{(b)} \quad \psi = C_2^* - KC_1^* = 0 \end{aligned}\right. \qquad \text{for} \begin{Bmatrix} t = 0, & x \geqslant 0 \\ t > 0, & x \to \infty \end{Bmatrix} \tag{206}$$

$$\left.\begin{aligned} &\text{(a)} \quad D\frac{\partial \varphi}{\partial x} = 0 \\ &\text{(b)} \quad \psi = \frac{\theta - K}{1 + \theta}\varphi \exp(-\lambda t) \end{aligned}\right\} \qquad \text{for } t > 0, \quad x = 0 \tag{207}$$

Equation (206b) states that before electrolysis, or also during electrolysis at a large distance from the electrode, reaction (199a) is at equilibrium. The boundary value problem of Eqs. (203) through (207) was solved by Koutecký (*23*) through integration in series, using the substitutions,

$$s = \frac{x}{\sqrt{12Dt/7}}; \qquad \xi = -\lambda t \tag{208}$$

where clearly s and ξ are dimensionless variables. The reader is referred to the appendix (Sec. VII.D, Example IV) for a solution of the above boundary value problem based on the unified procedure of Sec. VII.C. Noting that the current is proportional to the flux of A_1 at the electrode surface according to the equation,

$$i = -nFAD\left(\frac{\partial C_1}{\partial x}\right)_{x=0}$$

and taking Eqs. (202), (204), and (A115) into account, we obtain

$$i = \frac{nFAD}{1 + K}\exp(\lambda t)\left(\frac{\partial \psi}{\partial x}\right)_{x=0} = \sqrt{\frac{7D}{3t}}\,nFA\frac{K - \theta}{(1 + K)(1 + \theta)}\varphi^*$$

$$\times \exp(\lambda t)\sum_{i=0}^{\infty}\frac{(-\lambda)^i}{i!}\frac{\Gamma(\frac{3}{7}i + 1)}{\Gamma(\frac{3}{7}i + \frac{1}{2})}t^i \tag{209}$$

where the symbol $\Gamma(x)$ denotes a gamma function. If the rate k of reaction (199a) (regenerating the depolarizer) approaches zero, the polarographic current of Eq. (209) tends to the diffusion current i_∞ relative to the simple reversible process (199b). Under these conditions, which are encountered in the absence of the electroinactive substance B_3, K loses its significance and must be replaced by the ratio C_2^*/C_1^*, whereas λ equals zero. With the above substitutions, the only power of t in the series expansion of Eq. (209) that does not vanish is the zeroth power. Recalling that $\Gamma(1) = 1$ and $\Gamma(\frac{1}{2}) = \pi^{1/2}$, from Eq. (209) we immediately obtain the following expression for the diffusion current i_∞,

$$i_\infty = \sqrt{\frac{7D}{3\pi t}}\, nFA\, \frac{C_2^* - \theta C_1^*}{1 + \theta} \tag{210}$$

Equation (210) could have been obtained by using the Ilkovic equation (91), valid for a simple reversible process, and by noting that at the electrode surface we have $\bar{C}_2/\bar{C}_1 = \theta$, and $\bar{C}_1 + \bar{C}_2 = C_1^* + C_2^*$. On dividing the catalytic current i by i_∞ we obtain the simple expression,

$$\frac{i}{i_\infty} = \pi^{1/2} \exp(-\xi) \sum_i \frac{\Gamma(\frac{3}{7}i + 1)}{\Gamma(\frac{3}{7}i + \frac{1}{2})} \frac{\xi^i}{i!} \tag{211}$$

where ξ is defined in Eq. (208). Equation (211) shows that the ratio i/i_∞ depends on electrolysis time through the dimensionless parameter $\xi = -\lambda t$. Hence the above ratio has the same value either if we measure it at a given time t and at a given concentration of B_3 (and, consequently, at a given value of λ) or if we increase C_3 and we measure i/i_∞ at a shorter time, chosen in such a way that the product $\lambda t = -\xi$ remains constant.

In order to derive the mean current we recall that the area A of the dropping electrode varies with time according to Eq. (153). Hence, from Eq. (210) we have

$$i_\infty = \text{const} \times t^{1/6} \tag{212}$$

The ratio of the catalytic mean current $\bar{\imath}$ to the corresponding diffusion mean current $\bar{\imath}_\infty$ is therefore given by

$$\frac{\bar{\imath}}{\bar{\imath}_\infty} = \frac{\int_0^{t_d} \left(\frac{i}{i_\infty}\right) t^{1/6}\, dt}{\int_0^{t_d} t^{1/6}\, dt}$$

or also, on introducing the dimensionless variable ξ,

$$\frac{\bar{i}}{\bar{i}_\infty} = \frac{7}{6\xi_1^{7/6}} \int_0^{\xi_1} \frac{i}{i_\infty} \xi^{1/6}\, d\xi \tag{213}$$

where t_d is the drop time and ξ_1 equals $-\lambda t_d$. For $\xi_1 > 10$ the ratio $\bar{i}/\bar{i}_\infty$, as given by Eqs. (211) and (213), can be approximately expressed in closed form (*23*),

$$\frac{\bar{i}}{\bar{i}_\infty} = 0.812\xi_1^{1/2} + 1.92\xi_1^{-7/6} \cong 0.812\xi_1^{1/2} \tag{214}$$

In view of Eq. (212), $\bar{i}_\infty = 1/t_d \int_0^{t_d} i_\infty\, dt$ is proportional to $t_d^{1/6}$. Hence from Eq. (214) we can see that $\bar{i} \propto t_d^{2/3}$ and consequently the catalytic mean current is proportional to the maximum area $A_0 t_d^{2/3}$ of the drop surface. This result is readily understood if we consider that for $\xi_1 > 10$, the rate of the regeneration reaction is very high. Consequently the concentration gradient of the depolarizer A_1 at the electrode surface, which in the absence of B_3 would progressively decrease during the drop life, reaches its minimum value at a given time t_{min} much shorter than t_d. At t_{min} steady-state conditions are reached, so that the same amount of A_1 consumed at the electrode is simultaneously reproduced from the electrode product A_2 through reaction (199a) within a very thin solution layer adjacent to the electrode. From t_{min} up to t_d the current intensity increases as a consequence of the drop growth, but the current density $i/(A_0 t^{2/3})$ remains constant, owing to the constant value of $(\partial C_1/\partial x)_{x=0}$. Consequently the instantaneous catalytic current, as opposed to diffusion currents, increases with $t^{2/3}$ rather than with $t^{1/6}$, and since $t_{min} \ll t_d$, $\bar{i}$ is practically proportional to $t_d^{2/3}$. Figure 6 shows the ratio $\bar{i}/\bar{i}_\infty$ expressed by Eqs. (211) and (213) as a function of $\xi_1^{1/2}$. The asymptotic slope in Eq. (214) is represented by the straight line 2. The experimental value of $\bar{i}/\bar{i}_\infty$ can be readily obtained by polarographing a solution of A_1 both in the absence of B_3 and in its presence. If the A_1/A_2 redox couple behaves reversibly, any potential along the rising portion of the wave can be chosen for the computation of $\bar{i}/\bar{i}_\infty$ since Eq. (211) shows that this ratio does not depend on θ but only on $k(1 + K)$. In other words the presence of B_3 produces an increase in the polarographic wave of A_1 but does not cause it to shift along the potential axis. If the electrode process (199b) is irreversible, Eqs. (211) through (214) can still be applied to the limiting current. In fact, when the limiting current is reached, $\bar{C}_1$ is zero and this is exactly the boundary condition to which Nernst's equation (202) reduces when $\theta \to \infty$. Under these

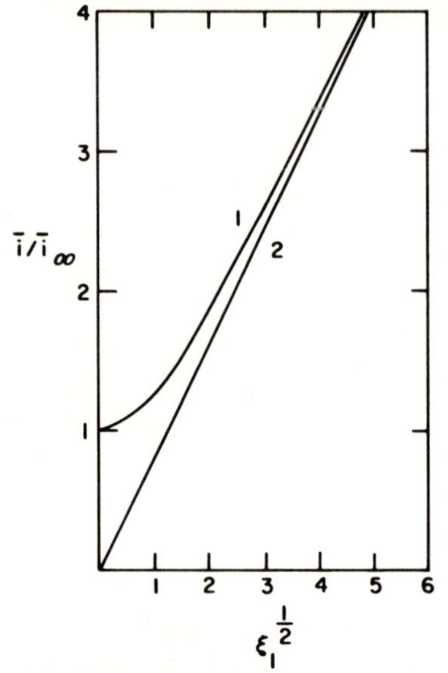

Fig. 6. Curve 1: Plot of $\bar{i}/\bar{i}_\infty$ against $\xi_1^{1/2}$, as derived from Eq. (211) and (213). Curve 2: Asymptotic behavior as expressed by Eq. (214) [From *23*] (By permission of the Czechoslovak Academy of Sciences.)

circumstances the boundary value problem in Eqs. (203) through (207) remains unchanged provided we set $(\theta - K)/(1 + \theta) = 1$. Once $\bar{i}/\bar{i}_\infty$ at a given drop time t_d is known, the diagram of Fig. 6 allows the value $-\lambda = k(1 + K)$ to be determined. Many examples of electrode processes of the type given in Eq. (199) are known. Some of them will be briefly considered in connection with the re-examination of scheme (199) by the approximate procedure.

2. Partial Regeneration of the Depolarizer by a First-Order Reaction

A further scheme of a catalytic process treated by following the rigorous procedure outlined in Sec. IV.A is

$$\left.\begin{aligned} &\text{(a) } A_2 \xrightarrow{k} \tfrac{1}{2}A_1 + \text{other products} \\ &\text{(b) } A_1 \rightleftharpoons A_2 + ne \end{aligned}\right\} \qquad (215)$$

Here the regeneration of the depolarizer A_1 from the electrode product A_2 by the agency of an electroinactive substance B_3 is partial. The regeneration reaction is assumed to be unidirectional. Hence the products of reaction

(215a), other than A_1, which diffuse away from the electrode and consequently contribute to diffusion overpotential, do not appear in the boundary value problem since their concentrations are not contained in the expression $\rho = kC_2$ for the rate of the regeneration reaction. In the present case the diffusion equations (187), the initial conditions (188), and the boundary conditions (191) and (192) read as follows:

$$\delta C_1 + \tfrac{1}{2}kC_2 = 0$$

$$\delta C_2 - kC_2 = 0$$

$$C_1 = C_1^*; \qquad C_2 = 0 \qquad \text{for}\begin{cases} t = 0, & x \geqslant 0 \\ t > 0, & x \to \infty \end{cases} \tag{216}$$

$$D\frac{\partial C_1}{\partial x} + D\frac{\partial C_2}{\partial x} = 0; \qquad \frac{C_2}{C_1} = \theta = \exp\left[\frac{nF}{RT}(E - E^\circ)\right] \tag{217}$$

The combination φ of Eq. (195) is now given by $\varphi = 2C_1 + C_2$, whereas the function ψ of Eq. (198) is expressed by $\psi = C_2 \exp(kt)$,

$$\delta\varphi = 0; \qquad \delta\psi = 0 \tag{218}$$

On introducing φ and ψ in the conditions of Eqs. (216) and (217), we obtain

$$\varphi = 2C_1^*; \qquad \psi = 0 \qquad \text{for} \quad t = 0, x \geqslant 0 \tag{219}$$

$$\left.\begin{aligned} &\frac{\partial\varphi}{\partial x} + \frac{\partial\psi}{\partial x}\exp(-kt) = 0 \\ &\varphi = \frac{2 + \theta}{\theta}\,\psi\exp(-kt) \end{aligned}\right\} \qquad \text{for } x = 0, \quad t > 0 \tag{220}$$

The boundary value problem in Eq. (218) through (220) was solved by Koutecký et al. (*24*) by the method of integration in series. The resulting expression for the polarographic current was used in the study of the reduction wave of O_2 to H_2O_2 in the presence of catalase. This enzyme decomposes H_2O_2 according to the reaction,

$$H_2O_2 \xrightarrow{\text{cat}} \tfrac{1}{2}O_2 + H_2O$$

first order with respect to H_2O_2, thus causing the partial regeneration of the depolarizer O_2.

3. Chemical Reaction Interposed between Two Charge-Transfer Processes

The depolarization scheme,

$$\left.\begin{array}{l}\text{(a) } A_3 \rightleftharpoons A_1 + n_3 e \\ \text{(b) } A_1 \xrightarrow{k} A_2 \\ \text{(c) } A_2 \rightleftharpoons A_4 + n_2 e\end{array}\right\} \qquad (221)$$

in which a chemical reaction (Eq. (221b)) intervenes between two electrochemical reactions (Eq. (221a) and (221c)) has received some attention recently. The polarographic theory for this reaction scheme, usually referred to as "ECE mechanism," can be developed along the lines of the general formulation of Sec. IV.A. If we assume equal diffusion coefficients for all the species, application of the diffusion equations (187) and of the initial conditions of Eq. (188) to the species A_3, A_1, and A_2 yields

$$\left.\begin{array}{r}\delta C_3 = 0 \\ \delta C_1 - kC_1 = 0 \\ \delta C_2 + kC_1 = 0\end{array}\right\} \qquad (222)$$

and

$$C_3 = C_3^*, \qquad C_1 = 0; \qquad C_2 = 0 \qquad \text{for} \begin{cases} t = 0; & x \geqslant 0 \\ t > 0; & x \to \infty \end{cases} \qquad (223)$$

The differential operator δ is defined in Eq. (83). In Eq. (223) we have assumed that only A_3 is present in the bulk of the solution. Under limiting current conditions the species A_3 and A_2 will be instantaneously electrooxidized as soon as they reach the electrode and consequently the following boundary conditions can be written,

$$C_3 = 0; \qquad C_2 = 0 \qquad \text{for} \qquad t > 0; \qquad x = 0 \qquad (224)$$

Application of Eq. (192) to the two species, A_3 and A_1, involved in the first charge-transfer reaction (221a) completes the set of boundary conditions

$$D\frac{\partial C_3}{\partial x} = -D\frac{\partial C_1}{\partial x} \quad \text{for } t > 0; \quad x = 0 \qquad (225)$$

In the present instance the combination φ introduced in Eq. (195) is given by

$$\varphi = C_1 + C_2 \qquad (226)$$

whereas the function ϕ obeying a diffusion equation of the type (197) is simply expressed by C_1. It follows that the function ψ introduced in Eq. (198) is now given by $\psi = C_1 \exp(kt)$. With the above simplifications the diffusional problem is reduced to the solution of the set of differential equations,

$$(a)\ \delta C_3 = 0; \qquad (b)\ \delta\varphi = 0; \qquad (c)\ \delta\psi = 0 \tag{227}$$

uncomplicated by kinetic terms, with the boundary conditions of Eqs. (224) and (225) and the initial conditions of Eq. (223). This diffusional problem can be solved by following the mathematical procedure outlined in the appendix. Thus Nicholson et al. (*24*$_{(1)}$) made use of the method of Laplace transforms by employing Eq. (A69); a very similar procedure had been previously followed by Kastening and Holleck (*24*$_{(2)}$).

It is interesting to note that the diffusion equation (227a) with the initial condition $C_3(x, t = 0) = C_3^*$ and the boundary condition $C_3(x = 0, t) = 0$ can be solved independently of the remaining diffusion equations of the set of Eq. (227), thus yielding an expression for the flux of A_3 at $x = 0$ which is given by the Ilkovic equation (91),

$$\left(\frac{\partial C_3}{\partial x}\right)_{x=0} = \frac{C_3^*}{\sqrt{\frac{3}{7}\pi D t}} \tag{228}$$

Clearly if the rate k for the chemical reaction is sufficiently low, the limiting current i_l is diffusion controlled and corresponds to the exchange of n_3 electrons according to the charge-transfer process (221a),

$$i_l = -n_3 FAD\left(\frac{\partial C_3}{\partial x}\right)_{x=0} = -\frac{n_3 FADC_3^*}{\sqrt{\frac{3}{7}\pi D t}} \equiv i_d \tag{229}$$

Conversely if the chemical reaction (221b) is sufficiently fast, the limiting current i_l corresponds to the exchange of $(n_3 + n_2)$ electrons, in accordance with the overall electrode process $A_3 \rightarrow A_4 + (n_3 + n_2)e$, and consequently is still diffusion controlled,

$$i_l = -(n_3 + n_2)FAD\left(\frac{\partial C_3}{\partial x}\right)_{x=0} = \frac{n_3 + n_2}{n_3}\, i_d$$

Kinetic complications due to the interposed chemical reaction (221b) are therefore to be expected only for an intermediate range of k values. A semi-rigorous approach to the present diffusional problem (*24*$_{(3)}$) will be considered in Sec. IV.F.4.

An example of ECE mechanism extensively studied both with the polarographic method ($24_{(1)}$) and with other related techniques ($24_{(4)}$, $24_{(5)}$) is represented by the reduction of *p*-nitrosophenol to *p*-aminophenol taking place according to the scheme,

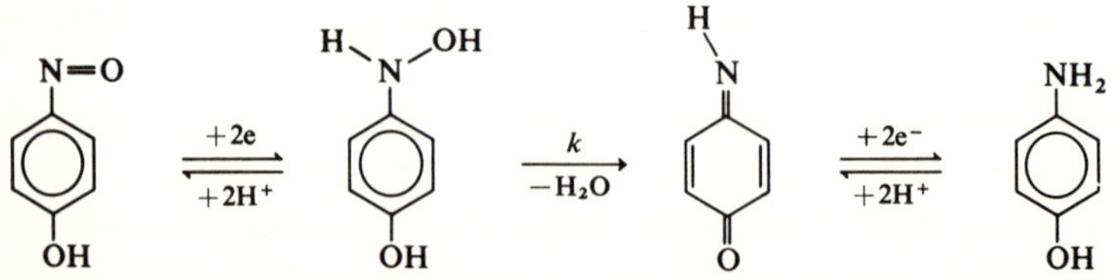

The chemical step consists in the dehydration of *p*-hydroxylaminophenol and is acid-base catalyzed.

C. Approximate Solution of the Diffusional Problem on the Basis of the Reaction Layer Concept

1. The Reaction-Layer Concept

We shall consider in general the electrode process (41) coupled with the homogeneous chemical reaction (185) and assume for simplicity that the electroinactive substances B_k taking part in the latter reaction are present in the solution in strong deficiency with respect to the supporting electrolyte. As already seen in Sec. I.B.2 for charge-transfer, penetration, diffusion, and ohmic overpotentials, in the approximate treatment of polarographic currents we try to separate the solution layer around the electrode into regions characterized by the exclusive presence of only one type of overpotential. When reaction overpotential comes into play, the region in which we imagine all and only the reaction overpotential to be located is called the "reaction layer." Limiting ourselves to the consideration of the solution layer bounded by the outer Helmholtz plane $x = a_2$ and a plane $x = c$ placed in the bulk of the solution, the total overpotential located in this layer consists of three contributions—namely, the ohmic, diffusion, and reaction overpotentials, as appears from the rigorous Eq. (36),

$$\eta^{a_2c} = -\frac{1}{I}\int_{a_2}^{c} \tilde{A}_\rho J_\rho \, dx - \frac{1}{I}\int_{a_2}^{c} \left(\sum_k |\mathbf{J}_k| \left|\frac{\partial \tilde{\mu}_k}{\partial x}\right| + \sum_i |\mathbf{J}_i| \left|\frac{\partial \tilde{\mu}_i}{\partial x}\right|\right) dx - \frac{1}{I}\int_{a_2}^{c} \sum_s |\mathbf{J}_s| \left|\frac{\partial \tilde{\mu}_s}{\partial x}\right| dx$$

$$\cong \eta_r^{a_2c} + \eta_d^{a_2c} + \eta_\Omega^{a_2c} \tag{230}$$

$\tilde{A}_\rho$ and J_ρ express the electrochemical affinity and the net rate of reaction (185), respectively. Any attempt to locate the various types of overpotential in different regions of the layer (a_2, c) requires extremely rough assumptions, unless reaction (185) is very fast. In the latter case we can make use of the concept of the Nernst diffusion layer as done in Sec. I.B.2, thus locating η_Ω exclusively in the zone (δ, ∞). In order to bring about the formal separation of the region (a_2, δ) into two zones where to concentrate the diffusion and reaction overpotentials, respectively, we may write the electrochemical affinity $\tilde{A}^{O\delta}$ of the electrode reaction (41) (see Eq. (55)) as follows (*5*):

$$\tilde{A}^{O\delta} = nF\eta^{O\delta} = \left(-n\tilde{\mu}_e - \sum_i \nu_i{}^{a_2}\tilde{\mu}_i\right) + \left(\sum_i \nu_i{}^{a_2}\tilde{\mu}_i - \sum_i \nu_i{}^r\tilde{\mu}_i\right)$$
$$+ \left(-\sum_i \nu_i'{}^r\tilde{\mu}_i - \sum_k \nu_k'{}^r\tilde{\mu}_k\right) + nF\eta^{r\delta} \tag{231}$$

Here $x = r$ is the plane separating the zone (a_2, r), characterized by the exclusive presence of reaction overpotential, from the zone (r, δ), where we shall imagine the sole diffusion overpotential is located. By comparing Eqs. (55) and (231) it is readily seen that η_d, which is formally located on the other side of the plane $x = r$ with respect to the electrode, is given by

$$nF\eta_d = nF\eta^{r\delta} = \sum_i \nu_i{}^r\tilde{\mu}_i - \sum_i \nu_i{}^\delta\tilde{\mu}_i + \sum_i \nu_i'{}^r\tilde{\mu}_i + \sum_k \nu_k'{}^r\tilde{\mu}_k \tag{232}$$

Since at $x = \delta$ the electrochemical affinity of the homogeneous chemical reaction (185) is practically zero, we have

$$\sum_i \nu_i'{}^\delta\tilde{\mu}_i + \sum_k \nu_k'{}^\delta\tilde{\mu}_k = 0 \tag{233}$$

Combining Eqs. (232) and (233) we obtain

$$\eta_d = \frac{1}{nF}\sum_i (\nu_i + \nu_i')({}^r\tilde{\mu}_i - {}^\delta\tilde{\mu}_i) + \frac{1}{nF}\sum_k \nu_k'({}^r\tilde{\mu}_k - {}^\delta\tilde{\mu}_k) \tag{234}$$

The first and the second term on the right-hand side of Eq. (234) clearly express the contributions to η_d due, respectively, to the electroactive substances and to the electroinactive species taking part in the chemical reaction (185). From Eq. (231) it is manifest that if we set

$$-n\tilde{\mu}_e - \sum_i \nu_i{}^{a_2}\tilde{\mu}_i = nF\eta_t$$

as in Sec. I.B.2, the reaction overpotential is formally expressed by

$$\eta_r = \eta^{a_2 r} = \frac{1}{nF}\left(\sum_i \nu_i\, {}^{a_2}\tilde{\mu}_i - \sum_i \nu_i\, {}^{r}\tilde{\mu}_i\right) + \frac{1}{nF}\left(-\sum_i \nu_i'\, {}^{r}\tilde{\mu}_i - \sum_k \nu_k'\, {}^{r}\tilde{\mu}_k\right) \quad (235)$$

Two methods can be employed in order to characterize the reaction layer (a_2, r) in a simple way. According to one of them it is assumed that the first term on the right-hand side of Eq. (235) is practically zero. The constancy of the electrochemical potentials of the electroactive species between $x = a_2$ and $x = r$ implies the uniformity of the local electrochemical affinity $\tilde{A}_\rho$ and of the local rate J_ρ of reaction (185) throughout the reaction layer. In order that the current which would flow through the idealized reaction layer previously described as well as through the adjacent Nernst diffusion layer be equal to the actual current, it is necessary to determine r in a way analogous to that employed by Brdička and Wiesner (*26*) for the particular case of a chemical reaction preceding the charge-transfer process. This particular model of reaction layer will be called the "Brdička-Wiesner reaction layer."

The second method employed in order to characterize the reaction layer in a simple way consists in assuming that the second term on the right-hand side of Eq. (235) is zero. This amounts to postulating that the homogeneous reaction (185) is in equilibrium at the boundary of the reaction layer. Under these assumptions, the whole reaction overpotential is given by the term $1/(nF) \sum_i \nu_i({}^{a_2}\tilde{\mu}_i - {}^{r}\tilde{\mu}_i)$, which has the form of a diffusion overpotential of the various electroactive species A_i between $x = a_2$ and $x = r$. This alternative way of idealizing the reaction layer leads to a value for r which is generally different from that previously considered in connection with the Brdička-Wiesner reaction layer. We shall designate this new value by r'. Hanuš (*27*) made an approximate calculation of r' for several homogeneous chemical reactions coupled with a polarographic electrode process, so that we shall refer to this model of reaction layer as to the "Hanuš reaction layer." According to this model the reaction overpotential is considered to be analogous to a type of diffusion overpotential of the various species A_i, produced by fictitious concentration gradients much larger than the actual ones. The two different ways of idealizing the reaction layer arise from the fact that both reaction and diffusion overpotentials contribute in comparable amounts to the total overpotential in the solution layer adjacent to the electrode as appears from the rigorous equation (230).

The formal separation of total overpotential previously outlined suggests the general procedure to be followed in order to obtain phenomenological relations between current and potential (5). In this connection it must be noted that under steady-state conditions, the rate v_t for the charge transfer, the rate v_r for the chemical reaction within the reaction layer, as well as the rate v_d at which diffusion proceeds within the diffusion layer are equal. Therefore the current intensity is given by

$$i = -nFAv_t = -nFAv_r = -nFAv_d \tag{236}$$

In the absence of specific adsorption, v_t can be expressed directly as a function of the concentrations of the electroactive species A_i at the outer Helmholtz plane and depends on the applied potential E and on the kinetic parameters that characterize the various steps of the electrode process (41),

$$v_t = v_t(\{^{a_2}C_i\}, E, \{k_h\}) \tag{237}$$

In Eq. (237) $\{k_h\}$ designates the set of kinetic parameters and $\{^{a_2}C_i\}$ the set of concentrations at $x = a_2$.

The rate v_r of the chemical reaction within the Brdička-Wiesner reaction layer is expressed by the number of $|\nu_i'|$ moles of A_i or $|\nu_k'|$ moles of B_k produced (ν_i' or $\nu_k' > 0$) or consumed (ν_i' or $\nu_k' < 0$) in a volume element of unit base normal to the electrode surface and bounded by the planes $x = a_2$ and $x = r$. Recalling the definition of ρ given in Eq. (186), it immediately follows that

$$v_r = (r - a_2)\rho \cong r\rho \tag{238}$$

owing to the postulated uniformity of the various concentrations in the Brdička-Wiesner reaction layer (a_2, r). As a matter of fact, on account of the large concentration gradients existing in the double layer, the above uniformity can be reasonably assumed only outside the reaction layer $(x > b)$. Consequently Eq. (238) is valid only if $r \gg b$, in which case the structure of the diffuse layer can be neglected. Under this assumption the reaction layer can be confined outside the diffuse layer† and ρ can be considered as a function of the concentrations of the various species A_1 and B_k at $x = b$, or also (which is the same in the Brdička-Wiesner approximation) at $x = r$,

$$\rho = \rho(\{^rC_i\}, \{^rC_k\}, \{k\}) \tag{239}$$

† If the chemical reaction (185) is very fast, r may become comparable with b, thus causing the reaction rate ρ to be influenced by the double-layer structure. This case will be considered in Sec. V.

The symbols $\{^rC_i\}$ and $\{^rC_k\}$ denote, respectively, the sets of concentrations of A_i and B_k at $x = r$, whereas $\{k\}$ designates the set of kinetic parameters of the homogeneous chemical reaction (185). From Eqs. (236), (238), and (239) we have

$$i = -nFAv_r = -nFAr\rho(\{^rC_i\}, \{^rC_k\}, \{k\}) \tag{240}$$

If ρ can be represented as the difference between a forward and a backward reaction rate, then when equilibrium is approached, the expressions for these rates must be such that on equating them the expression for the equilibrium constant of reaction (185) is obtained. Under the assumption that $r \gg b$, the probability that an electroactive species A_i present in the diffuse double layer and diffusing toward the electrode will be converted into an electroinactive species as a consequence of the homogeneous chemical reaction is very low. Consequently we may express the concentrations $^{a_2}C_i$ of the various species A_i at $x = a_2$ in terms of the corresponding concentrations ^bC_i at $x = b$ by making use of Eq. (54), which expresses a Boltzmann distribution of ionic species in the double layer (penetration overpotential $\eta_p = 0$). By this procedure, which neglects the effects of reaction (185) within the diffuse layer, v_t assumes the form,

$$v_t = -\frac{i}{nFA} = v_t'(\{^bC_i\}, E, \{k_h\}) \tag{241}$$

which is identical with the form (157) used in Sec. III.A. If the electrode process (41) is so fast that its electrochemical affinity can be considered practically zero, Eq. (241) is replaced by Nernst's equation applied to the concentrations ^bC_i,

$$\theta = \exp\left[\frac{nF}{RT}(E - E^\circ)\right] = \prod_i {}^bC_i^{\nu_i} \tag{242}$$

The rate v_d of diffusion within the layer (r, δ) is expressed by the number of $|\nu_i + \nu_i'|$ moles of A_i, or by the number of $|\nu_k'|$ moles of B_k, which diffuse through the plane $x = r$ per unit time per unit area. If the influence of the electric field strength upon the species A_i and B_k is negligible for $x \geq r$, in view of Fick's first law, we may write,

$$i = -nFAv_d = \frac{nFA}{\nu_i + \nu_i'} D_i \left(\frac{\partial C_i}{\partial x}\right)_{x=r} = \frac{nFA}{\nu_k'} D_k \left(\frac{\partial C_k}{\partial x}\right)_{x=r} \qquad \text{for all } i \text{ and } k \tag{243}$$

D_i and D_k are the diffusion coefficients of the species A_i and B_k respectively. Taking into account that in the ideal model of diffusion layer (r, δ), the concentrations of all the species other than the ions composing the supporting electrolyte change linearly with the distance x from the electrode, we have

$$v_d = -\left(\frac{D_i}{v_i + v_i'}\right)\left(\frac{{}^{\delta}C_i - {}^{r}C_i}{\delta - r}\right) = -\left(\frac{D_k}{v_k'}\right)\left(\frac{{}^{\delta}C_k - {}^{r}C_k}{\delta - r}\right) \tag{244}$$

where the concentrations ${}^{\delta}C_i$ and ${}^{\delta}C_k$ at $x = \delta$ coincide with the corresponding concentrations in the bulk of the solution. Eq. (244) holds for any value of i and k.

If we choose the Hanuš model of reaction layer, thus considering reaction overpotential as a particular type of diffusion overpotential, the rate v_r for the homogeneous reaction (185) within the layer (b, r') is expressed by a relation analogous to Eq. (244) as

$$v_r = -\frac{i}{nFA} = -\frac{D_i}{v_i}\frac{{}^{r}C_i - {}^{b}C_i}{r' - b} \cong -\frac{D_i}{v_i}\frac{{}^{r}C_i - {}^{b}C_i}{r'} \quad \text{for all the species } A_i \tag{245}$$

According to the Brdička-Wiesner model of reaction layer, the polarographic current-potential curve is obtained by combining Eq. (241), or alternatively (242), with Eqs. (240) and (244). As in the approximate treatment of slow charge-transfer processes based on the diffusion layer concept, here too the concentrations ${}^{b}C_i = {}^{r}C_i$ and ${}^{b}C_k = {}^{r}C_k$ will be assumed constant during the drop life. Noting that in the most general case r is a function of the applied potential E, of the kinetic parameters $\{k\}$ and $\{k_h\}$, as well as of the concentrations of some of the species A_i and B_k at $x = r$, r is also assumed independent of electrolysis time. In view of Eqs. (236), (241), (240), and (244) and recalling that $A = A_O t^{2/3}$ we then have, for the mean polarographic current $\bar{i}$,

$$\bar{i} = -\frac{nFA_O v_t}{t_d}\int_0^{t_d} t^{2/3}\,dt = -\frac{3}{5}\,nFA_O t_d^{2/3} v_t'(\{{}^{b}C_i\}, E, \{k_h\}) \tag{246}$$

$$\bar{i} = -\frac{nFA_O v_r}{t_d}\int_0^{t_d} t^{2/3}\,dt = -\frac{3}{5}\,nFA_O t_d^{2/3} r\rho(\{{}^{r}C_i\}, \{{}^{r}C_k\}, \{k\}) \tag{247}$$

$$\bar{\imath} = -\frac{nFA_O}{t_d}\int_0^{t_d} v_d t^{2/3}\,dt = \begin{cases} \dfrac{nFA_O D_i({}^{\delta}C_i - {}^{r}C_i)}{(v_i + v_i')t_d}\displaystyle\int_0^{t_d}\frac{t^{2/3}}{\delta - r}\,dt \\ \qquad \text{for all } A_i \text{ with } v_i + v_i' \neq 0 \\ \dfrac{nFA_O D_k({}^{\delta}C_k - {}^{r}C_k)}{v_k' t_d}\displaystyle\int_0^{t_d}\frac{t^{2/3}}{\delta - r}\,dt \\ \qquad \text{for all } B_k \end{cases} \tag{248}$$

Noting that in the present case the diffusion layer thickness is given by $\delta - r$, in view of Eq. (110) we have,

$$\delta - r = \sqrt{\frac{3\pi Dt}{7}}$$

On substituting the value of $\delta - r$ in Eq. (248) and integrating over t_d, we obtain,

$$\left.\begin{aligned} \bar{\imath} &= \sqrt{\frac{12D_i}{7\pi}}\,\frac{nFA_O t_d^{1/6}}{v_i + v_i'}({}^{\delta}C_i - {}^{r}C_i) \qquad \text{for all } A_i \text{ with } v_i + v_i' \neq 0 \\ \bar{\imath} &= \sqrt{\frac{12D_k}{7\pi}}\,\frac{nFA_O t_d^{1/6}}{v_k'}({}^{\delta}C_k - {}^{r}C_k) \qquad \text{for all } B_k \end{aligned}\right\} \tag{249}$$

Let us designate by p the number of the species A_i characterized by a zero value of $v_i + v_i'$. Equations (246), (247), and (249) express $n' - p + 2$ relations between the applied potential E and the $n' + 1$ potential-dependent variables ${}^{r}C_i = {}^{b}C_i$, ${}^{r}C_k$, and $\bar{\imath}$. The concentrations of A_i and B_k at $x = \delta$ are known, since according to the diffusion layer thickness approximation they coincide with the corresponding bulk concentrations. It follows that a simple algebraic manipulation of Eqs. (246), (247), and (249) leads directly to the current-potential characteristic $\bar{\imath} = \bar{\imath}(E, \{k\}, \{k_h\})$, provided the expression for r is known and $p = 1$. This latter condition is always satisfied, with the exclusion of the case of total regeneration of the depolarizer. Here, as appears from the depolarization scheme of Eqs. (199a–b), both the depolarizer and the product of the electrode reaction are characterized by $v_i + v_i' = 0$ and consequently p equals 2. Owing to this peculiarity, the total regeneration of the depolarizer will be considered separately. If the electrode process (41) is polarographically reversible, Eq. (246) is replaced by Nernst's equation

(242) and the theoretical current-potential characteristic has the form $\bar{i} = \bar{i}(E, E^\circ, \{k\})$.

When choosing the Hanuš model of reaction layer, we shall still make use of Eq. (246), or alternatively Eq. (242), and Eq. (249), as in the preceding approach, but we shall replace Eq. (240) by Eq. (245). The Hanuš reaction layer thickness r', like r, is also independent of electrolysis time. Hence, on integrating Eq. (245) over the drop life, we have

$$\bar{i} = \frac{3}{5} \frac{nFA_O D_i t_d^{2/3}}{v_i r'} ({}^rC_i - {}^bC_i) \qquad \text{for all } A_i \tag{250}$$

If we substitute the various ${}^rC_i$ from Eq. (250) into Eq. (246), after rearrangement we obtain

$$\bar{i} = \bar{i}(\{{}^rC_i\}, E, \{k_h\}, \{k\}) \tag{251}$$

provided the expression for r' is known. A further relation among the concentrations ${}^rC_i$ is obtained by recalling that according to the Hanuš model, the chemical reaction (185) is at equilibrium at $x = r'$. Thus denoting by K the equilibrium constant for the above reaction, we have

$$K = \prod_i {}^rC_i^{v_i'} \prod_k {}^rC_k^{v'_k} \tag{252}$$

Equations (251), (252), and (249) represent $n' - p + 2$ algebraic relations between E and the $n' + 1$ potential dependent variables ${}^rC_i$, ${}^rC_k$, and $\bar{i}$. Hence on combining the above equations, the current-potential characteristic $\bar{i} = \bar{i}(E, \{k_h\}, \{k\})$ is obtained, provided $p = 1$. Here, too, if the electrode process is polarographically reversible, Eq. (246) is replaced by Eq. (242). Obviously, in this latter case the final expression for the polarographic current will not contain the kinetic parameters $\{k_h\}$. If the expressions for r and r' are properly chosen, the theoretical current-potential characteristics obtained both according to the Brdička-Wiesner model and to the Hanuš model of reaction layer are entirely identical.

2. The Determination of the Reaction-Layer Thickness

Approximate expressions for the reaction layer thickness have been determined by different ways. Thus Wiesner (*28*) related r to the average distance traveled by a molecule of the depolarizer or of the electrode product in its average lifetime. For a preceding reaction r defines a reaction layer such that a molecule of the depolarizer produced at any point within it has a good chance of reaching the electrode surface before disappearing by the backward reaction. The method proposed by Wiesner allows r to be determined apart from a constant factor. Here we shall describe the procedure employed independently by Koutecký (*22–24*) and by Hans and Hencke (*29*) and subsequently generalized by Koutecký, Brdička, Hanuš (*30*), and by Koutecký and Koryta (*14*). In applying this procedure to the general chemical reaction (185) coupled with the general electrode process (41), we shall not follow the unified method proposed by Koutecký and coworkers and based on the use of suitable combinations of the concentrations of the diffusing species. Rather, a slightly different method based on the previously outlined formal separation of the solution layer around the electrode into regions characterized by the presence of only one type of overpotential will be described. The results obtained by applying this method to specific problems are similar to those obtained by Koutecký's method. Furthermore these results come to coincide when some simplifying assumptions are made, accounting for the fact that only very fast chemical reactions can be treated by both the above methods. With respect to Koutecký's procedure, the method described here is more intuitive and helps to penetrate into the intricacies of diffusional problems in a more direct way.

Let us focus our attention on the substance taking part both in the electrode reaction (41) and in the chemical reaction (185) and characterized by $\nu_i + \nu_i' = 0$. We shall designate it by A_1. The species A_1 may be either one of the reactants of the electrode process, as in the case of a preceding chemical reaction, or one of the products, as in the case of a subsequent chemical reaction. A_1 is the only species which, according to the formal separation of overpotential outlined in Sec. I.B.2, does not contribute to diffusion overpotential in the diffusion layer. This circumstance is encountered since the amount of A_1 produced (consumed) at the electrode surface per unit time is practically consumed (produced) within the reaction layer in the same unit time. In this respect A_1 is also the only species for which the concentration gradient between the electrode surface and the bulk of the solution is almost entirely concentrated in the reaction layer (b, r). Taking into account that the concentration gradients of the remaining species are distributed over the whole diffusion layer δ and that $\delta \gg r$, it follows that the concentration gradient of A_1 in the reaction layer is much larger than that of the other species. In this regard the concentrations of all the diffusing species, other than A_1, can be taken as uniform throughout the reaction layer. In view of the rigorous definition of overpotential expressed by Eq. (230) this amounts to saying that A_1 is the only species which contributes appreciably to the diffusion overpotential η_d within the reaction layer. Analogously, if we consider the approximate expression (235) for the reaction overpotential η_r, we can state that A_1 is the only species contributing to the first term on the right-hand side of Eq. (235) in a significant way. Let us temporarily disregard the identification of reaction overpotential either with the first term (Hanuš model) or with the second term (Brdička-Wiesner model) on the right-hand side of Eq. (235).

On the basis of the previous assumptions, if we apply the law of conservation of mass to A_1 in the reaction layer,

$$\frac{\partial C_1}{\partial t} = D_1 \frac{\partial^2 C_1}{\partial x^2} + \frac{2x}{3t} \frac{\partial C_1}{\partial x} + \nu_1' \rho(\{C_i\}, \{C_k\}, \{k\}) \tag{253}$$

all C_i and C_k, other than C_1, can be taken as uniform throughout this layer. Furthermore since A_1 is simultaneously produced and consumed in the very thin reaction layer, we may assume that after a very short time from the beginning of the drop life, steady-state conditions are reached, so that the change of C_1 with time can be neglected ($\partial C_1/\partial t = 0$). Since the reaction layer is very thin with respect to the diffusion layer, we may further postulate that the effects of the convective motions caused by the growth of the drop and expressed by the term $2x/3t\ \partial C_1/\partial x$ in Eq. (253) are not appreciably felt for $x < r$. Hence Eq. (253) takes the following simplified form,

$$D_1 \frac{\partial^2 C_1}{\partial x^2} = -\nu_1' \rho \tag{254}$$

which in general may be easily integrated. Thus if $\nu_1' \rho = d_i C_1/dt$ is the difference between a forward rate ρ_f, zero order with respect to A_1, and a backward rate m order with respect to A_1, we have

$$D_1 \frac{\partial^2 C_1}{\partial x^2} = k_b C_1^m - \rho_f \tag{255}$$

In the most general case k_b and ρ_f depend on all C_i and C_k, other than C_1. Noting that $\partial/\partial x(\partial C_1/\partial x)^2 = 2(\partial C_1/\partial x)(\partial^2 C_1/\partial x^2)$, Eq. (255) can be written as

$$\frac{\partial}{\partial x}\left(\frac{\partial C_1}{\partial x}\right)^2 = \frac{2}{D_1}(k_b C_1^m - \rho_f)\frac{\partial C_1}{\partial x}$$

Upon integrating over x between $x = b$ and $x = r$, we have

$$\left(\frac{\partial C_1}{\partial x}\right)^2_{x=b} - \left(\frac{\partial C_1}{\partial x}\right)^2_{x=r} - \left[\frac{2}{D_1}\left(\frac{k_b C_1^{m+1}}{m+1} \quad \rho_f C_1\right)\right]^{{}^bC_1}_{{}^rC_1} \tag{256}$$

Recalling that the change in the concentration of A_1 between the electrode surface and the bulk of the solution is almost entirely concentrated in the reaction layer (b, r) and noting that the absolute value $|\partial C_1/\partial x|$ of the concentration gradient of A_1 must necessarily decrease in a gradual way in passing from $x = b$ to $x = r$, we may approximately neglect $(\partial C_1/\partial x)^2_{x=r}$ with respect to $(\partial C_1/\partial x)^2_{x=b}$. Hence

$$\left(\frac{\partial C_1}{\partial x}\right)_{x=b} = \pm\left\{\left[\frac{2}{D_1}\left(\frac{k_b C_1^{m+1}}{m+1} - \rho_f C_1\right)\right]^{{}^bC_1}_{{}^rC_1}\right\}^{1/2} \tag{257}$$

Because the chemical reaction (185) attains equilibrium at the outer boundary $x = r$ of the reaction layer, Eq. (252) allows ${}^rC_1$ to be readily obtained as a function of all the other concentrations at $x = r$,

$${}^rC_1 = \left[\frac{K}{(\prod_{i\neq 1} {}^rC_i^{\nu_i'} \prod_k {}^rC_k^{\nu k'})}\right]^{1/\nu_1'} \tag{258}$$

Taking into consideration Eqs. (241) and (189), and recalling from Sec. II.A.2 that the notation $x = 0$ in the latter equation indicates the outer boundary $x = b$ of the diffuse layer, we have

$$\frac{i}{nFA} = -v_t'(\{{}^bC_i\}, E, \{k_h\}) = \frac{D_1}{\nu_1}\left(\frac{\partial C_1}{\partial x}\right)_{x=b} \tag{259}$$

Substituting $(\partial C_1/\partial x)_{x=b}$ from Eq. (259) into Eq. (257) and rearranging terms, it is possible to derive ${}^bC_1$ as a function of ${}^rC_1$, $\underset{i\neq 1}{{}^rC_i} = \underset{i\neq 1}{{}^bC_i}$, and ${}^rC_k$. Since the expression for ${}^rC_1$ in terms of $\underset{i\neq 1}{{}^rC_i}$ and ${}^rC_k$ is known (cf. Eq. 258), the combination of Eqs. (257), (258), and (259) allows $(\partial C_1/\partial x)_{x=b}$ to be obtained as a function of ${}^rC_i$ and ${}^rC_k$,

$$\left(\frac{\partial C_1}{\partial x}\right)_{x=b} = \varphi\left(\left\{\underset{i\neq 1}{{}^rC_i}\right\}, \{{}^rC_k\}, E, \{k\}, \{k_h\}\right) \tag{260}$$

† Clearly the sign on the right-hand side of Eq. (257) must be chosen in such a way that $(\partial C_1/\partial x)_{x=b}$ be positive if A_1 diffuses towards the electrode and negative if A_1 diffuses away from the electrode.

If the charge-transfer reaction is polarographically reversible, the expression for ${}^bC_1$ as a function of ${}^rC_{\substack{i \\ i\neq 1}} = {}^bC_{\substack{i \\ i\neq 1}}$ is very easily obtained by replacing Eq. (259) with Nernst's equation (242), written in the form,

$$ {}^bC_1 = \left(\frac{\theta}{\prod_{i\neq 1} {}^rC_i^{\nu_i}} \right)^{1/\nu_1} \tag{261} $$

A particularly simple situation is encountered when the limiting current for an electrode process controlled by a preceding chemical reaction is considered. In this case the electroactive species A_1 is instantaneously consumed by the charge-transfer reaction as soon as it reaches the electrode surface and ${}^bC_1$ equals zero. In the preceding derivation we have assumed that within the reaction layer both the chemical overpotential as defined in the rigorous equation (36) and the diffusion overpotential η_d due to the sole species A_1 contribute to the total overpotential. With these assumptions, the following expression for the current is obtained,

$$ i = nFA \frac{D_1}{\nu_1} \left(\frac{\partial C_1}{\partial x} \right)_{x=b} \equiv nFA \frac{D_1}{\nu_1} \varphi\left(\left\{ {}^rC_{\substack{i \\ i\neq 1}} \right\}, \{{}^rC_k\} \right) \tag{262} $$

According to the Brdička-Wiesner model, the thickness of the reaction layer must be chosen in such a way that, assuming that only reaction overpotential is located in this layer, the current flowing through it equals the current expressed by Eq. (262). If we postulate the uniformity of C_1 throughout the reaction layer and we give C_1 a value ${}^{\bar{r}}C_1$ included between the values ${}^bC_1$ and ${}^rC_1$, in view of Eqs. (240) and (262) we can write

$$ \frac{D_1}{\nu_1} \varphi\left(\left\{ {}^rC_{\substack{i \\ i\neq 1}} \right\}, \{{}^rC_k\} \right) = -r\rho\left(\left\{ {}^rC_{\substack{i \\ i\neq 1}} \right\}, {}^{\bar{r}}C_1, \{{}^rC_k\} \right) \tag{263} $$

where r is the Brdička-Wiesner reaction layer thickness. The value for r determined from Eq. (263) depends on the arbitrary choice of ${}^{\bar{r}}C_1$, but this fact does not alter the final results. In fact, whatever the value for ${}^{\bar{r}}C_1$, this value must also be used in the expression for ρ in Eq. (247). Thus, on deriving r from Eq. (263) and replacing it in Eq. (247), we obtain

$$ \bar{i} = \frac{3}{5} nFA_0 t_d^{2/3} \frac{D_1}{\nu_1} \varphi\left(\left\{ {}^rC_{\substack{i \\ i\neq 1}} \right\}, \{{}^rC_k\} \right) \tag{264} $$

Equation (264) could also be derived from Eq. (262) by integrating the current over the drop life, under the assumption, commonly made in approximate treatments, that ${}^{r}C_{\substack{i \\ i \neq 1}}$ and ${}^{r}C_k$ are constant with electrolysis time. It is manifest that the derivation of r represents an unnecessary step in the determination of the approximate current-potential characteristic. This result is to be expected since the knowledge of r or r' is exclusively required in Eq. (247) or (250) in order to determine the dependence of $\bar{\imath}$ on ${}^{r}C_{\substack{i \\ i \neq 1}}$ and ${}^{r}C_k$. Once this dependence, which is directly expressed by Eq. (264), is known, Eqs. (249) based on the concept of diffusion layer thickness allow the final expression for the theoretical $i - E$ curve to be readily obtained.

The same reasoning holds for the Hanuš model of reaction layer. According to this model the reaction layer thickness must be determined in such a way that on assuming that only reaction overpotential in the form of a type of diffusion overpotential of the various electroactive species (the first term on the right-hand side of Eq. (235)) is located in the reaction layer, then the current flowing through this layer is equal to that expressed by Eq. (262). Taking into account the assumptions on which Eq. (262) is based, we must postulate that only A_1 contributes to reaction overpotential; that is, $\eta_r = \eta^{br}$ in Eq. (235) is simply given by $(v_1{}^{b}\tilde{\mu}_1 - v_1{}^{r}\tilde{\mu}_1)/nF$. The reaction layer thickness r', analogously to the diffusion layer thickness δ, is determined by the intersection of the tangent to the $C_1 - x$ profile at $x = b$ with the horizontal line expressing the concentration of A_1 outside the reaction layer. The combination of Eqs. (245) and (262) then yields

$$\frac{i}{nFA} = \frac{D_1}{v_1}\frac{{}^{r}C_1 - {}^{b}C_1}{r'} = \frac{D_1}{v_1}\varphi\left(\left\{{}^{r}C_{\substack{i \\ i\neq 1}}\right\}, \{{}^{r}C_k\}\right) \tag{265}$$

If the reaction layer thickness r' derived from Eq. (265) as a function of ${}^{b}C_1$ and of all the concentrations ${}^{r}C_i$ at $x = r$ is replaced in Eq. (250), we immediately obtain Eq. (264).

We can conclude that the direct determination of r or r' is not necessary for the derivation of the $i - E$ curve according to the approximate procedure. In spite of this the concept of reaction layer thickness retains an outstanding importance not only for the understanding of the developments of the theory of kinetic currents, but also because it represents a successful attempt to carry out an ideal separation of the total overpotential.

D. The Semirigorous Procedure

A more rigorous expression for the current-potential characteristic is obtained if we now consider the various concentrations at $x = r$ as time dependent, and we use Eq. (262) as a boundary condition for the diffusion equations of all the species other than A_1. These differential equations are applied outside the reaction layer and do not contain the source term,

$$\frac{\partial C_\alpha}{\partial t} = D_\alpha \frac{\partial^2 C_\alpha}{\partial x^2} + \frac{2x}{3t}\frac{\partial C_\alpha}{\partial x} \qquad x > r \tag{266}$$

In Eq. (266), α denotes both i (with the exclusion of $i = 1$) and k. Consequently the diffusion equations (266) are $n' - 1$ in number and correspond to the $n' - 1$ species contributing to the diffusion overpotential in the diffusion layer. The source term is lacking in Eq. (266) since, according to the ideal separation of overpotential previously performed, the homogeneous reaction (185) takes place only in the reaction layer (b, r). Outside the reaction layer the chemical reaction is at equilibrium and is shifted in one direction in such a complete way that the diffusion of the various species toward or from the electrode does not cause any appreciable interconversion of these species at $x > r$. Besides Eq. (262), the remaining $n' - 2$ boundary conditions for the diffusion equations (266) are obtained from Eqs (243), which express the proportionality between all possible pairs of concentration gradients $(\partial C_\alpha/\partial x)_{x=r}$ of the $n' - 1$ diffusing species. If we represent the concentration gradients at $x = r$ of all these species but one (for example, A_2) in terms of the concentration gradient of this latter, then we can write

$$\underset{i \neq 1,2}{\frac{D_i}{\nu_i + \nu_i'}} \left(\frac{\partial C_i}{\partial x}\right)_{x=r} = \frac{D_k}{\nu_k'} \left(\frac{\partial C_k}{\partial x}\right)_{x=r} = \frac{D_2}{\nu_2 + \nu_2'} \left(\frac{\partial C_2}{\partial x}\right)_{x=r}$$

The initial conditions for the differential equations (266) are

$$C_\alpha = {}^{\delta}C_\alpha \qquad \text{for} \begin{cases} t = 0, \ x \geqslant r \\ t > 0, \ x \to \infty \end{cases}$$

where ${}^{\delta}C_\alpha$ is the bulk concentration of the αth species. By a procedure entirely analogous to that followed in Sec. II.B.2c[Eqs. (123) through (127)], we obtain

$$\frac{D_2^{1/2}}{\nu_2 + \nu_2'}({}^{\delta}C_2 - {}^{r}C_2) = \underset{i \neq 1,2}{\frac{D_i^{1/2}}{\nu_i + \nu_i'}}({}^{\delta}C_i - {}^{r}C_i) = \frac{D_k^{1/2}}{\nu_k'}({}^{\delta}C_k - {}^{r}C_k) \qquad (267)$$

Equations (267) express the concentrations at $x = r$ of all the species, except A_1 and A_2, as functions of ${}^{r}C_2$. On combining Eqs. (243), (262), and (267) we have

$$i = nFA\frac{D_1}{\nu_1}\left(\frac{\partial C_1}{\partial x}\right)_{x=b} = nFA\frac{D_2}{\nu_2 + \nu_2'}\left(\frac{\partial C_2}{\partial x}\right)_{x=r} \qquad (268)$$

and

$$nFA\frac{D_1}{\nu_1}\varphi\left(\left\{{}^{r}C_i\right\}_{i\neq 1}, \{{}^{r}C_k\}\right) = nFA\frac{D_1}{\nu_1}\varphi'({}^{r}C_2) = nFA\frac{D_2}{\nu_2 + \nu_2'}\left(\frac{\partial C_2}{\partial x}\right)_{x=r} \qquad (269)$$

Hence the diffusion problem is reduced to the solution of the single differential equation (266), where $\alpha = 2$, with the boundary condition (269) and the initial condition,

$$C_2 = {}^{\delta}C_2 \qquad \text{for} \quad t > 0, x \to \infty \qquad (270)$$

The boundary condition in Eq. (269) is usually rather involved since part of the complications inherent in the rigorous formulation of the diffusional problem (cf. Sec. IV.A) are now transferred to this condition. Nevertheless the solution of the boundary value problem of Eqs. (266), (269), (270) can be suitably tackled by the mathematical procedure outlined in the appendix.

It must be emphasized that the "semirigorous" method previously described is perfectly equivalent to the approximate procedure of Sec. IV.C, based on the use of the reaction layer concept, as far as concerns the formal separation of the total overpotential, since the assumptions as to the nature of overpotential in the reaction layer are identical. The only difference consists in the fact that with the semirigorous procedure the diffusion layer thickness δ is calculated in a more precise way. Thus if we write the equation

$$\left(\frac{\partial C_2}{\partial x}\right)_{x=r} = \frac{{}^{\delta}C_2 - {}^{r}C_2}{\delta}$$

and we replace $(\partial C_2/\partial x)_{x=r}$ and ${}^rC_2$ by their time-dependent expressions, as derived by the semirigorous procedure, the diffusion layer thickness δ so obtained takes into better account the diffusion phenomena occurring around the electrode than does the expression $(\frac{3}{7}\pi Dt)^{1/2}$, which holds rigorously only for a diffusion-controlled electrode process. As already anticipated the semirigorous procedure described here differs from the procedure employed by Koutecký and coworkers (*14, 30*) mainly in that recourse to linear combinations of the concentrations of the diffusing species is avoided. From the foregoing treatment it is manifest that the connections between the approximate and the semirigorous treatment are relatively strict. Thus it is not surprising that the discrepancies between the results obtained through the approximate and the semirigorous procedures, respectively, are relatively small. The approximate procedure can usefully be employed whenever it is desired to ascertain the correctness of a postulated mechanism through the functional relation between mean current and potential. When the nature of the mechanism has been clearly elucidated, the values for the kinetic parameters of the chemical reaction are more satisfactorily derived by the semirigorous procedure. In this way the complex and time-consuming numerical solution of the boundary value problem of Eq. (269), on which the semirigorous method is based, is performed only once in connection with the definitively ascertained mechanism. In any case, when applied to sufficiently fast reactions, the approximate procedure yields kinetic parameters that are of the correct order of magnitude and that can deviate from their true value by no more than a few tenths of a percent.

E. Two Species Characterized by $\nu_i + \nu_i' = 0$. Total Regeneration of the Depolarizer

Thus far we have limited ourselves to the examination of cases in which only one species, namely A_1, is characterized by $\nu_i + \nu_i' = 0$. We shall now consider the depolarization scheme,

$$A_2 \rightleftharpoons A_1 + ne; \qquad A_1 \underset{kK}{\overset{k}{\rightleftharpoons}} A_2 \tag{271}$$

where the depolarizer A_2 is totally regenerated from the electrode product A_1 by the agency of an electroinactive species B_3. This is the only case in which two electroactive species—that is, A_1 and A_2—have $\nu_i + \nu_i' = 0$. In other words A_2 may be considered to be produced through the chemical reaction within the reaction layer and consumed at the electrode surface, while A_1 is produced at the electrode and thoroughly consumed within the reaction layer. Neither A_1 nor A_2 contribute appreciably to diffusion overpotential for $x > r$, and, furthermore, both A_1 and A_2 exhibit a high concentration gradient within the reaction layer. Usually the molecules of A_1 and A_2 have comparable sizes, so that we may assume approximately equal diffusion coefficients for the two species,

$$D_1 = D_2 = D$$

Under the above assumptions, from the rigorous equations (202) and (203) we see that the combination $(C_1 + C_2)$ satisfies the diffusion equation $\delta(C_1 + C_2) = 0$ uncomplicated by the source term, with the boundary condition $D\, \partial/\partial x_{x=0}(C_1 + C_2) = 0$. By considerations analogous to those leading to Eq. (86) we then have

$$C_1(x) + C_2(x) = {}^{\delta}C_1 + {}^{\delta}C_2 \qquad (272)$$

which holds at any distance x from the electrode. Equation (272) allows the present problem to be reduced to that of a single species with $\nu_i + \nu_i' = 0$. In fact, by the use of Eq. (272) we may express C_2 as a linear function of C_1 within the reaction layer. The subsequent calculations are carried out as in Eqs. (255) through (260) and the final expression is analogous to Eq. (262). The only difference is represented by the fact that in the present case the set of concentrations $\{{}^{r}C_i\}$ does not contain either C_1 or C_2.

It must be noted that if the regeneration reaction is very fast, even a relatively high excess of the electroinactive substance B_3 may not be sufficient to avoid the formation of an appreciable concentration gradient of B_3 near the electrode. Hence a finite contribution of B_3 to diffusion overpotential may arise. In this case the concentration of B_3 must be included in the diffusional problem.

F. The Various Types of Kinetic Currents

We shall now examine some schemes of electrode processes coupled with preceding, parallel, and subsequent reactions, together with some examples of polarographic kinetic currents illustrating the various schemes. In order to emphasize the separation of overpotential carried out both in the approximate and semirigorous procedures, depolarization schemes will be depicted schematically by representing the electrode surface on the right and by writing, from right to left in the following order, the electrode process (41), the chemical reaction (185), and the various species diffusing toward and from the reaction layer.

1. Prekinetic Currents

a. A General Depolarization Scheme. We shall first consider the case of an electroinactive substance B_3 in chemical equilibrium with an electroactive species A_1, which is oxidized at the electrode giving rise to a species A_2.

$$B_3 \underset{kK}{\overset{k}{\rightleftharpoons}} A_1; \qquad A_1 \rightarrow A_2 + ne \tag{273a}$$

k, kK, and K are, respectively, the forward and backward rate constants and the equilibrium constant. We shall assume that the chemical equilibrium is strongly shifted toward B_3 in the bulk of the solution, in order to exclude the contribution of A_1 to the diffusion overpotential in the diffusion layer ($K = {}^{\delta}C_3/{}^{\delta}C_1 \gg 1$). Obviously if the equilibrium were strongly shifted in the opposite direction, the current would be practically controlled by the diffusion of A_1 toward the electrode and the chemical reaction would not play any significant role in the overall process. The reaction scheme is represented as follows:

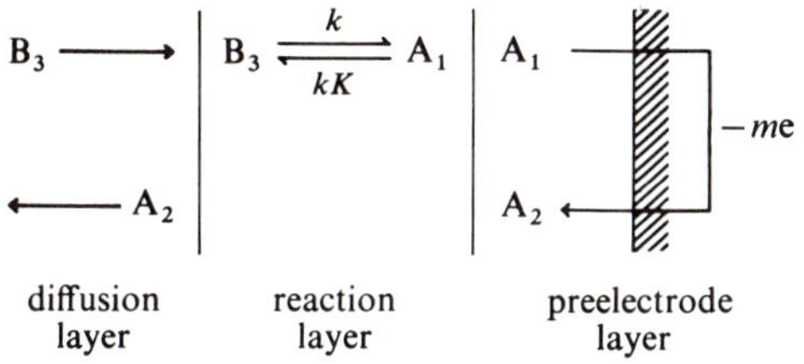

It is readily seen that it is actually A_1, the species for which $v_i + v_i' = 0$. From Eq. (255) we have

$$D_1 \frac{\partial^2 C_1}{\partial x^2} = kKC_1 - kC_3 \qquad \text{for } x < r \tag{274}$$

On integrating once and taking Eq. (257) into account, we obtain

$$\left(\frac{\partial C_1}{\partial x}\right)_{x=b} = \left\{\left[\frac{kKC_1^2 - 2k\,{}^rC_3\,C_1}{D_1}\right]_{{}^rC_1}^{{}^bC_1}\right\}^{1/2} \tag{275}$$

If we confine ourselves to the limiting current, ${}^{b}C_1$ equals zero. Since at $x = r$ equilibrium conditions with respect to the homogeneous reaction are attained, ${}^{r}C_1 = {}^{r}C_3/K$. Substitution of ${}^{b}C_1$ and ${}^{r}C_1$ in Eq. (275) results in

$$\left(\frac{\partial C_1}{\partial x}\right)_{x=b} = \sqrt{\frac{k}{KD_1}}\,{}^{r}C_3 = \frac{D_3}{D_1}\left(\frac{\partial C_3}{\partial x}\right)_{x=r} \tag{276}$$

where use has been made of Eq. (268). Under the assumption that the rate for the backward reaction $A_1 \xrightarrow{kK} B_3$ can be neglected, the Brdička-Wiesner reaction layer thickness is obtained by applying Eq. (263),

$$D_1\left(\frac{\partial C_1}{\partial x}\right)_{x=b} = \sqrt{\frac{kD_1}{K}}\,{}^{r}C_3 = rk\,{}^{r}C_3 \tag{277}$$

From Eq. (277) it immediately follows that

$$r = \sqrt{\frac{D_1}{kK}} \tag{278}$$

As a matter of fact, for the derivation of the current-potential characteristic according to the approximate method, the explicit expression for the reaction layer thickness r is not required. In fact the replacement of $(\partial C_1/\partial x)_{x=b}$ from Eq. (276) into Eq. (264) yields the expression,

$$\bar{i}_1 = -\frac{3}{5}\,nFA_{\mathrm{O}}\,t_{\mathrm{d}}^{2/3}\sqrt{\frac{kD_1}{K}}\,{}^{r}C_3 \tag{279}$$

On deriving ${}^{r}C_3$ from Eq. (249) and substituting it in Eq. (279), after simple passages, we obtain,

$$\frac{\bar{i}_1}{\bar{i}_{\mathrm{d}} - \bar{i}_1} = \frac{3}{5}\sqrt{\frac{7\pi}{12}\frac{D_1kt_{\mathrm{d}}}{D_3K}} = 0.81\sqrt{\frac{D_1kt_{\mathrm{d}}}{D_3K}} \tag{280}$$

where $\bar{i}_{\mathrm{d}} = -\sqrt{12D_3/(7\pi)}\,nFA_{\mathrm{O}}\,t_{\mathrm{d}}^{1/6}\,{}^{\delta}C_3$ expresses the diffusion limiting current which would be attained if B_3 were electroactive or also if the homogeneous reaction $B_3 \xrightarrow{k} A_1$ were infinitely fast ($k \to \infty$).

According to the semirigorous procedure outlined in Sec. IV.D, the recourse to the diffusion layer thickness approximation is avoided and

Eq. (276), written in the form,

$$D_3\left(\frac{\partial C_3}{\partial x}\right)_{x=r} = \sqrt{\frac{kD_1}{K}}\,{}^rC_3$$

is employed as the boundary condition at $x = r$ for the diffusion equation of the species B_3,

$$\frac{\partial C_3}{\partial t} = D_3 \frac{\partial^2 C_3}{\partial x^2} + \frac{2x}{3t}\frac{\partial C_3}{\partial x} \qquad \text{for } x > r \tag{281}$$

The initial condition is

$$C_3 = {}^{\delta}C_3 \qquad \text{for} \begin{cases} t = 0, & x \geqslant r \\ t > 0, & x \to \infty \end{cases} \tag{282}$$

Although A_2 contributes to diffusion overpotential for $x > r$, as appears from the scheme of Eq. (273b), there is no need for considering the diffusion equation of A_2 since this species does not affect the diffusion of B_3 towards the electrode under limiting current conditions. According to the semirigorous method proposed by Koutecký (*31*), the limiting current is obtained by solving the following boundary value problem,

$$\left.\begin{aligned} \frac{\partial \psi}{\partial t} &= D \frac{\partial^2 \psi}{\partial x^2} + \frac{2x}{3t}\frac{\partial \psi}{\partial x} && \text{for } x > r \\ \psi &= \frac{D_1\,{}^{\delta}C_1 + D_3\,{}^{\delta}C_3}{D} && \text{for } t = 0, \quad x \geqslant r \\ \frac{\partial \psi}{\partial x} &= \frac{D_1}{D_3}\sqrt{\frac{l}{K}}\,\psi && \text{for } t > 0, \quad x = r \end{aligned}\right\} \tag{283}$$

where

$$D = \frac{D_1 + KD_3}{1 + K}; \qquad \psi = \frac{D_1 C_1 + D_3 C_3}{D}; \qquad l = k(1 + K)\frac{D}{D_1 D_3}$$

We shall note that the reaction layer approximation, on which both the approximate and the semirigorous procedures are based, can be applied with sufficient accuracy only if the equilibrium $B_3 \rightleftharpoons A_1$ is strongly shifted towards B_3, in which case K is $\gg 1$. Consequently,

$$D \cong D_3, \quad \psi_{x>r} \cong C_3, \quad l \cong \frac{kK}{D_1} \tag{284}$$

and the boundary value problem Eq. (283) becomes identical with the other Eqs. (276), (281), and (282). With the simplifications of Eq. (284), the solution of the diffusional problem (283) carried out by Koutecký (*31*) leads to an expression for the mean limiting current of the form,

$$\bar{i}_l = \bar{i}_d \bar{F}(\xi_1) \tag{285}$$

where

$$\xi_1 = \sqrt{\frac{3D_1 k t_d}{7D_3 K}}$$

$\bar{F}(\xi_1)$ is a power series in ξ_1. The reader is referred to the mathematical appendix for the derivation of Eq. (285). $\bar{F}(\xi_1)$ can be approximately expressed in closed form,

$$\bar{F}(\xi_1) = \frac{\xi_1}{1.5 + \xi_1} \tag{286}$$

Combination of Eqs. (285) and (286) yields

$$\frac{\bar{i}_l}{\bar{i}_d - \bar{i}_l} = 0.886\sqrt{\frac{D_1 k t_d}{D_3 K}} \tag{287}$$

Equation (287) is identical with the approximate expression given in Eq. (280) apart from the numerical factor.

Some interesting conclusions can be drawn from the approximate equation (280), where for simplicity D_1 is set equal to D_3. If the antecedent reaction is slow, $\bar{i}_l \ll \bar{i}_d$ and, in view of the definition of $\bar{i}_d$, Eq. (280) can be written as

$$\bar{i}_l = \frac{3}{5}\sqrt{\frac{7\pi}{12}\frac{kt_d}{K}}\,\bar{i}_d = -\frac{3}{5}\,nFA_O\sqrt{\frac{Dk}{K}}\,t_d^{2/3}\,{}^{\delta}C_3 \tag{288}$$

where $D = D_1 = D_3$. Comparing Eq. (288) with the general equation (247) and recalling the expression for the reaction layer thickness r given in Eq. (278), we readily see that the concentration of B_3 at $x = r$ is practically equal to the bulk concentration ${}^{\delta}C_3$. This situation is encountered when the rate of the antecedent chemical reaction is so slow that no appreciable depletion of B_3 around the electrode is observed during the flow of current. Hence diffusion overpotential is practically zero and the current is completely kinetic in nature. In view of Eq. (153), the area of the electrode surface at $t = 1$ sec equals $0.85\, m^{2/3}$, where m is the flowrate of mercury. Since m is directly proportional and the drop time t_d is inversely proportional to the height h of the mercury head, $\bar{\imath}_l$ does not depend on h. This represents an important criterion of distinction between purely prekinetic currents and diffusion currents, which latter are proportional to $h^{1/2}$. It is worth recalling that purely catalytic instantaneous currents are also proportional to the area of the instantaneous electrode surface, the instantaneous current density remaining constant during the drop growth. If the rate of the preceding reaction is sufficiently high, $\bar{\imath}_l$ cannot be neglected with respect to $\bar{\imath}_d$. In actuality Eqs. (280) and (287) show that when k increases beyond a certain limit, $\bar{\imath}_l$ coincides with $\bar{\imath}_d$ and is exclusively controlled by the diffusion of B_3 towards the electrode. At this point things proceed as if B_3 were directly discharged at the electrode. The transition from a purely kinetic current to the corresponding diffusion current is characterized by an increasing dependence of the current on the height of the mercury head.

Let us now examine a reaction scheme somewhat more general than that previously considered. We shall assume that the electroinactive species B_3 and B_4 interact according to the reversible reaction,

$$B_3 + B_4 \underset{kK}{\overset{k}{\rightleftharpoons}} A_1 + B_5 \tag{289a}$$

originating another electroinactive species B_5 and the electroactive species A_1. Here A_1 is oxidized to A_2 at the electrode surface. Depicting the reaction scheme as usual,

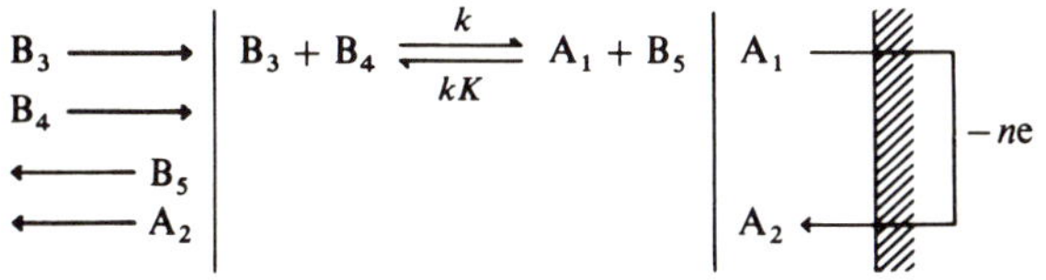

we see that all the species but A_1 contribute to diffusion overpotential. According to the general procedure of Sec. IV.C.2, the diffusion equation for

A_1 in the reaction layer can be written (see Eqs. (255) and (274)) as follows:

$$D_1 \frac{\partial^2 C_1}{\partial x^2} - kKC_1C_5 - kC_3C_4 \qquad \text{for } x < r \tag{290}$$

where C_3, C_4, and C_5 must be considered uniform within the reaction layer. Integration of Eq. (290) yields

$$\left(\frac{\partial C_1}{\partial x}\right)_{x=b} = \left\{\left[\frac{kKC_5C_1^2 - 2kC_3C_4C_1}{D_1}\right]_{^rC_1}^{^bC_1}\right\}^{1/2} \tag{291}$$

Under limiting current conditions $^bC_1 = 0$. Recalling that $^rC_1 = {^rC_3}{^rC_4}/(K{^rC_5})$, Eq. (291) becomes

$$\left(\frac{\partial C_1}{\partial x}\right)_{x=b} = {^rC_3}{^rC_4}\sqrt{\frac{k}{D_1K{^rC_5}}} \tag{292}$$

According to the Brdička-Wiesner model if the rate for the backward reaction may be neglected, the reaction layer thickness is obtained by equating $D_1(\partial C_1/\partial x)_{x=b}$ to $rk{^rC_3}{^rC_4}$. Hence $r = \sqrt{D_1/(kK{^rC_5})}$. From Eqs. (264) and (292) it follows that the approximate expression for the mean limiting current $\bar{\imath}_1$ is given by

$$\bar{\imath}_1 = -\frac{3}{5}nFA_0t_d^{2/3}\,{^rC_3}{^rC_4}\sqrt{\frac{D_1k}{K{^rC_5}}} \tag{293}$$

where $^rC_3$, $^rC_4$, and $^rC_5$ must be expressed in terms of $\bar{\imath}_1$ and of their respective bulk concentrations by the use of Eq. (249).

According to the semirigorous treatment we have in view of Eq. (267),

$$D_5^{1/2}({^\delta C_5} - {^rC_5}) = D_3^{1/2}({^rC_3} - {^\delta C_3}) = D_4^{1/2}({^rC_4} - {^\delta C_4}) \tag{294}$$

Equation (294) allows $^rC_5$ and $^rC_4$ to be expressed as functions of $^rC_3$ in Eq. (292),

$$\frac{D_3}{D_1}\left(\frac{\partial C_3}{\partial x}\right)_{x=r} = {^rC_3}\left[\frac{D_3^{1/2}}{D_4^{1/2}}({^rC_3} - {^\delta C_3}) + {^\delta C_4}\right] \times \sqrt{k\Big/\left\{D_1K\left[\frac{D_3^{1/2}}{D_5^{1/2}}({^\delta C_3} - {^rC_3}) + {^\delta C_5}\right]\right\}} \tag{295}$$

Therefore the diffusional problem reduces to the solution of the differential equation,

$$\frac{\partial C_3}{\partial t} = D_3 \frac{\partial^2 C_3}{\partial x^2} + \frac{2x}{3t}\frac{\partial C_3}{\partial x} \qquad x > r$$

with the initial condition $C_3 = {}^{\delta}C_3$ for $t = 0$, $x \geqslant r$ and the boundary condition (295) (*32*).

Let us consider an important particular case of the general scheme of Eq. (289),

$$\left.\begin{array}{l} \text{(a)}\ \mathrm{B_3} \underset{kK}{\overset{k}{\rightleftharpoons}} \mathrm{A_1} + \mathrm{B_5} \\ \text{(b)}\ \mathrm{A_1} \rightarrow \mathrm{A_2} + ne \end{array}\right\} \qquad (296)$$

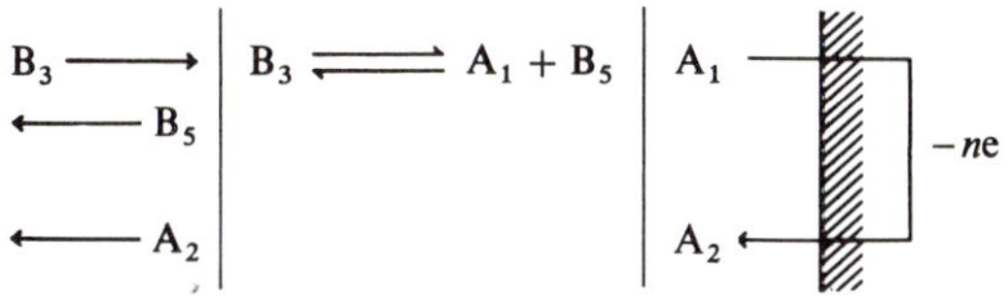

Here the electroinactive species B_3 dissociates into another electroinactive species B_5 and the electroactive species A_1. Equations (290) through (293) can be applied to the scheme of Eq. (296) by setting $C_4 \equiv 1$—that is, by ignoring the presence of C_4 in the expression for the rate of the forward reaction (289a). We shall assume that only C_3 is different from zero in the bulk of the solution (${}^{\delta}C_5 = 0$) and that equilibrium of Eq. (296a) is strongly shifted to the left. In view of Eqs. (249) and (293) the approximate expression for the limiting current is obtained by combining the following equations:

$$\bar{i}_l = -\frac{3}{5} nFA_0 t_d^{2/3}\, {}^{r}C_3 \sqrt{\frac{D_1 k}{K\, {}^{r}C_5}} \qquad (297)$$

$$\bar{i}_l = -\sqrt{\frac{12 D_3}{7\pi}}\, nFA_0 t_d^{1/6}({}^{\delta}C_3 - {}^{r}C_3) \qquad (298)$$

$$\bar{i}_l = -\sqrt{\frac{12 D_5}{7\pi}}\, nFA_0 t_d^{1/6}\, {}^{r}C_5 \qquad (299)$$

On setting $-\sqrt{12D_3/(7\pi)}\,nFA_O t_d^{1/6}\,{}^{\delta}C_3 \equiv \bar{i}_d$, where $\bar{i}_d$ expresses the value that the limiting current would attain if B_3 were electroactive, Eqs. (297) through (299) yield, after simple passages (*27*),

$$\frac{\bar{i}_1}{\bar{i}_d} = 0.815^{2/3}\beta^{1/3}\left(1 - \frac{\bar{i}_1}{\bar{i}_d}\right)^{2/3} \tag{300}$$

where

$$\beta = \frac{D_1}{D_3}\left(\frac{D_5}{D_1}\right)^{1/2} \frac{k}{K}\frac{t_d}{{}^{\delta}C_3}$$

b. Some Examples of Prekinetic Currents. Many organic compounds containing a carbonyl group are present in aqueous solution predominantly under their hydrated form. Usually this form is electroinactive and the rate of the dehydration reaction,

$$R_1R_2C(OH)_2 \underset{kK}{\overset{k}{\rightleftharpoons}} R_1R_2C{=}O + H_2O \tag{301}$$

partially controls the limiting current due to the electroreduction of the electroactive form $R_1R_2C{=}O$. Formaldehyde was the first compound exhibiting the above behavior to be studied by the polarographic technique. The height of the reduction wave of formaldehyde in buffer solutions is much lower than would be expected for a diffusion controlled current and is strongly influenced by temperature changes and by the pH of the solution (*33–35*). Temperature and pH do not appreciably affect the ratio of the dehydrated to the hydrated form in the bulk of the solution, which ratio is always $\ll 1$, but rather the rate at which the free aldehydic form consumed at the electrode is produced by dehydration of methyleneglycol (*34*). The rate constant k depends in general on the concentrations of all the acids A and bases B present in the solution, according to the relation,

$$k = k_0 + k_{H_3O^+}[H_3O^+] + k_{OH^-}[OH^-] + \sum k_A[A] + \sum k_B[B] \tag{302}$$

Here k_0 represents the contribution to k due to the solvent H_2O, which is itself a proton donor and a proton acceptor. In contrast the constant K

for the hydration equilibrium, which for formaldehyde amounts to $2.3\ 10^3$ (*36*), is independent of pH as well as of the nature and concentration of the buffer components. The electroreduction of formaldehyde in buffer solutions falls within the depolarization scheme of Eq. (273). By using the expression (280) or (287) for the limiting current, k can be determined for different pH values and buffer compositions (*37*). The values for k_0, $k_{H_3O^+}$, k_{OH^-}, k_A, k_B can be subsequently obtained from Eq. (302). Thus $k_{OH^-} = 1.3\ 10^3$ and $k_0 = 3.4\ 10^{-2}$ mole^{-1} l sec^{-1}. The high value for k_{OH^-} explains why the limiting current of formaldehyde increases sharply with increasing pH, reaching its maximum value at pH 13.15. For higher pH values the limiting current decreases again, due to the acid dissociation of formaldehyde and the consequent formation of its electroinactive anionic form (*34*).

Prekinetic waves analogous to that of formaldehyde have been observed for several other aldehydes and ketones (*38–43*).

Another very important class of prekinetic waves falling within the scheme of Eq. (273) is represented by the cathodic waves of the anions of many weak organic acids in buffered solutions. Here the current is controlled by the rate of protonation of the anions,

$$\begin{aligned} &\text{(a)}\ A^- + H^+ \rightleftharpoons HA \\ &\text{(b)}\ HA + e \rightarrow \text{products} \end{aligned} \qquad (303)$$

The fact that the undissociated acid HA is more easily reduced than the corresponding anion can be explained by the influence of the double layer structure on charge-transfer processes (*6*). The electroreduction usually occurs at potentials situated on the negative side of the electrocapillary maximum, and consequently the transfer of electrons from the electrode to the depolarizer is facilitated if this latter is neutral rather than negatively charged. Furthermore, the concentration of HA is higher than that of A^- in the pre-electrode layer. The prekinetic wave due to the reduction of the undissociated acid is usually followed at more negative potentials by the diffusion wave of the A^- ions, which are directly reduced at the electrode surface owing to the increased negative potential difference between the electrode surface and the plane of closest approach. The sum of the two waves remains practically constant with changing pH and corresponds to the current controlled by the diffusion of the acid, both in its dissociated and undissociated form, toward the electrode.

In order to relate the mechanism of Eq. (303) with the general depolarization scheme (273), let us set

$$K \equiv \frac{[A^-]}{[HA]} = \frac{K_a}{[H^+]} \tag{304}$$

where

$$K_a = \frac{[A^-][H^+]}{[HA]} \tag{305}$$

in the dissociation constant of the acid. A^- and HA correspond, respectively, to the electroinactive species B_3 and to the electroactive one A_1 of scheme (273). The hydrogen ion concentration is practically constant throughout the solution up to the electrode, provided the solution is well buffered. Consequently the rate constant k of scheme (273) is proportional to $[H^+]$,

$$k = k_H[H^+] \tag{306}$$

In view of Eq. (305) we have

$$\mathrm{pH} = \mathrm{p}K_a - \log \frac{[HA]}{[A^-]} = \mathrm{p}K_a + \log\left[\frac{[HA] + [A^-]}{[HA]} - 1\right] \tag{307}$$

If the pH is lower by one or more units than the pK_a of the acid, the analytical concentration of the acid practically coincides with the concentration of its undissociated form. Under these circumstances only one cathodic wave is observed, falling within the potential range where only HA is electroreducible and controlled by the diffusion of the undissociated acid. When pH $\cong$ pK_a, the bulk concentrations of HA and A^- are comparable, and the limiting current of the first cathodic wave is partially controlled by the diffusion of HA from the bulk of the solution and partially by the rate of reaction (303a), generating HA from the electroinactive anion. In general, under these pH conditions, the rate k is so high (cf. Eq. 306) that the first wave is still diffusion controlled and no second wave is observed. Usually the height of the first wave begins to decrease with respect to its maximum value only for pH values higher than the pK_a of the acid by one or two units—that is, when the bulk concentration of the undissociated molecules is much smaller than that of the anions. This represents a quite favorable circumstance, because only when $K \gg 1$ is the semirigorous treatment of the reaction scheme (273) applicable. With increasing pH the rate constant k, and consequently the height of the first wave, gradually decrease. At the same time the height of the second wave, occurring at more negative potentials, increases, so that the sum of the two waves remains constant. For sufficiently high pH values the rate of HA formation is so low that the first wave is no longer detected and only one wave is observed. This is controlled by the diffusion of the anions and falls within the more negative potential range where A^- is directly electroreduced.

On application of Eq. (287) to the present case we have, in view of Eqs. (304) and (306),

$$\frac{\bar{i}_l}{\bar{i}_d - \bar{i}_l} = \frac{\bar{i}_1}{\bar{i}_2} = 0.886[H^+]\sqrt{\frac{k_H t_d}{K_a}} \tag{308}$$

where the diffusion coefficients of A^- and HA have been set equal. The height of the second wave, denoted by $\bar{i}_2$, is equal to $\bar{i}_d - \bar{i}_1$, where $\bar{i}_1$, also designated by $\bar{i}_l$, is the kinetic limiting current expressing the height of the first wave. Rearranging Eq. (308), we obtain

$$pH = pK_a' - \log\frac{\bar{i}_1}{\bar{i}_2} = pK_a' + \log\left(\frac{\bar{i}_d}{\bar{i}_1} - 1\right) \tag{309}$$

where $pK_a' \equiv \log 0.886\sqrt{k_H t_d/K_a}$ expresses the pH at which the two waves have equal heights. The comparison of Eqs. (307) and (309) shows that on plotting $\bar{i}_1/\bar{i}_d$ against pH, an S-shaped curve is obtained that strictly resembles the dissociation curve—that is, the $[HA]/([HA] + [A^-])$ versus pH plot. This is the reason why K_a' is called the "apparent polarographic" dissociation constant. As a matter of fact the $\bar{i}_1/\bar{i}_d$ versus pH curve is strongly shifted towards higher pH values with respect to the corresponding dissociation curve (*26*, *44*). The identity between the two plots would be realized only if the attainment of the protonation equilibrium of Eq. (303a) were slow in the time scale of polarographic measurements. Under these circumstances, the first limiting current would be exclusively controlled by the diffusion of undissociated molecules ($\bar{i}_1 = \bar{K}[HA]^*$ and $\bar{i}_2 = \bar{K}[A^-]^*$, where $\bar{K} = \sqrt{12D/7\pi}\, nFA_0\, t_d^{1/6}$). The relation between the true pK_a of the acid and its apparent polarographic value pK_a',

$$pK_a' = \log 0.886\sqrt{k_H t_d} + \tfrac{1}{2}pK_a$$

allows the rate constant for the protonation reaction to be readily determined

(*22*, *26*, *45*). It must be noted that, in general, the rate constant k for the protonation of anions depends not only on the concentration of hydroxonium ions, but also on that of the other proton donors HA_i present in the solution as buffer components,

$$k = k_0 + k_H[H_3O^+] + \sum_i k_{HA_i}[HA_i] \tag{310}$$

k_0 is the contribution to k due to water molecules. Usually k_0 and $\sum_i k_{HA_i}[HA_i]$ are negligible with respect to $k_H[H_3O^+]$, but for relatively high buffer concentrations the contribution of the various buffer components can be detected (*46*, *47*). In this case the more general relation,

$$\frac{\bar{i}_l}{\bar{i}_d - \bar{i}_l} = 0.886\sqrt{\frac{t_d[H^+]}{K_a}\left(k_0 + k_H[H_3O^+] + \sum_i k_{HA_i}[HA_i]\right)}$$

derived from Eqs. (287), (304), and (310) must be used. By employing solutions of different buffer capacity, the partial protonation constants k_{HA_i} of the various proton donors can be determined (*46*, *48*, *49*).

Kinetic currents controlled by the rate of an antecedent protonation reaction are furnished by pyruvic acid, glyoxalic acid (*50*), terephtalic acid (*51*), nitrophtalic and nitroterephtalic acids (*52*), phenylglyoxalic acid (*44*), pyridinecarboxylic acid (*53*), quinolinecarboxylic acid (*54*), *p*-hydroxybenzaldehyde (*55*), and many other organic acids. Several nitrogen-containing substances forming onium compounds, such as cupferron (*56*), hydrazines, semicarbazones (*57*), and oximes (*58*) also give kinetic currents controlled by a protonation reaction. Maleic acid yields two consecutive kinetic waves (*59*). The first, more positive wave corresponds to the electroreduction of the neutral dibasic acid H_2A and its limiting current is controlled by the two consecutive reactions,

$$A^{2-} + H^+ \rightleftharpoons HA^- + H^+ \rightleftharpoons H_2A; \qquad H_2A \rightarrow \text{products}$$

The other wave is due to the direct electroreduction of the monovalent anion HA^-, and its kinetic nature is explained with the antecedent protonation reaction: $A^{2-} + H^+ \rightleftharpoons HA^-$. This latter wave falls within the scheme of Eq. (303) and its limiting current is adequately expressed by Eq. (308). The expression for the kinetic limiting current of the first wave furnished by dibasic acids has been derived both by the approximate (*59*) and the semirigorous method (*60*)).

An unbuffered solution of boric acid yields a cathodic current due to the discharge of the hydrogen ions originated from the dissociation of the acid (*61*)),

$$\left.\begin{array}{l}\text{(a) } HA \underset{kK}{\overset{k}{\rightleftharpoons}} H^+ + A^- \\ \text{(b) } H^+ + e \rightarrow H\end{array}\right\} \tag{311}$$

The limiting current for the foregoing process is partially controlled by the rate of reaction (311a). Such a current does not fall within the reaction scheme (303), or more generally Eq. (273), since the backward reaction is bimolecular and cannot be treated as a pseudomonomolecular reaction, as is done for the forward reaction (303a). In other words the contribution of A^- to diffusion overpotential cannot be disregarded. The process (311) is a particular case of the general scheme (296) and, in fact, the cathodic limiting current of boric acid obeys Eq. (300) (*61*). Another case in which Eq. (300) is employed is the electroreduction of an undissociated weak acid HA, following the protonation of the anion A^- in an unbuffered solution of a salt of the weak acid with a strong base. If the acid is only moderately weak ($pK < 8$), the hydrolysis equilibrium,

$$A^- + H_2O \underset{Kk_0}{\overset{k_0}{\rightleftharpoons}} HA + OH^- \tag{312}$$

is shifted to the left to such an extent that the bulk concentration of HA is negligible compared to that of A^-. On the other hand the concentration of hydrogen ions is so low that even if the rate of the protonation reaction within the reaction layer were controlled by the diffusion of H^+ ions toward the electrode, the calculated limiting current, equal to the diffusion current of hydrogen ions, would be practically negligible with respect to that actually observable. Hence, under the present conditions the only effective proton donor is water, and hydroxyl ions are formed as products of the preceding chemical reaction. These tend to shift the equilibrium of Eq. (312) to the left and consequently their contribution to diffusion overpotential in the diffusion layer must be taken into account. Clearly the electroreduction of HA coupled with the preceding reaction (312) is a particular case of the depolarization scheme of Eq. (296). The corresponding kinetic current is usefully employed for the determination of the rate constant k_0 for protonation by water, (*62*), which would be hardly obtained in buffer solutions, where the contributions of buffer components and of H_3O^+ ions to the actual protonation rate constant (310) are predominant.

c. Decomposition of an Electroinactive Dimer into an Electroactive Monomer. A few cases are known (*63–65*) in which the preceding reaction controlling the overall electrode process consists in the decomposition of an electroinactive dimer yielding an electroactive monomer. The general reaction scheme is

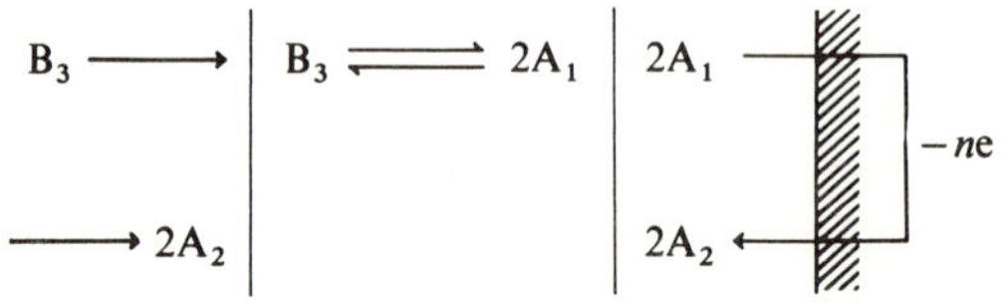

In order to derive the reaction layer thickness, we shall consider the diffusion equation,

$$D_1 \frac{\partial^2 C_1}{\partial x^2} = (kKC_1^2 - kC_3) \qquad x \le r \tag{314}$$

valid under steady-state conditions (cf. Eq. (255)). In view of Eq. (257) the integration of Eq. (314) yields

$$\left(\frac{\partial C_1}{\partial x}\right)_{x=b} = \left\{\left[\frac{2}{D_1}\left(\frac{kK}{3}C_1^3 - k^rC_3C_1\right)\right]_{^rC_1}^{^bC_1}\right\}^{1/2} \tag{315}$$

Under limiting current conditions $^bC_1 = 0$. $^rC_1$ is derived from the expression for the law of mass action, $K = {}^rC_3/{}^rC_1^2$, which holds at the outer boundary of the reaction layer. Substituting $^bC_1$ and $^rC_1$ in Eq. (315) and rearranging terms, we obtain

$$\left(\frac{\partial C_1}{\partial x}\right)_{x=b} = \frac{2D_3}{D_1}\left(\frac{\partial C_3}{\partial x}\right)_{x=r} = 2\sqrt{\frac{k}{3D_1K^{1/2}}}\,{}^rC_3^{3/4} \tag{316}$$

The equality between the first two members of Eq. (316) follows immediately from Eq. (268). The reaction layer thickness r' according to the Hanuš model is obtained from Eq. (316) by placing $(\partial C_1/\partial x)_{x=b} = {}^rC_1/r' = {}^rC_3^{1/2}/(K^{1/2}r')$. Hence

$$r' = \frac{1}{2}\sqrt{\frac{3D_1}{k(K\,{}^rC_3)^{1/2}}} \tag{317}$$

On the basis of Wiesner's original concept of reaction layer, Hanuš (*27*) writes instead†,

$$r' \geqslant \frac{1}{2}\sqrt{\frac{D_1}{k(K\,{}^rC_3)^{1/2}}} \tag{318}$$

thus neglecting the numerical factor $\sqrt{3}$. This amounts to setting the flux of the electroactive species A_1 at $x = b$ equal to

$$D_1\left(\frac{\partial C_1}{\partial x}\right)_{x=b} = 2\sqrt{D_1 \frac{k}{K^{1/2}}}\,{}^rC_3^{3/4} \tag{319}$$

Writing

$$\bar{i}_1 = -\left(\frac{3}{10}\right) nFA_0\, t_d^{2/3} D_1\left(\frac{\partial C_1}{\partial x}\right)_{x=b} \tag{320}$$

in accordance with Eq. (264), and recalling in view of Eq. (249) that

$$\bar{i}_1 = -\sqrt{\frac{12D_3}{7\pi}}\, nFA_0\, t_d^{1/6}({}^\delta C_3 - {}^rC_3) \tag{321}$$

we have, after combination of Eqs. (319) through (321),

$$\frac{\bar{i}_1}{\bar{i}_d} = 0.81\sqrt{\frac{D_1 k t_d}{D_3(K\,{}^\delta C_3)^{1/2}}}\left(1 - \frac{\bar{i}_1}{\bar{i}_d}\right)^{3/4} \tag{322}$$

Here $\bar{i}_d = -\sqrt{12D_3/(7\pi)}\, nFA_0\, t_d^{1/6}\, {}^\delta C_3$ is the limiting current that would be observed were B_3 electroactive. A more correct approximate expression for the current-potential characteristic can be derived if Eq. (316) is used in place of Eq. (319).

The last two members of Eq. (316) constitute the boundary condition at $x = r$ for the diffusion equation,

$$\frac{\partial C_3}{\partial t} = D_3 \frac{\partial^2 C_3}{\partial x^2} + \frac{2x}{3t}\frac{\partial C_3}{\partial x} \qquad x > r \tag{323}$$

† In reality, Hanuš writes $r' \geqslant \sqrt{D_1/[k(K\,{}^rC_3)^{1/2}]}$, but since he improperly assumes that $\bar{i}_l = -\frac{3}{5}\, nFA_0\, t_d^{2/3}\, D_1\,(\partial C_1/\partial x)_{x=b}$, the final formula (322), which he derives, is identical with that which would be obtained by the correct position (320) and by giving r' the value in Eq. (318).

according to the semirigorous procedure (cf. Eq. (266)). The boundary-value problem of Eqs. (316) and (323), apart from minor differences, has been solved by Koutecký and Hanuš (*66*) by following the mathematical procedure outlined in the appendix. The above authors set $\psi = (2C_3 + C_1)/2$ instead of C_3 in Eqs. (316) and (323), but since equilibrium must be strongly shifted towards the left for the semirigorous procedure to be consistently applied, we have $\psi \cong C_3$ at $x \geqslant r$.

The electrooxidation of dithionite in neutral and alkaline media offers an example of the depolarization scheme of Eq. (313) (*63*, *64*). At low temperatures ($\cong$0°C) only an anodic wave due to the direct oxidation of the dimer $S_2O_4^{2-}$ is observed. With increasing temperature, the rate of the monomer formation reaction,

$$S_2O_4^{2-} \underset{kK}{\overset{k}{\rightleftharpoons}} 2SO_2^- \tag{324}$$

increases and two new waves of equal heights appear. These waves, one anodic and the other cathodic, are not diffusion controlled and fall at potentials more negative than those corresponding to the direct electrooxidation of $S_2O_4^{2-}$. They are attributed, respectively, to the electrooxidation and to the electroreduction of the monomer SO_2^- formed through reaction (324). A further increase in temperature causes the anodic wave of SO_2^- to grow at the expense of the more positive anodic wave of $S_2O_4^{2-}$, the sum of these two waves varying with temperature as an ordinary diffusion-controlled wave.

2. Catalytic Currents

a. A General Depolarization Scheme. In Sec. IV.B we have already examined parallel reactions in connection with the rigorous procedure. Here we shall consider the application of the approximate and semirigorous procedures to the more general reaction scheme,

$$\left.\begin{array}{l} \text{(a)}\ A_2 + B_3 \underset{kK}{\overset{k}{\rightleftharpoons}} A_1 \\ \text{(b)}\ A_1 \rightarrow A_2 + ne \end{array}\right\} \tag{325}$$

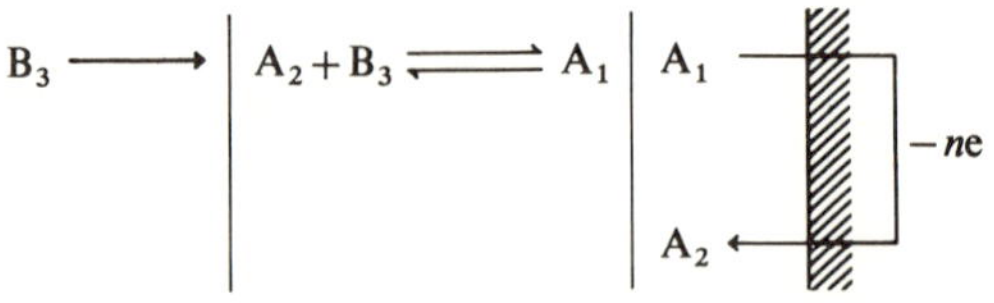

B_3 is an electroinactive substance reacting with the product A_2 of the charge-transfer process and regenerating the depolarizer A_1. We shall assume that B_3 is not present in excess with respect to the other diffusing species. Consequently, the contribution of B_3 to diffusion overpotential must be taken into account.

We must note that while the occurrence of prekinetic currents depends on the nature of the depolarizer (weak acids, carbonyl compounds), kinetic currents characterized by a parallel reaction are artificially produced by introducing into the solution of an ordinary depolarizer A_1 an electro-inactive substance B_3 with a strong oxidizing or reducing power. The direct electroreduction (or oxidation) of B_3 must require a high overvoltage in order that the depolarization process (325) may take place. In fact, if both A_1 and B_3 are reversibly reduced (or oxidized) at the electrode, it is evident by simple thermodynamic considerations that no potential range can exist where the depolarization process of Eq. (325) is feasible. Usually a low concentration of A_1 is sufficient to cause a sort of "indirect" electroreduction (or oxidation) of B_3, which otherwise would not take place. This is the reason why A_1 is called "catalyst" and kinetic currents of the present type are termed "catalytic." The possibility of "creating" catalytic currents by a proper choice of A_1 and B_3 allows the polarographic method to be advantageously employed in the study of fast redox reactions. Unfortunately the limitations imposed on this method by the necessity of disposing of a substance B_3 that is electroreduced (or oxidized) with a high overvoltage are severe.

In the case under study neither A_1 nor A_2 contribute to diffusion overpotential outside the reaction layer and consequently the simplified diffusion equation (254), valid under steady-state conditions, can be applied to both these species. Thus for A_1 we write

$$D_1\left(\frac{\partial^2 C_1}{\partial x^2}\right) = kKC_1 - kC_2 C_3 \tag{326}$$

If we assume that the diffusion coefficients of A_1 and A_2 are equal, we have, in view of Eq. (272),

$$C_1 + C_2 = C^* = {}^{\delta}C_1 1 + \frac{K}{{}^{\delta}C_3} \tag{327}$$

where C^* is the sum of the bulk concentrations of A_1 and A_2. On substituting

C_2 from Eq. (327) into Eq. (326) and integrating this latter equation once, we obtain

$$\left(\frac{\partial C_1}{\partial x}\right)_{x=b} = \left\{\left[\frac{k}{D_1}[(K + {}^rC_3)C_1^2 - 2C^* {}^rC_3 C_1]\right]_{^rC_1}^{^bC_1}\right\}^{1/2} \tag{328}$$

At $x = r$ reaction (325a) attains equilibrium. Hence, taking Eq. (327) into account, we have ${}^rC_1 = C^* {}^rC_3/(K + {}^rC_3)$. Noting that under limiting current conditions ${}^bC_1 = 0$, and substituting ${}^rC_1$ and ${}^bC_1$ in Eq. (328) yields

$$\left(\frac{\partial C_1}{\partial x}\right)_{x=b} = \sqrt{\frac{k}{D_1(K + {}^rC_3)}}\, C^* {}^rC_3 \tag{329}$$

Usually the equilibrium of Eq. (325a) is strongly shifted in favor of A_1 in the bulk of the solution: that is, $K/{}^\delta C_3 = {}^\delta C_2/{}^\delta C_1 \ll 1$. Hence we have $C^* \cong {}^\delta C_1$ and Eq. (329) takes the simplified form,

$$\left(\frac{\partial C_1}{\partial x}\right)_{x=b} = \sqrt{\frac{k\, {}^rC_3}{D_1}}\, C^* \tag{330}$$

It is readily seen that Eq. (330) could also be obtained by disregarding the rate of the backward reaction $A_1 \xrightarrow{kK} A_2 + B_3$ in the diffusion equation (326),

$$D_1\left(\frac{\partial^2 C_1}{\partial x^2}\right) = -kC_2 C_3 \tag{331}$$

Integration of Eq. (331) yields

$$\left(\frac{\partial C_1}{\partial x}\right)_{x=b} = \left\{\left[\frac{k}{D_1}({}^rC_3 C_1^2 - 2C^* {}^rC_3 C_1)\right]_{^rC_1}^{^bC_1}\right\}^{1/2} \tag{332}$$

Noting that at $x = r$ practically all the molecules of A_2 produced at the electrode are transformed again into A_1 due to the action of B_3, from Eq. (327) it follows that ${}^rC_1 \cong C^*$. On substituting ${}^rC_1$ in Eq. (332) and setting ${}^bC_1 = 0$, Eq. (330) is immediately obtained. The possibility of neglecting the backward reaction is typical of catalytic and postkinetic currents with respect to prekinetic currents. The consideration of unidirectional chemical reactions results in the advantage that the equilibrium constant K is eliminated from the diffusional problem (cf. Eqs. (329) and (330)). If the chemical reaction precedes the charge-transfer proper, as in the reaction scheme (273), the rate for the backward reaction cannot be ignored because this would amount to assuming that the electroinactive species B_3 is transformed into the electroactive one A_1 by a rapid unidirectional reaction in the very bulk of the solution. The immediate consequence of this state of affairs is that with prekinetic currents the knowledge of the rate constant k for the forward reaction $B_3 \rightarrow A_1$ is subordinated to the previous knowledge of the equilibrium constant $K = {}^{\delta}C_3/{}^{\delta}C_1$.

Returning to reaction scheme (325), the expression for the current according to the approximate procedure is obtained by making use of Eqs. (249) and (264), which are now written as

$$\bar{i}_1 = -\sqrt{\frac{12D_3}{7\pi}}\, nFA_0\, t_d^{1/6}({}^{\delta}C_3 - {}^{r}C_3) = \bar{i}_{d,3}\left(1 - \frac{{}^{r}C_3}{{}^{\delta}C_3}\right) \tag{333}$$

$$\bar{i}_1 = -\frac{3}{5}\, nFA_0\, t_d^{2/3}\sqrt{D_1 k\, {}^{r}C_3\, C^*} \tag{334}$$

In Eq. (333)

$$\bar{i}_{d,3} = -\sqrt{\frac{12D_3}{(7\pi)}}\, nFA_0\, t_d^{1/6}\, {}^{\delta}C_3$$

is the diffusion limiting current furnished by B_3 if it were electroactive and A_1 were missing. Noting that the diffusion limiting current $\bar{i}_d$ of A_1 in the absence of B_3 is given by

$$\bar{i}_d = -\sqrt{\frac{12D_1}{7\pi}}\, nFA_0\, t_d^{1/6} C^* \tag{335}$$

and combining Eqs. (333) through (335), we obtain the final expression (*67*),

$$\frac{\bar{i}_1}{\bar{i}_d} = 0.81\sqrt{kt_d\, {}^{\delta}C_3\left(1 - \frac{\bar{i}_1}{\bar{i}_{d,3}}\right)} \tag{336}$$

If the semirigorous procedure is followed, on noting from Eq. (268) that $D_3(\partial C_3/\partial x)_{x=r} = D_1(\partial C_1/\partial x)_{x=b}$ and making use of Eq. (330), we have

$$D_3\left(\frac{\partial C_3}{\partial x}\right)_{x=r} = \sqrt{D_1 k\, {}^{r}C_3\, C^*} \tag{337}$$

Equation (337) is the boundary condition at $x = r$ for the diffusion equation,

$$\frac{\partial C_3}{\partial x} = D_3 \frac{\partial^2 C_3}{\partial x^2} + \frac{2x}{3t}\frac{\partial C_3}{\partial x} \qquad x > r$$

Koutecký, in considering the present problem (*68*), made some erroneous assumptions in the derivation of the boundary condition. From the general treatment of kinetic currents in polarography formulated by the above author et alii (*30*), a boundary condition practically identical with Eq. (337) can be readily deduced.

If B_3 is present in the solution in strong excess with respect to A_1, $\bar{i}_{d,3} \gg \bar{i}_1$ and Eq. (336) simplifies as

$$\frac{\bar{i}_1}{\bar{i}_d} = 0.81\sqrt{kt_d{}^{\delta}C_3} \tag{338}$$

It should be noted that the validity of Eq. (338) does not depend exclusively on the value for the ratio of the bulk concentrations of B_3 and A_1, it also depends on the value for the rate of the regeneration reaction. The higher this rate, the larger the ratio ${}^{\delta}C_3/{}^{\delta}C_1$ required for a proper application of Eq. (338). This equation corresponds to a pseudo-first-order regeneration reaction of the type,

$$A_2 \xrightarrow{k^{\delta}C_3} A_1; \qquad A_1 \rightarrow A_2 + ne \tag{339}$$

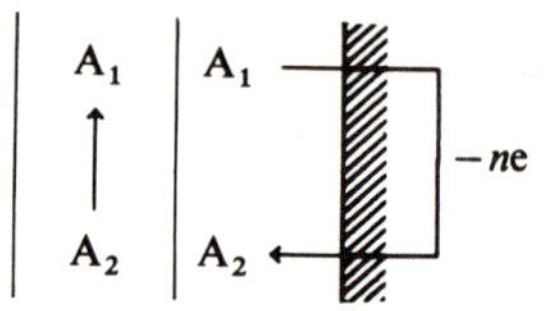

Here no substance contributes appreciably to diffusion overpotential outside the reaction layer. Scheme (339) was already considered in Sec. IV.B.1 in connection with the rigorous procedure. The approximate expression (338) practically coincides with the rigorous Eq. (214), valid for very high reaction rates, provided we note that with the present notations ξ_1 equals $k^{\delta}C_3(1 + K/{}^{\delta}C_3)t_d$ and that $K/{}^{\delta}C_3$ is $\ll 1$.

b. Some Examples of Catalytic Currents. Although many examples of catalytic currents are reported in the literature, the number of electroinactive additives of the B_3 type employed—usually characterized by a strong oxidizing power—is relatively minor. The electroinactive additive more widely used is undoubtedly hydrogen peroxide, although hydroxylamine, chloric, perchloric, and nitric acid have also been frequently employed. It is the great variety of catalysts (namely, oxidants that are electroreduced reversibly on mercury, giving rise to species that are chemically oxidized in the solution by the electroinactive substance of the B_3 type) which is responsible for the large number of catalytic currents. Ferric ion, either free (*69*) or bound in a complex, as in hemoglobin, hematin (*70*, *71*) myoglobine (*72*), ferri-triethanolamine (*73*), and Fe^{III} (EDTA) (*74*), is an excellent catalyst towards H_2O_2 reduction. The depolarization scheme is as follows,

$$Fe^{II} + 1/2\,H_2O_2 + H^+ \xrightarrow{k} Fe^{III} + H_2O \qquad (340)$$
$$Fe^{III} + e \rightleftharpoons Fe^{II}$$

Since the parallel reaction is first order both with respect to Fe^{II} and to H_2O_2, it seems to be composed of the two consecutive elementary steps,

$$Fe^{II} + H_2O_2 \xrightarrow{k/2} Fe^{III} + OH^- + \dot{O}H \qquad (341)$$
$$Fe^{II} + \dot{O}H \rightleftharpoons Fe^{III} + OH^- \qquad (342)$$

of which only the first one is rate determining. The second step, which is practically at equilibrium, consumes a further ferrous ion. Hence the rate $-d[Fe^{II}]/dt$ of the overall chemical reaction (340) is twice that of the rate-determining step (341). The mechanism (341)–(342) seems to be confirmed by the experimental observation that substances capable of reacting with $\dot{O}H$ radicals but not with H_2O_2 molecules (e.g., alcohols, acrylonitrile, acetic acid) lower the catalytic wave of Fe^{III} (*69*). In general, the catalytic waves of iron (III) complexes in the presence of H_2O_2 decrease in strongly alkaline media, owing to the reduced oxidizing power of HO_2^- with respect to that of undissociated H_2O_2 (*70*, *71*). Besides Fe^{III}, molybdenum, tungsten, and vanadium catalyze the reduction of H_2O_2 on mercury (*75*). In an analogous way the exavalent osmium deriving from the electroreduction of osmium tetroxide on mercury is reoxidized to OsO_4 by the agency of H_2O_2, giving rise to a catalytic current (*76*). A catalytic current is also encountered in the electroreduction of aquoamine complexes of Co^{III} to Co^{II} in the presence of hydrogen peroxide due to the regeneration of Co^{III} from Co^{II} (*77*). Besides H_2O_2, hydroxylamine is also electroreduced with a large overvoltage, so that several reversible redox couples may be employed for catalyzing the reduction of NH_2OH at more positive potentials. Thus in the presence of the triethanolamine complex of Fe^{III}, the reduction of NH_2OH occurs at the potential of electroreduction of Fe^{III} to Fe^{II} according to the mechanism (*78*),

$$Fe^{II} + \tfrac{1}{2}NH_2OH + \tfrac{1}{2}H_2O \xrightarrow{k} Fe^{III} + \tfrac{1}{2}NH_3 + OH^-$$
$$Fe^{III} + e \rightarrow Fe^{II}$$

Here, too, as in the $Fe^{II} - H_2O_2$ system, the chemical reaction is first order both with respect to Fe^{II} and to NH_2OH, so that it seems to proceed through the two consecutive steps,

$$Fe^{II} + NH_2OH \xrightarrow{k/2} Fe^{III} + OH^- + \dot{N}H_2$$
$$Fe^{II} + \dot{N}H_2 + H_2O \rightleftharpoons Fe^{III} + NH_3 + OH^-$$

The first step is rate determining, whereas the second, involving the highly reactive $\dot{N}H_2$ radical, is practically at equilibrium. The reduction of hydroxylamine is also catalyzed by the oxalate complex of Ti^{IV} (*79, 80*). In this case Ti^{III}, deriving from the electroreduction of Ti^{IV}, is reoxidized to Ti^{IV} through the elementary reaction,

$$Ti^{III} + NH_2OH \xrightarrow{k} Ti^{IV} + OH^- + \dot{N}H_2$$

The $\dot{N}H_2$ radical so formed seems to react more rapidly with the excess of oxalic acid than with Ti^{III}. This may explain why, according to the stoichiometry of the overall regeneration reaction, one mole of NH_2OH reacts with only one mole of Ti^{III}. Other catalytic currents are encountered in the electroreduction of the oxalate complex of Ti^{IV} to Ti^{III} (*81*), of Cr^{III} to Cr^{II} (*82*), or Mo^{VI} to Mo^{V} (*83*), and of W^{VI} to W^{V} (*84, 85*) in the presence of chloric acid, which reoxidizes the product of the electroreduction thus regenerating the depolarizer. In an analogous way the reduction of perchloric acid is catalyzed by the electroreduction of W^{VI} to W^{V} (*86*), and of Mo^{VI} to Mo^{V} (*87–91*), whereas nitric acid reoxidizes chemically the product of the electroreduction of W^{VI} to W^{V}, of Mo^{VI} to Mo^{V} (*92*), and of U^{VI} to U^{III} (*93*). The regeneration of the depolarizer from the electrode product can also occur by the agency of suspensions. Thus the quinone produced in the electrooxidation of hydroquinone at a dropping mercury electrode is reduced again to hydroquinone by the hydrogen absorbed in a suspension of colloidal palladium (*94*). Analogously, during the electroreduction of $Ag(CN)_2^-$ in the presence of colloidal AgBr (*95*), the cyanide ions liberated at the electrode regenerate $Ag(CN)_2^-$ from AgBr according to the reaction,

$$AgBr + 2CN^- \rightarrow Ag(CN)_2^- + Br^-$$

Many of the regeneration reactions previously described are so fast they can be treated as pseudo-first order only if the electroinactive additive is in strong excess with respect to the catalyst. With increasing concentration of the catalyst, the catalytic limiting current tends to coincide with the diffusion limiting current of the electroinactive additive. At the same time the instantaneous limiting current, which at low concentrations of the catalyst is proportional to $t^{2/3}$, tends to become proportional to $t^{1/6}$.

3. Postkinetic Currents

Postkinetic currents are originated from a chemical reaction inactivating the product of the charge-transfer step. The number of postkinetic currents reported in the literature is much less than that of prekinetic and catalytic currents. This is partly due to the fact that the requirements for the detection of subsequent reactions are more stringent than those for preceding and parallel reactions. Thus prekinetic and catalytic currents are usually studied under limiting conditions because the complications deriving from a nonzero charge-transfer overpotential are avoided. This is not feasible with postkinetic currents because under limiting conditions these currents are exclusively controlled by the diffusion of the depolarizer towards the electrode. Consider for example the reaction scheme,

$$\left.\begin{array}{l}\text{(a)}\ A_2 \underset{k_b}{\overset{k_f}{\rightleftharpoons}} A_1 + ne \\ \text{(b)}\ A_1 \underset{kK}{\overset{k}{\rightleftharpoons}} B_3\end{array}\right\} \qquad (343)$$

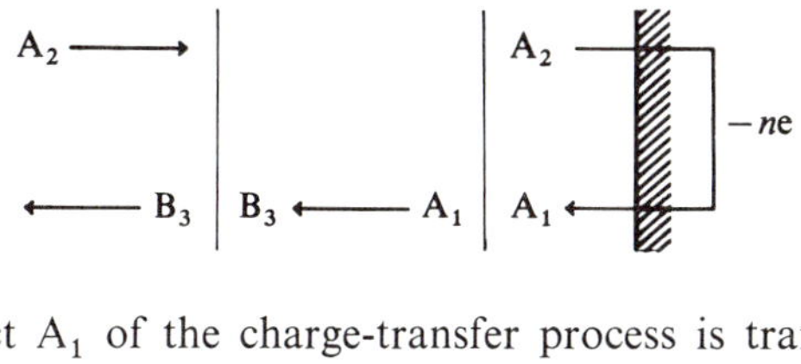

where the product A_1 of the charge-transfer process is transformed to the electroinactive species B_3. It is intuitive that the fate of A_1, whether it diffuses away from the electrode or is involved in the inactivation reaction (343b), does not exert any influence on the limiting current. This situation is clearly reflected by the boundary value problem, which is expressed by the diffusion equation relative to A_2 uncomplicated by kinetic terms,

$$\frac{\partial C_2}{\partial t} = D_2 \frac{\partial^2 C_2}{\partial x^2} + \frac{2x}{3t}\frac{\partial C_2}{\partial x}$$

with the boundary condition

$$^{b}C_2 = 0 \tag{344}$$

and the initial condition $C_2 = {}^{\delta}C_2$ for $t = 0$, $x \geqslant b$. Noting that $\bar{\imath} = -nFA\, D_2 \times (\partial C_2/\partial x)_{x=b}$, the solution of the above boundary value problem leads directly to the well-known Ilkovic equation,

$$\bar{\imath} = -\sqrt{\frac{12D_2}{7\pi}}\, nFA_0\, t_{\mathrm{d}}^{1/6}\, {}^{\delta}C_2$$

An analogous situation is encountered along the rising portion of the polarographic wave if the charge-transfer step is so slow that the rate for the backward electrode process can be neglected with respect to that for the forward. In this case the boundary condition (344) is replaced by the other,

$$D_2\left(\frac{\partial C_2}{\partial x}\right)_{x=b} = k_{\mathrm{f}}\, {}^{b}C_2 \tag{345}$$

which, like Eq. (344), does not contain the concentration of A_1. Consequently, the diffusional problem can be solved without taking into account the diffusion equations of A_1 and B_3, which contain the kinetic term. In conclusion, only if the charge-transfer process is quasi reversible, or, still better, reversible, can a subsequent reaction be detected through the investigation of the rising portion of the polarographic wave.

The recourse to relaxation techniques does not offer particular advantages over polarography. On the contrary the use of the potentiostatic technique, which is advantageous in the study of very fast preceding reactions, would increase the contribution of charge-transfer overpotential with respect to that of reaction overpotential, thus obscuring the features of postkinetic currents. An ingenious procedure for the study of postkinetic currents consists in keeping the applied potential at a value corresponding to the diffusion limiting current of A_2 for a short time, during which the electrode product A_1 together with its electroinactive form B_3 accumulate around the electrode (*96*). Subsequently the potential is caused to jump to a value at which A_1 is instantaneously retransformed into A_2 as soon as it reaches the electrode surface. This way the postkinetic current is converted into a prekinetic one.

Let us derive the expression of the polarographic current for the reaction scheme (343). The application of Eq. (254) to the substance not contributing to diffusion overpotential, namely A_1, results in

$$D_1 \frac{\partial^2 C_1}{\partial x^2} = kC_1 \tag{346}$$

where the rate of the backward reaction $B_3 \xrightarrow{kK} A_1$ has been neglected. On integrating Eq. (346) once through the procedure outlined in Eqs. (255) through (257), we obtain

$$\left(\frac{\partial C_1}{\partial x}\right)_{x=b} = -\left\{\left[\frac{k}{D_1} C_1^2\right]_{^rC_1}^{^bC_1}\right\}^{1/2} \tag{347}$$

Owing to the postulated unidirectional character of reaction (343b), $^rC_1$ is practically zero. If the charge-transfer process (343a) is quasi reversible, application of Eq. (259) yields

$$-D_1\left(\frac{\partial C_1}{\partial x}\right)_{x=b} = k_f{}^bC_2 - k_b{}^bC_1 \tag{348}$$

where k_f and k_b are the potential dependent rate constants for the forward and backward electrode processes. Combination of Eqs. (347) and (348) results in

$$\sqrt{D_1 k}\,{}^bC_1 = k_f{}^bC_2 - k_b{}^bC_1 \tag{349}$$

From Eq. (349) $^bC_1$ is obtained as a function of the concentration of A_2, which is taken as constant throughout the reaction layer ($^bC_2 \cong {}^rC_2$). On substituting $^bC_1$ in Eq. (347) we have

$$-D_1\left(\frac{\partial C_1}{\partial x}\right)_{x=b} = \frac{\sqrt{D_1 k}\, k_f{}^rC_2}{k_b + \sqrt{D_1 k}} \tag{350}$$

Confining ourselves to the case of a perfectly reversible charge-transfer process, let k_f and k_b tend to infinity while keeping their ratio equal to

$$\frac{k_f}{k_b} = \frac{{}^bC_1}{{}^bC_2} = \theta = \exp\left[\frac{nF}{RT}(E - E^{\circ})\right] \tag{351}$$

E° is the standard potential of the A_1/A_2 couple. Hence Eq. (350) becomes

$$-D_1\left(\frac{\partial C_1}{\partial x}\right)_{x=b} = \sqrt{D_1 k}\,\theta\,{}^rC_2 \tag{352}$$

The expression for the thickness r' of the Hanuš reaction layer is readily obtained by equating the right-hand side of Eq. (352) to $D_1\,{}^bC_1/r'$ and noting that $\theta\,{}^rC_2 \cong \theta\,{}^bC_2 = {}^bC_1$,

$$r' = \sqrt{\frac{D_1}{k}}$$

The expression for the polarographic current according to the approximate procedure is directly obtainable from Eq. (352) by making use of Eqs. (264) and (249), which for the present case are written as,

$$\bar{\imath} = \frac{3}{5}\,nFA_0\,t_d^{2/3}D_1\left(\frac{\partial C_1}{\partial x}\right)_{x=b} \tag{353}$$

$$\bar{\imath} = -\sqrt{\frac{12D_2}{7\pi}}\,nFA_0\,t_d^{1/6}({}^{\delta}C_2 - {}^rC_2) \tag{354}$$

Substituting ${}^rC_2$ from Eq. (354) into Eq. (352) and $D_1(\partial C_1/\partial x)_{x=b}$ from Eq. (352) into Eq. (353) we have, after rearrangement,

$$\frac{\bar{\imath}}{\bar{\imath}_d} = 0.81\sqrt{\frac{D_1}{D_2}kt_d}\left(1 - \frac{\bar{\imath}}{\bar{\imath}_d}\right)\theta \tag{355}$$

where $\bar{\imath}_d = -\sqrt{12D_2/(7\pi)}\,nFA_0\,t_d^{1/6}\,{}^{\delta}C_2$ is the diffusion limiting current that is actually reached when $\theta \to \infty$. In view of Eq. (351), Eq. (355) can also be written in the more convenient form (97),

$$E = E^0 - \frac{RT}{nF}\ln\frac{\bar{\imath}_d - \bar{\imath}}{\bar{\imath}} - \frac{RT}{nF}\ln 0.81\sqrt{\frac{D_1}{D_2}kt_d} \tag{356}$$

From Eq. (356) it is manifest that the shape of the wave is identical with that of a simple reversible wave in the absence of the inactivation reaction. The shape of this latter wave is described by the well-known equation,

$$E = E^{\circ} - \frac{RT}{nF} \ln \frac{\bar{i}_{\mathrm{d}} - \bar{i}}{\bar{i}}$$

which is readily obtained from Eq. (356) by letting k tend to zero. One of the main features of postkinetic waves is that their half-wave potential

$$E_{1/2} = E^{\circ} - \frac{RT}{nF} \ln 0.81 \sqrt{\frac{D_1}{D_2} k t_{\mathrm{d}}}$$

is shifted in the direction opposite to that of increasing overvoltages with respect to the standard potential E°. This shift, which increases slightly with an increase in the drop time, is due to the rapid subtraction of A_1 from the charge-transfer equilibrium $A_2 \rightleftharpoons A_1 + ne$, as a consequence of B_3 formation.

According to the semirigorous procedure, the polarographic current for the quasi-reversible case is obtained by solving the diffusion equation,

$$\frac{\partial C_2}{\partial t} = D_2 \frac{\partial^2 C_2}{\partial x^2} + \frac{2x}{3t} \frac{\partial C_2}{\partial x} \qquad \text{for } x > r \tag{357}$$

The boundary condition at $x = r$ for Eq. (357) is derived from Eq. (350) by noting that in view of Eq. (268),

$$D_1 \left(\frac{\partial C_1}{\partial x} \right)_{x=b} = -D_2 \left(\frac{\partial C_2}{\partial x} \right)_{x=r} \tag{358}$$

Hence

$$D_2 \left(\frac{\partial C_2}{\partial x} \right)_{x=r} = \frac{\sqrt{D_1 k}\, k_{\mathrm{f}}\, {}^{r}C_2}{k_{\mathrm{b}} + \sqrt{D_1 k}} \tag{359}$$

Koutecký (*98*) solved a boundary value problem that can be reduced to the present, expressed by Eqs. (357) and (359), by setting the equilibrium constant K for the subsequent reaction $A_1 \underset{kK}{\overset{k}{\rightleftharpoons}} B_3$ much lower than unity. On combining Eqs. (352) and (358), we obtain the boundary condition for the diffusion equation (357) in the case of a reversible charge-transfer step,

$$D_2 \left(\frac{\partial C_2}{\partial x} \right)_{x=r} = \sqrt{D_1 k}\, \theta\, {}^{r}C_2 = \sqrt{D_1 k} \exp\left[\frac{nF}{RT} (E - E^{\circ}) \right] {}^{r}C_2 \tag{360}$$

It is interesting to note that Eq. (360) is perfectly analogous to the boundary condition relative to the totally irreversible discharge $A_2 \xrightarrow{k_f} A_1 + ne$ of the species A_2 at the electrode (cf. Eq. (315) in which k_f is set equal to k_s exp $[\alpha nF(E - E°)/RT]$). The only difference is represented by the fact that in the case under study the standard rate constant k_s for the charge-transfer step is replaced by $\sqrt{D_1 k}$ and the transfer coefficient α by unity. Consequently the theoretical current-potential characteristic as obtained by solving the boundary value problem of Eqs. (357) and (360) is entirely analogous to that for a totally irreversible discharge of A_2 (*22*) and is expressed by the equation,

$$E = E° - \frac{RT}{nF} \ln \frac{\bar{\imath}_d - \bar{\imath}}{\bar{\imath}} - \frac{RT}{nF} \ln 0.886 \sqrt{\frac{D_1 k t_d}{D_2}} \tag{361}$$

The semirigorous equation (361) is identical with the approximate equation (356), apart from the numerical factor. The current-time curves along the rising portion of a prekinetic wave are also identical in shape with those observed along an irreversible wave. Hence, in the lower portion of the wave expressed by Eq. (361) the instantaneous current is proportional to $t^{2/3}$ and not to $t^{1/6}$ as could be inferred from the reversible slope ($\Delta E/\Delta[\ln \bar{\imath}/(\bar{\imath}_d - \bar{\imath})] = RT/nF$). This is a further interesting feature of postkinetic currents.

A typical example of a postkinetic wave is provided by the reversible electrooxidation of ascorbic acid, which leads to an unstable intermediate product (*97*, *99*). This is subsequently transformed into the electroinactive dehydroascorbic acid. Other cases of postkinetic currents are encountered when a reversible electrode process—which is preceded by a chemical reaction when carried out in a given direction—is made to proceed in the opposite direction (*100*).

4. The ECE Mechanism

Although a chemical reaction intervening between two charge-transfer steps (cf. Eq. (221), Sec. IV.B.3) does not fall in any of the three categories of chemical reactions previously considered, it can still be treated with some minor modifications according to the semirigorous procedure of Sec. IV.D. ($24_{(3)}$). Taking into account the depolarization scheme of Eq. (221) and

depicting it as usual,

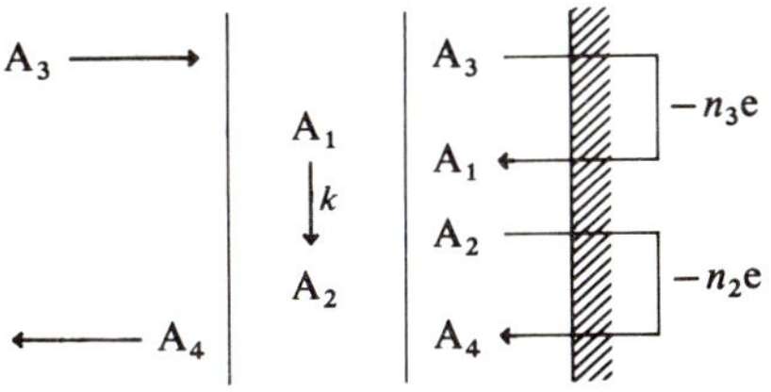

we immediately see that the species A_1 does not contribute to diffusion overpotential within the diffusion layer. Hence the simplified diffusion equation can be applied to this species, yielding

$$D \frac{\partial^2 C_1}{\partial x^2} = kC_1 \tag{362}$$

On integrating Eq. (362) through the procedure outlined in Eqs. (255) through (257), we obtain

$$\left(\frac{\partial C_1}{\partial x}\right)_{x=b} = -\left\{\left[\frac{k}{D} C_1^2\right]_{{}^rC_1}^{{}^bC_1}\right\}^{1/2}$$

In view of the postulated unidirectional character of the chemical reaction $A_1 \xrightarrow{k} A_2$, ${}^rC_1$ is practically zero. Hence

$$\left(\frac{\partial C_1}{\partial x}\right)_{x=b} = -\sqrt{\frac{k}{D}}\, {}^bC_1 \tag{363}$$

Combination of Eq. (363) with the boundary condition in Eq. (225) and with the expression (228) for the flux of A_3 at $x = 0$, valid under limiting current conditions, yields

$${}^bC_1 = \frac{C_3^*}{\sqrt{\frac{3}{7}\pi k t}} \tag{364}$$

In view of Eq. (364) and of the initial and boundary conditions of Eqs. (223) and (224), the linear combination φ, defined in Eq. (226), obeys the diffusion equation,

$$\delta\varphi = 0 \tag{365}$$

with the initial condition:

$$\varphi = 0 \qquad \text{for} \qquad t = 0,\ x \geqslant 0 \tag{366}$$

and the boundary condition,

$$\varphi = \frac{C_3^*}{\sqrt{\frac{3}{7}\pi k t}} \qquad \text{for } t > 0,\ x = 0 \tag{367}$$

The reader is referred to the appendix (Sec. VII.D. Example V) for a solution of the boundary value problem of Eqs. (365) through (367) based on the unified procedure of Sec. VII.C. The resulting expression for the flux of φ at $x = 0$ (Eq. A117) is

$$\left(\frac{\partial \varphi}{\partial x}\right)_{x=0} \equiv \left(\frac{\partial C_1}{\partial x}\right)_{x=0} + \left(\frac{\partial C_2}{\partial x}\right)_{x=0} = -\frac{C_3^*}{\sqrt{\frac{3}{7}\pi D t}}\frac{0.573}{\sqrt{kt}} = -\left(\frac{\partial C_3}{\partial x}\right)_{x=0}\frac{0.573}{\sqrt{kt}} \tag{368}$$

Clearly the limiting current i_l is proportional to the sum of the fluxes of A_3 and A_2 at $x = 0$, each flux being multiplied by the number of electrons involved in the corresponding charge-transfer step

$$i_l = -n_3 FAD\left(\frac{\partial C_3}{\partial x}\right)_{x=0} - n_2 FAD\left(\frac{\partial C_2}{\partial x}\right)_{x=0}$$

Taking into account Eqs. (368) and (225), the limiting current i_l can be readily expressed in the form,

$$i_l = -\left(n_3 + n_2 - n_2\frac{0.573}{\sqrt{kt}}\right)FAD\left(\frac{\partial C_3}{\partial x}\right)_{x=0}$$

Denoting by i_d the contribution $-n_3 FAD(\partial C_3/\partial x)_{x=0}$ to the total limiting current due to the flux of A_3 at $x = 0$ (contribution that actually coincides with i_l if the rate constant k is sufficiently low) we have

$$\frac{i_l - i_d}{i_d} = \frac{n_2}{n_3}\left(1 - \frac{0.573}{\sqrt{kt}}\right) \tag{369}$$

The approximate expression (369) holds to a good approximation for $kt \geqslant 4$. For lower values of the dimensionless parameter kt, the ratio $(i_l - i_d)/i_d$ is expressed by a more involved function of kt, whose tabulated value can be found in References ($24_{(1)}$) and ($24_{(2)}$).

G. Diagnostic Criteria for the Distinction of Polarographic Currents Characterized by Comparable Contributions of Reaction and Diffusion Overpotentials to Total Overpotential (Kinetic Currents)

Under limiting current conditions prekinetic and catalytic currents, as opposed to postkinetic currents, are not diffusion controlled. Prekinetic limiting currents ($i_{l,p}$) are lower than the corresponding diffusion limiting currents (i_d) which would be observed if the rate of the preceding reaction were infinitely fast. Conversely, catalytic limiting currents ($i_{l,c}$) are higher than the corresponding diffusion limiting currents (i_d) which would be observed in the absence of the electroinactive additive. Under the most favorable experimental conditions, preceding reactions in prekinetic currents as well as parallel reactions in catalytic currents are usually pseudo-first order. Consequently it is possible to change at will the "apparent rate constant" for the chemical reaction (e.g., by varying pH or buffer capacity in the case of preceding protonation or dehydration reactions or the concentration of the electroinactive additive in the case of parallel reactions). By doing so the heights $i_{l,p}$ of prekinetic currents may be caused to vary from values much less than i_d to values practically identical with i_d by increasing the apparent rate constant for the preceding reaction. Analogously, the heights $i_{l,c}$ of catalytic currents can be made to vary from i_d to values much greater than i_d by increasing the apparent rate constant for the parallel reaction. The more prekinetic and catalytic currents are made to approach i_d, the more closely the dependence of the instantaneous current intensity on electrolysis time is expressed by the proportionality relation,

$$i \propto m^{2/3} t^{1/6} \tag{370}$$

and the plot of log i against log t approaches $\frac{1}{6}$. Conversely, the more prekinetic currents decrease with respect to i_d or catalytic currents grow with respect to i_d, the more closely the dependence of i on t is expressed by the relation,

$$i \propto m^{2/3} t^{2/3} \tag{371}$$

In practice Eq. (371) describes the constancy of the current density $I = i/A$ with varying electrolysis time. This constancy is due to somewhat different causes according to whether prekinetic or catalytic currents are considered.

With prekinetic currents, the slow chemical reaction preceding the charge-transfer proper causes the current density i/A to change very slightly with t from the very beginning of electrolysis, owing to the negligible depletion of the electroinactive form of the depolarizer near the electrode. For very high values of t, the contribution of η_d to total overpotential begins to be felt, so that for $t \to \infty$ the prekinetic current would tend to coincide with i_d. This situation is exactly analogous to that encountered with a current controlled by the rate of the charge-transfer step. In the case of catalytic currents, for very low values of t the current density is extremely high because the concentration gradient of the depolarizer at the electrode surface is very large and diffusion is practically the only controlling factor. With increasing t, the concentration gradient of the depolarizer at $x = b$ does not decrease progressively, as occurs with diffusion limiting currents, but rapidly reaches a constant value as a consequence of the regeneration reaction. This constant value is obviously larger, the greater the rate constant for the regeneration reaction. At this point the catalytic current density ceases decreasing with t. The typical variation of prekinetic and catalytic current densities with time is represented schematically in Fig. (7). In practice, in the time scale of

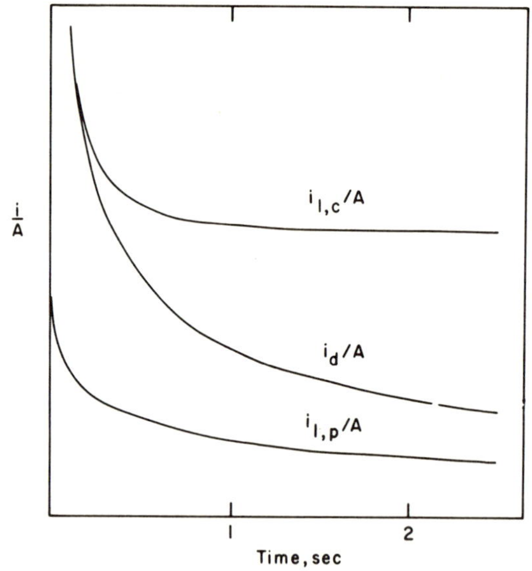

Fig. 7. Qualitative behavior of a typical pre-kinetic limiting current density ($i_{1,p}$/A) and of a typical catalytic limiting current density ($i_{1,c}$/A) as opposed to that of the corresponding diffusion current density i_d/A.

polarographic measurements both the prekinetic and catalytic current densities of Fig. (7) can be taken as time independent with the consequence that Eq. (371) may be applied. The gradual passage from the time dependence of Eq. (371) to the time dependence of Eq. (370) with an increase in the ratio $i_{l,p}/i_d < 1$ in the case of prekinetic currents, or with a decrease in the ratio $i_{l,c}/i_d > 1$ in the case of catalytic currents, is a valuable criterion for clarifying their nature. The passage from the proportionality relation of Eq. (371) to relation of Eq. (370) can be verified either directly from current-time curves or indirectly from the dependence of mean currents on the height of the mercury head. In fact it follows from Eq. (152) that when Eq. (370) applies, we also have $\bar{i} \propto h^{1/2}$ whereas, in the case of validity of Eq. (371), $\bar{i}$ is independent of h.

A particular situation is encountered in the case of partial regeneration of depolarizer by the action of an electroinactive additive (e.g., the decomposition of H_2O_2 by catalase (*24*)). On increasing the concentration of the additive the mean limiting current, which is originally proportional to $h^{1/2}$, tends to become independent of h and then subsequently to return to the original proportionality to $h^{1/2}$.

In general, when the forward and backward components of the antecedent chemical reaction controlling a prekinetic current are first order or pseudo-first order, the prekinetic current increases linearly with increasing concentration C_3^* of the electroinactive form of the depolarizer. Hence the ratio $\bar{i}_{l,p}/\bar{i}_d$ is independent of C_3^*. If, however, the backward reaction is second or higher order (e.g., in the decomposition of an electroinactive dimer to active monomer (*63–65*) or in the dissociation of a metal complex in the absence of an excess of the complex-forming agent (*101*)), an increase in C_3^* causes the ratio $\bar{i}_{l,p}/\bar{i}_d$ to decrease. In other words $\bar{i}_{l,p}$ increases "less than proportionally" to C_3^*. The decrease of $\bar{i}_{l,p}/\bar{i}_d$ with C_3^* is particularly pronounced if the limiting current is "purely kinetic," that is, if the ratio $\bar{i}_{l,p}/\bar{i}_d$ is $\ll 1$ and relation of Eq. (371) is closely approached.

An analogous situation is encountered with catalytic currents. As an example let us consider the electroreduction of uranyl ions in acid media, yielding pentavalent uranium UO_2^+ (*25*). This latter ion disproportionates into UO_2^{2+} and $UOOH^+$ according to a reaction second order with respect to UO_2^+. The ratio of the catalytic limiting current $\bar{i}_{l,c}$, controlled by the partial regeneration of UO_2^{2+}, to the diffusion limiting current $\bar{i}_d$ of UO_2^{2+} increases with the bulk concentration of uranyl ions. In this case the rate of the regeneration reaction, due to its second-order character, increases with an increase in the depolarizer concentration more than if it were first order.

Postkinetic currents are easily distinguished from kinetic and catalytic currents because under limiting conditions they are diffusion controlled. The distinctive features of postkinetic waves as opposed to reversible and irreversible waves have already been examined in Sec. IV.F.3. Suffice it here to say that the half-wave potential of postkinetic waves is independent of the concentration of depolarizer only if the inactivation reaction is first order with respect to the electrode product. Thus in the reduction of some aromatic carbonyl compounds;

$$RR'C{=}O + H^+ + e \rightleftharpoons RR'\dot{C}OH$$

followed by the second-order dimerization of the radical $RR'\dot{C}OH$ to the corresponding pinacol,

$$2RR'\dot{C}OH \rightarrow \underset{\substack{|\\OH}}{RR'C}{-}\underset{\substack{|\\OH}}{C}{-}RR'$$

the half-wave potential of the wave shifts in the direction of positive potentials with increasing concentration of the depolarizer (*102*).

V. SLOW HOMOGENEOUS CHEMICAL REACTIONS INFLUENCED BY THE DIFFUSE LAYER STRUCTURE

So far we have assumed that the reaction layer thickness r is much larger than the "effective thickness" b of the diffuse double layer. In this case the rate of the chemical reaction (185) is not appreciably influenced by the double layer structure. In fact the large concentration gradients of the ionic species in the diffuse double layer due to the high electric field barely affect the mean concentrations of these species in the whole reaction layer. Recalling that the reaction layer thickness is a measure of the mean displacement of the electroactive particle A_1 during its mean life in the direction normal to the electrode surface (*28*), the probability that a molecule of A_1 moving from $x = b$ to the pre-electrode layer will be converted into electroinactive particles is very low. Consequently the correction for the diffuse layer structure can be made directly on A_1 as well as on the other species A_i taking part in the electrode process (41). On assuming a zero contribution of the penetration overpotential η_r to the total overpotential, the relation between the concentration

${}^{b}C_i$ of A_i at the boundary of the diffuse layer and the corresponding concentration ${}^{a_2}C_i$ at the outer Helmholtz plane is (cf. Eq. (54))

$$ {}^{a_2}C_i = {}^{b}C_i \exp\left(-\frac{z_i F}{RT}\phi_2\right) \qquad \text{for all } A_i \tag{372} $$

where $\phi_2 = \varphi^{a_2} - \varphi^{b}$ is practically equal to the potential difference between $x = a_2$ and the bulk of the solution. Equation (372), expressing a Boltzmann distribution of electric charges, can be extended in an approximate manner to any distance x from the electrode, within the diffuse double layer (a_2, b)

$$ C_i(x) = {}^{b}C_i \exp\left[-\frac{z_i F}{RT}\phi(x)\right] \tag{373} $$

Here $\phi(x)$ is the potential at x as measured with respect to the potential in the bulk of the solution.

A. The Inner Potential at the Outer Helmholtz Plane

According to the Gouy-Chapman theory (*103*), the value for ϕ_2 can be calculated by applying the Poisson equation,

$$ \frac{d^2\phi(x)}{dx^2} = -\frac{4\pi\rho(x)}{\varepsilon} \tag{374} $$

$\rho(x)$ is the charge density at x and ε is the dielectric constant, taken as uniform up to the electrode surface. Obviously,

$$ \rho(x) = \sum_i z_i F C_i(x) \tag{375} $$

where the summation is extended to all the charged species present in the solution. If the supporting electrolyte is present in strong excess, its contribution to the structure of the diffuse layer is preponderant with respect to that of the other species contained in the solution, provided none of these latter species is specifically adsorbed.

We recall that a molecule is to be considered as specifically adsorbed at an electrode when its interaction with the metal surface cannot be accounted for by ascribing it exclusively to purely coulombic forces. In this case it is necessary to appeal to specific forces due to a sort of covalent bonding. According to Grahame's model of double layer (*104*), specifically adsorbed species can approach the electrode more closely than species exclusively subject to coulombic forces. Thus the electrical centers of these latter species occupy the outer Helmholtz plane $x = a_2$, whereas specifically adsorbed species form a monolayer with their electrical centers in a plane $x = a_1$ (the inner Helmholtz plane) located at a smaller distance from the electrode.

Taking into account only the supporting electrolyte, which for simplicity is assumed z-z valent, the combination of Eqs. (373), (374), and (375) yields

$$\frac{d^2\phi}{dx^2} = -\frac{4\pi |z| FC_s^*}{\varepsilon}\left[\exp\left(-\frac{|z| F}{RT}\phi\right) - \exp\left(\frac{|z| F}{RT}\phi\right)\right] \tag{376}$$

C_s^* is the bulk concentration of the supporting electrolyte, practically coincident with the corresponding concentration ${}^bC_s$ immediately outside the diffuse layer. A first integration between $x = a_2$ and $x = b$ is performed by multiplying both members of Eq. (376) by $2d\phi/dx$ and noting that $2(d\phi/dx)(d^2\phi/dx^2) = d/dx(d\phi/dx)^2$. Since both ϕ and $d\phi/dx$ tend to zero for $x \to \infty$, we find

$$\left(\frac{d\phi}{dx}\right)_{x=a_2} = \pm\left(\frac{8\pi RTC_s^*}{\varepsilon}\right)^{1/2}\left[\exp\left(\frac{|z| F}{RT}\phi_2\right) + \exp\left(-\frac{|z| F}{RT}\phi_2\right) - 2\right]^{1/2}$$

$$= -\left(\frac{8\pi RTC_s^*}{\varepsilon}\right)^{1/2}\left[\exp\left(\frac{|z| F}{2RT}\phi_2\right) - \exp\left(-\frac{|z| F}{2RT}\phi_2\right)\right] \tag{377}$$

The choice of the minus sign in Eq. (377) is justified by the fact that when ϕ_2 is positive, $\phi(x)$ decreases with increasing x at $x \geqslant a_2$. According to Gauss' theorem, the electric field strength $d\phi/dx$ produced by a planar layer of charges with an uniform charge density q is given by

$$\frac{d\phi}{dx} = -\frac{4\pi q}{\varepsilon} \tag{378}$$

On applying Eq. (378) at $x = a_2$ and combining it with Eq. (377) we obtain

$$q = +\left(\frac{RT\varepsilon C_s^*}{2\pi}\right)^{1/2}\left[\exp\left(\frac{|z| F}{2RT}\phi_2\right) - \exp\left(-\frac{|z| F}{2RT}\phi_2\right)\right] \tag{379}$$

where q is the sum of the charge density q_M on the metal surface plus the charge density q_1 due to the ions of the supporting electrolyte eventually adsorbed at $x = a_1$.

If the supporting electrolyte is nonspecifically adsorbed, q is simply given by q_M, which can be obtained as a function of the applied potential E either from electrocapillary curves or from differential capacity measurements. Consequently, by the use of Eq. (379) the value of ϕ_2 at any given potential can be deduced from the corresponding value of q_M. In using a solution of a 1-1 valent supporting electrolyte for which there is no adsorption of any constituent ion, the accurate values of q_M as a function of E obtained by Grahame (*105*) for aqueous NaF in contact with mercury can be profitably employed. Figure 8 shows the potential drop ϕ_2 across the diffuse layer for

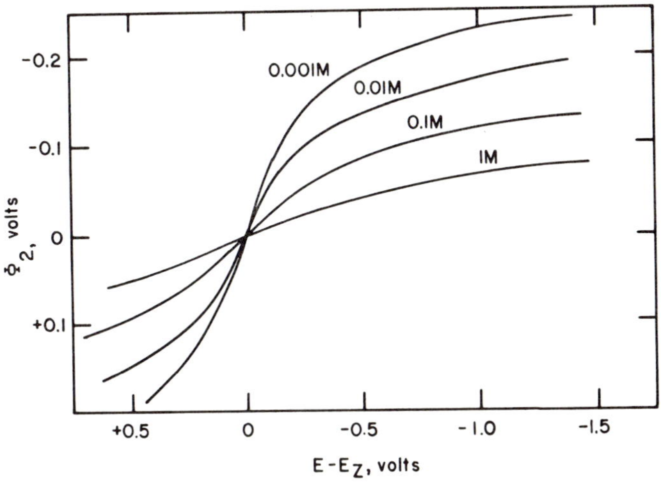

Fig. 8. Plot of ϕ_2 against $E - E_z$ for NaF solutions, as calculated from the data of Grahame (*105*). (From *106*) (By permission of Wiley.)

solutions of nonspecifically adsorbed NaF at different concentrations (*106*). At the potential of zero charge, E_z, ϕ_2 is obviously zero. For $(E - E_z) \neq 0$, ϕ_2 has the same sign as $(E - E_z)$. With increasing $|E - E_z|$, the rate of change of ϕ_2 with potential decreases. Thus for $E - E_z < -1$ V, ϕ_2 does not change appreciably with E. The effect of a change in C_s^* on ϕ_2 at constant E is remarkable. This can be better seen if we note that for $\phi_2 < -0.1$ V, the first exponential within the braces in Eq. (379) can be neglected with respect to the second. Conversely, $\exp(-|z|F\phi_2/2RT)$ in Eq. (379) can be dropped for $\phi_2 > 0.1$ V. Denoting by C' the integral capacity of the inner part of the double layer, bounded by the planes $x = 0$ and $x = a_2$, q_M is given by

$C'(E - E_z - \phi_2)$. In fact $E - E_z$ measures the potential difference across the whole double layer and consequently $E - E_z - \phi_2$ is the potential difference across the inner layer. For $|\phi_2| > 0.1$ V, Eq. (379) can then be written in the form,

$$C'(E - E_z - \phi_2) = \pm \left[\frac{RT\varepsilon C_s^*}{2\pi} \exp\left(\pm \frac{|z|F}{RT} \phi_2 \right) \right]^{1/2} \tag{380}$$

where the $\pm$ sign holds for $\phi_2 \gtrless 0$, respectively. Although C' varies somewhat with potential and electrolyte concentration, it can be considered roughly constant and equal to 18 μF/cm^2 at the water-mercury interphase (*107*). If we take into account that the rate of change of $\exp(\pm|z|F\phi_2/2RT)$ with varying ϕ_2 at constant $(E - E_z)$—that is, with varying C_s^*—is much higher than the rate of change of the left-hand side of Eq. (380), as a first approximation we can consider the right-hand side of this equation as constant with ϕ_2. Hence Eq. (380) reads as

$$\phi_2 = \text{const} \pm \frac{RT}{|z|F} \ln C_s^* \tag{381}$$

where now the $\pm$ sign holds for $\phi_2 \lessgtr 0$. Equation (381), which has been widely applied by Frumkin and other Russian authors, shows that a tenfold increase in C_s^* causes a $60/|z|$ mV decrease in the absolute value of ϕ_2.

If any of the constituent ions of the supporting electrolyte are specifically adsorbed, q_1 is $\neq 0$ and must be taken into due account. q_1 can be computed from electrocapillary curves according to a procedure described by Grahame and Soderberg (*108*) provided we can assume that only the cation or the anion of the electrolyte is specifically adsorbed. The determination of q_M, again from electrocapillary or differential capacity data, allows $q = q_M + q_1$ to be obtained as a function of potential. At this point Eq. (379) permits the calculation of ϕ_2 for different values of E. In general the specific adsorption of cations (anions) is the more pronounced, the higher the negative (positive) value of the charge density q_M on the metal surface. It follows that the net effect of specific adsorption upon ϕ_2 consists in decreasing the rate of change of $q = q_M + q_1$, and consequently of ϕ_2, with applied potential E. Thus ϕ_2 changes with E by about 0.2 V at the interface between mercury and 0.1 M KF, while it changes by only 0.03 V when F^- is replaced by I^-, which is specifically adsorbed.

It can be concluded that both the nature and the concentration of the supporting electrolyte exert a strong influence on the double layer structure and consequently on slow charge-transfer processes and on those chemical reactions taking place within the double layer.

B. Homogeneous Chemical Reactions Taking Place within the Diffuse Layer and Heterogeneous Chemical Reactions

The double layer structure is varied either by changing the nature and concentration of the supporting electrolyte or by changing the applied potential *E*. In connection with the effects of a change in *E*, a distinction between charge-transfer processes and purely chemical reactions occurring in the diffuse layer seems necessary. The rate of a charge-transfer step is directly influenced by the applied potential, independent of the double layer structure. In fact, even if no double layer existed and if a purely geometrical plane separated the metallic from the solution phase, the potential difference between these phases would still affect the potential energy barrier opposing the passage of electrons to or from the electrode. A progressive shift of *E* toward negative values increases the rate of a cathodic electrochemical step, such that at a certain point some nonelectrochemical step must become rate determining. Usually it is diffusion that becomes rate controlling, so that the diffusion limiting current is reached. If, however, the overall cathodic process comprises a purely chemical step, taking place either at the metal-solution interphase or in the solution phase, at sufficiently negative potentials the overall process may become simultaneously controlled by the rate of diffusion and by that of the chemical step. In this case a limiting current of a kinetic nature is attained.

As opposed to diffusion limiting currents, kinetic limiting currents may be altered by a change in the applied potential as well as by a change in the nature and concentration of the supporting electrolyte. This occurs when the chemical reaction controlling the kinetic current is heterogeneous or when, being homogeneous, takes place to a great extent in the diffuse double layer. Under these circumstances the chemical reaction is influenced by the double layer structure, which on its turn is affected by a change in the applied potential as well as by a variation in the nature and concentration of the supporting electrolyte. Thus the first cathodic wave of phenylglyoxylic acid (*44*), due to the reduction of the undissociated acid formed by protonation of the corresponding anion, exhibits a limiting current which, after having reached a maximum value, falls again at sufficiently negative potentials (*109*). Analogously, the cadmium cyanide complex for cyanide concentrations higher than 0.05 *M* gives a limiting current showing a depression. According to several authors (*110–112*) $Cd(CN)_4^{2-}$ is reduced at the dropping mercury electrode via the antecedent chemical reaction,

$$Cd(CN)_4^{2-} \underset{Kk}{\overset{k}{\rightleftharpoons}} Cd(CN)_3^- + CN^-$$

which originates the electroactive species $Cd(CN)_3^-$. The depression of the limiting current is reduced with increasing concentration of a nonspecifically adsorbed supporting electrolyte, like $NaNO_3$, and practically eliminated with a sufficiently high concentration of specifically adsorbed cations, like tetraalkylammonium ions (*113*).

Heterogeneous chemical reactions—that is, reactions taking place among substances specifically adsorbed—are not easily distinguishable from homogeneous chemical reactions occurring mainly in the diffuse double layer. Thus some cases of chemical reactions taking place simultaneously at the electrode surface and in a very thin layer of solution around the electrode (bulk-surface reactions) are reported in the literature (*114*, *115*). It should be mentioned that for many protonation reactions preceding the charge-transfer proper, it is still controversial whether they are controlled by a heterogeneous or a homogeneous reaction (*116*). In general we can say that homogeneous reactions occurring in the diffuse layer are particularly sensitive to a change in the concentration of a nonspecifically adsorbed supporting electrolyte and to the consequent alteration of the ionic atmosphere. Such a change affects only indirectly the surface concentrations of specifically adsorbed substances. Consequently heterogeneous reactions are less influenced by a variation in the ionic strength of the solution. This is particularly true for heterogeneous reactions occurring among neutral specifically adsorbed species. In this case the rate of the chemical reaction at constant E can be changed by altering the state of affairs in the inner double layer—for example, by adding to the solution a specifically adsorbed supporting electrolyte or a neutral surface-active electroinactive substance. Since by doing so the ionic atmosphere in the diffuse layer is also perturbed, it is advisable to start checking the behavior of a kinetic current by varying the concentration of a nonspecifically adsorbed supporting electrolyte. If the kinetic current is not affected by such a change of concentration, but is instead influenced by the addition of a specifically adsorbed electrolyte (e.g., a tetraalkylammonium salt) and by a change in the applied potential, we can reasonably infer that the current is controlled by a heterogeneous reaction. If, however, a change in the concentration of a nonspecifically adsorbed supporting electrolyte is effective, the heterogeneous nature of the chemical reaction cannot be excluded categorically. In fact, some of the substances participating in a heterogeneous reaction may react while remaining in the outer Helmholtz plane without being specifically adsorbed. This situation is probably encountered in many antecedent protonation reactions in which the organic base is specifically adsorbed whereas hydroxonium ions are not. We shall first consider the effect of the diffuse layer structure on homogeneous reactions taking place within the double layer. Section VI will deal with some distinctive features of kinetic currents controlled by heterogeneous chemical reactions.

C. Treatment of Polarographic Currents Controlled by Homogeneous Chemical Reactions Influenced by the Diffuse Layer Structure

We have seen that homogeneous reactions coupled with a charge-transfer process can be thought of as taking place in a liquid layer around the electrode, termed the "reaction layer." Usually a chemical reaction will be pseudo-first order or even second or higher order and, consequently, the reaction layer thickness r is a function of the concentration of one or more reaction partners in the reaction layer. If the reaction layer thickness, as computed in the absence of double layer effects, is of the same order of magnitude as the diffuse layer thickness $1/\kappa$, the concentration distributions of the ionic reaction partners are strongly affected by the electric field existing in the double layer, and the same must be said for the "actual" reaction layer thickness. We recall that the diffuse layer thickness $1/\kappa$ measures the distance $x - a_2$ from the outer Helmholtz plane at which $\phi(x)$ equals ϕ_2/e (cf. Sec. I.B.2). For a z-z valent electrolyte, $1/\kappa$ is given by the equation,

$$\frac{1}{\kappa} = \left(\frac{RT\varepsilon}{8\pi z^2 F^2 C_S^*}\right)^{1/2} \tag{382}$$

1. A Semirigorous Treatment of Prekinetic Currents

Let us consider the influence of the diffuse layer structure on the electrode process,

$$A_1 \rightarrow A_2 + ne \tag{383a}$$

preceded by the chemical reaction

$$B_3^{z_3} \underset{kK}{\overset{k}{\rightleftharpoons}} A_1^{z_1} + \nu B_5^{z_5} \qquad (z_3 = z_1 + \nu z_5) \tag{383b}$$

Here the nonreducible substance B_3 is transformed into the electroreducible one A_1 in presence of a large excess of the electroinactive species B_5. The following derivation is inspired by Matsuda's treatment (*117*). The same problem was considered for conditions somewhat less general by Hurwitz (*118*). Assume that the solution contains a large excess of an indifferent z-z valent electrolyte, for which there is no adsorption of any constituent ion. In view of Eq. (377), which can be applied at any distance $x > a_2$, the potential gradient within the diffuse layer is given by

$$\frac{d\phi}{dx} = -\frac{RT}{|z|F}\kappa\left[\exp\left(\frac{|z|F}{2RT}\phi\right) - \exp\left(-\frac{|z|F}{2RT}\phi\right)\right] \tag{384}$$

where the diffuse layer thickness $1/\kappa$ is defined by Eq. (382). Owing to the large excess of B_5, we can reasonably assume that the concentration of this species does not vary with time during electrolysis. Hence the concentration distribution of B_5 is expressed by Eq. (373)

$$C_5 = {}^{\delta}C_5 \exp\left(-\frac{z_5 F}{RT}\phi\right) \tag{385}$$

in which the concentration ${}^{b}C_5$ of B_5 just outside the diffuse layer ($\phi = 0$) is made to coincide with the corresponding bulk concentration ${}^{\delta}C_5$. Carrying out the usual separation of the solution layer around the electrode into different regions characterized by the presence of only one type of overpotential, we shall continue locating diffusion overpotential outside the reaction layer r. Hence, in view of Eq. (230), the driving force of the field-assisted diffusion for the electroinactive species B_3, that is, $\partial\tilde{\mu}_3/\partial x$, will be equated to zero within the reaction layer,

$$\frac{\partial\tilde{\mu}_3}{\partial x} = \frac{\partial}{\partial x}(z_3 F\phi + RT \ln C_3) = 0 \tag{386}$$

If the major part of the reaction layer falls outside the diffuse layer, $\partial\phi/\partial x$ is practically zero at $x \leqslant r$. Consequently we can write $\partial\tilde{\mu}_3/\partial x \cong \partial\mu_3/\partial x = 0$, which implies the uniformity of C_3 in the reaction layer. This was exactly the simplifying assumption made in Sec. IV.C.1. Under the present conditions no such simplification is feasible and the integration of Eq. (386) over x between any given distance x and $x = r$, or alternatively $x = b$, according to whether r is $>$ or $<b$, yields

$$C_3(x) = {}^{b,r}C_3 \exp\left(-\frac{z_3 F\phi(x)}{RT}\right) \tag{387}$$

which expresses the usual Boltzmann distribution. ${}^{b,r}C_3$ is the concentration of B_3 immediately outside the diffuse and reaction layers. Unlike ${}^{b}C_5$, ${}^{b,r}C_3$ depends on electrolysis time owing to the depletion of B_3 around the electrode.

In order to obtain the expression for the actual reaction layer thickness, we shall follow a procedure analogous to that outlined in Sec. IV.C.2. Thus on temporarily disregarding the assumption that no diffusion overpotential is located in the reaction layer, we shall focus our attention on the species A_1, which, being characterized by $v_i + v_i' = 0$, does not contribute to diffusion overpotential in the diffusion layer; we shall apply to it, for $0 \leqslant x \leqslant r$, the diffusion equation complicated with the kinetic term. In the case of field-assisted diffusion, Eq. (253) is not longer valid because it has been derived from the general diffusion equation (62) under the assumption that the flux of the particular diffusing species under study is expressed by Fick's first law (Eq. (61)). According to the Onsager theory, the diffusion flow of a given species A_1 is an irreversible phenomenon that can be considered as proportional to the cause giving rise to its occurrencc. We have seen in Sec. I.A that this cause, termed "driving force of diffusion," is expressed by grad μ_1. Thus in the case of linear diffusion we have

$$J_{1,x} = C_1 v_{1,x} = l\frac{\partial RT \ln C_1}{\partial x} = \frac{RTl}{C_1}\frac{\partial C_1}{\partial x} \tag{388}$$

where l is a proportionality coefficient. Upon comparing Eq. (61) and (388) we see that l is proportional to C_1 according to the equation,

$$l = -\frac{D_1 C_1}{RT} \tag{389}$$

In the case of field-assisted diffusion, the driving force is more generally expressed by $\partial\tilde{\mu}_1/\partial x$. Hence the proportionality relation between the flux $J_{1,x}$ and the corresponding force reads as

$$J_{1,x} = l\left(\frac{RT}{C_1}\frac{\partial C_1}{\partial x} + z_1 F \frac{\partial \phi}{\partial x}\right) \tag{390}$$

where l is still given by Eq. (389), as appears from Eqs. (388) and (390) on letting $\partial\phi/\partial x$ tend to zero. Combination of Eqs. (389) and (390) yields

$$J_{1,x} = -D_1\left(\frac{\partial C_1}{\partial x} + \frac{z_1 F C_1}{RT}\frac{\partial \phi}{\partial x}\right) \tag{391}$$

On substituting the flux $J_{1,x}$ from Eq. (391) into Eq. (62) and assuming, as was done in Sec. IV.C.2, that steady-state conditions are rapidly attained ($\partial C_1/\partial t = 0$) and that for $x < r$ and b the correction term for convection can be neglected, we obtain the following diffusion equation:

$$D_1 \frac{\partial}{\partial x}\left(\frac{\partial C_1}{\partial x} + \frac{z_1 F}{RT} C_1 \frac{\partial \phi}{\partial x}\right)$$
$$= kKC_1{}^{\delta}C_5^{\nu} \exp\left(-\frac{\nu z_5 F}{RT}\phi\right) - k^{b,r}C_3 \exp\left(-\frac{z_3 F}{RT}\phi\right) \qquad x < b \text{ and } r \tag{392}$$

Here ϕ depends on the distance x according to Eq. (384). The current is proportional to the flux of A_1 at $x = a_2$ and, therefore, from Eq. (391) we have

$$i = -nFAD_1\left(\frac{\partial C_1}{\partial x} + \frac{z_1 F}{RT} C_1 \frac{\partial \phi}{\partial x}\right)_{x=a_2}$$

where A is the instantaneous area of the electrode. The solution of the differential equation (392) with the boundary condition ${}^{a_2}C_1 = 0$ is quite complicated and leads to the following expression of i

$$-\frac{i}{nFA} = \sqrt{\frac{D_1 k}{K^{\delta}C_5^{\nu}}}\, {}^{b,r}C_3\, G_{\mathrm{I}}^{-1} \tag{393}$$

where G_{I} is an involved function of the potential ϕ_2 at the outer Helmholtz plane, of the charges z_1, z_3, and z_5 of the reacting species, and of the parameter $\mu = \sqrt{D_1/(kK^{\delta}C_5^{\nu})}$. For the expression of G_{I} the reader is referred to the original paper (*117*).

A comparison of the depolarization schemes of Eqs. (273) and (383) shows that they are identical in the absence of diffuse layer effects. In fact the backward reaction (383b) is pseudo-first order with respect to A_1, with a rate constant equal to $kK^{\delta}C_5^{\nu}$. Hence, in view of Eq. (278), μ expresses the reaction layer thickness in the absence of diffuse layer effects. According to the Brdička-Wiesner model, the actual thickness of the reaction layer is readily obtained by equating the right-hand side of Eq. (393) to $rk^{b,r}C_3$ as done in Eq. (277). It follows that

$$r = \sqrt{\frac{D_1}{kK^{\delta}C_5^{\nu}}}\, G_{\mathrm{I}}^{-1} = \frac{\mu}{G_{\mathrm{I}}}$$

and consequently G_{I} expresses the ratio of the reaction layer thickness in the absence of diffuse layer effects to its actual thickness. The last step for the derivation of the limiting current consists, as usual, in solving the diffusion equation for B_3 uncomplicated by kinetic terms,

$$\frac{\partial C_3}{\partial t} = D_3 \frac{\partial^2 C_3}{\partial x^2} + \frac{2x}{3t}\frac{\partial C_3}{\partial x} \qquad x \geq b, r \tag{394}$$

with the initial condition

$$C_3 = {}^{\delta}C_3 \qquad \text{for } x \geqslant b,\ t = 0 \tag{395}$$

and the boundary condition

$$D_1\left(\frac{\partial C_1}{\partial x} + \frac{z_1 F}{RT} C_1 \frac{\partial \phi}{\partial x}\right)_{x=a_2} = D_3\left(\frac{\partial C_3}{\partial x}\right)_{x=b,r} = \sqrt{\frac{D_1 k}{K^{\delta}C_5^{\nu}}}\ {}^{b,r}C_3\, G_{\mathrm{I}}^{-1} \tag{396}$$

The boundary value problem of Eqs. (394) through (396) is practically identical with that of Eqs. (281), (282), and (276). Hence, in view of Eq. (287), the final result is

$$\frac{\bar{\imath}_1}{\bar{\imath}_{\mathrm{d}} - \bar{\imath}_1} = 0.886 \sqrt{\frac{D_1 k t_{\mathrm{d}}}{D_3 K^{\delta} C_5^{\nu}}}\, G_{\mathrm{I}}^{-1} \tag{397}$$

Figure (9) shows $\log G_{\mathrm{I}}$ as a function of $-F\phi_2/(2 \times 2.3RT)$ for different values of the ratio $1/\mu\kappa$ and for $|z| = 1$, $z_3 = -2$, and $z_1 = -1$. It is evident that for values of $1/\mu\kappa \ll 1$, G_{I} is practically equal to one even if ϕ_2 is relatively high, since under these conditions the electric field, no matter how strong,

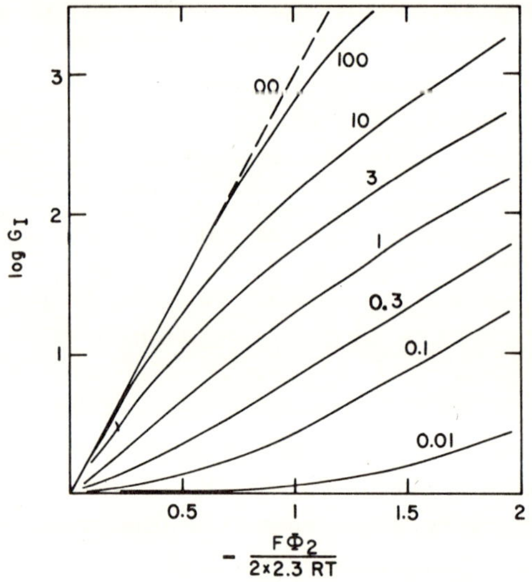

Fig. 9. Plot of log G_I as a function of $-F\phi_2/(2 \times 2.3RT)$ for $|z| = 1$, $z_3 = -2$, and $z_1 = -1$. The numbers on each curve refer to the value of $1/\mu\kappa$. (From *117*) (By permission of the American Chemical Society.)

does not appreciably alter the concentration distributions in the reaction layer. For a given ϕ_2, log G_I increases with increasing $1/\mu\kappa$ until for $1/\mu\kappa > 100$, it becomes practically proportional to $-F\phi_2/(2 \times 2.3RT)$ according to the proportionality constant 3. In the general case the limiting value of G_I for high values of $1/\mu\kappa$ is given by

$$G_I = \exp\left[\frac{(z_1 + z_3)F}{2RT}\phi_2\right] \tag{398}$$

provided the supporting electrolyte is 1-1 valent ($|z| = 1$).

2. The Case of the Reaction Layer Thickness much less than the Diffuse Layer Thickness

The limiting value (398) of G_I for $1/\mu\kappa \gg 1$ can easily be obtained from Eq. (392) by assuming that the chemical reaction occurs in a solution layer $a_2 \leqslant x \leqslant a_2 + dx$ that is so thin the changes in the inner potential ϕ within this layer can be ignored. Setting $\phi \cong \phi_2$ and neglecting $z_1 FC_1/(RT)\, \partial\phi/\partial x$ with respect to $\partial C_1/\partial x$, Eq. (392) takes the simplified form,

$$D_1 \frac{\partial^2 C_1}{\partial x^2} = kKC_1\, {}^{\delta}C_5^{\nu} \exp\left(-\frac{\nu z_5 F}{RT}\phi_2\right) - k\, {}^{b}C_3 \exp\left(-\frac{z_3 F}{RT}\phi_2\right) \quad \text{for } a_2 \le x \le a_2 + dx \quad (399)$$

Here the concentration at $x = a_2$ of the species B_5, present in strong excess, is expressed as a function of the corresponding bulk concentration ${}^{\delta}C_5$, whereas the concentration ${}^{a_2}C_3$ of the electroinactive species B_3 contributing to diffusion overpotential is expressed in terms of the corresponding concentration immediately outside the diffuse double layer ${}^{b}C_3$. The integration of Eq. (399) between $x = a_2$ and $x = a_2 + dx$, performed in a way entirely analogous to that employed in connection with scheme of Eq. (273), yields

$$\left(\frac{\partial C_1}{\partial x}\right)^2_{x=a_2} - \left(\frac{\partial C_1}{\partial x}\right)^2_{x=a_2+dx} = \left[\left\{kK\, {}^{\delta}C_5^{\nu} \exp\left(-\frac{\nu z_5 F}{RT}\phi_2\right)C_1^2 - 2k\, {}^{b}C_3 \exp\left(-\frac{z_3 F}{RT}\phi_2\right)C_1\right\}\Big/ D_1\right]^{a_2 C_1}_{a_2+dx C_1} \quad (400)$$

At $x = a_2 + dx$, reaction (383b) is practically at equilibrium and the application of the law of mass action yields

$$K = \frac{C_3}{C_1 C_5^{\nu}} = \frac{{}^{b}C_3}{{}^{a_2+dx}C_1\, {}^{\delta}C_5^{\nu}} \exp\left[-(z_3 - \nu z_5)\frac{F}{RT}\phi_2\right] \quad (401)$$

Noting that under the previously made assumption of a practically uniform inner potential in the region $a_2 \leqslant x \leqslant a_2 + dx$ the concentration gradient of A_1 at $x = a_2 + dx$ can be neglected with respect to that at $x = a_2$, substituting ${}^{a_2+dx}C_1$ from Eq. (401) into Eq. (400), and recalling that under limiting current conditions ${}^{a_2}C_1$ equals zero, we obtain

$$D_1\left(\frac{\partial C_1}{\partial x}\right)_{x=a_2} = \sqrt{\frac{D_1 k}{K\, {}^{\delta}C_5^{\nu}}}\, {}^{b}C_3 \exp\left[-\left(z_3 - \frac{\nu z_5}{2}\right)\frac{F}{RT}\phi_2\right] \quad (402)$$

Since $z_3 = z_1 + \nu z_5$, the comparison of Eq. (402) with Eq. (393) shows that the factor G_I obtained by the present approximate procedure is identical with that expressed by Eq. (398).

Assume that the homogeneous chemical reaction preceding the charge-transfer process,

$$\mathrm{A}_1 \rightarrow \mathrm{A}_2 + ne$$

has the form,

$$\mathrm{B}_3^{z_3} + \nu \mathrm{B}_5^{z_5} \underset{kK}{\overset{k}{\rightleftharpoons}} \mathrm{A}_1^{z_1} \tag{403}$$

In the present case the expression for the concentration gradient of the electroactive species A_1 at $x = a_2$, as obtained by solving a differential equation analogous to Eq. (399), is given by

$$D_1\left(\frac{\partial C_1}{\partial x}\right)_{x=a_2} = D_3\left(\frac{\partial C_3}{\partial x}\right)_{x=b} = \sqrt{\frac{D_1 k}{K}}\, {}^{\delta}C_5^{\nu}\, {}^{b}C_3 \exp\left(-\frac{z_1 F}{RT}\phi_2\right) \tag{404}$$

This equation, which holds for $\mu \ll 1/\kappa$, is identical with that derived by Matsuda under the same simplifying assumptions (*117*). Using Eq. (404) as the boundary condition for the diffusion equation (394), we immediately obtain the equation,

$$\frac{\bar{i}_1}{\bar{i}_\mathrm{d} - \bar{i}_1} = 0.886\sqrt{\frac{D_1}{D_3}\frac{k t_\mathrm{d}}{K}}\, {}^{\delta}C_5^{\nu} \exp\left(-\frac{z_1 F}{RT}\phi_2\right) \tag{405}$$

which is analogous to Eq. (397). A relation similar to Eq. (405) was employed by Grabowski and Bartel (*119*) in studying the influence of ionic strength, and consequently of ϕ_2, on the kinetic current of *p*-dimethylaminobenzaldehyde. This current is controlled by the rate of the antecedent protonation reaction,

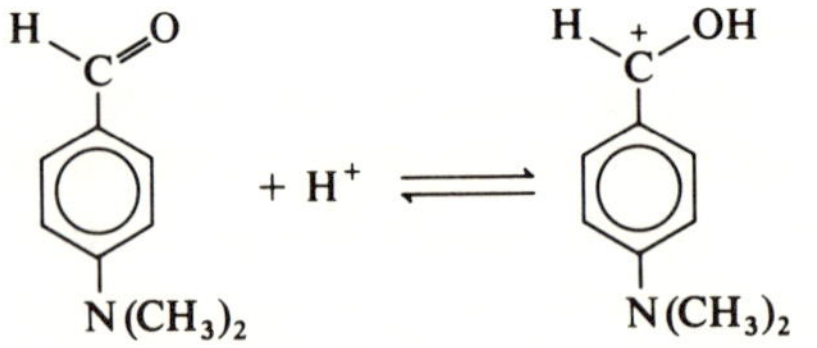

In the present case $B_5 = H^+$ and $z_1 = +1$.

Some authors (*111*, *112*, *119*) when correcting the reaction layer thickness for the diffuse layer structure under the assumption that $\mu \ll 1/\kappa$, start directly from Eq. (287), valid for an antecedent first-order or pseudo-first-order reaction in the absence of diffuse layer effects, and correct the rate constants of the forward and backward reactions. It should be noted that, in general, the final expression derived by this procedure is in disagreement with that obtained by calculating directly the actual reaction layer thickness r. In this connection consider the depolarization scheme,

$$B_3^{z_3} \underset{kK}{\overset{k}{\rightleftharpoons}} A_1^{z_1} \qquad (z_3 = z_1)$$

$$A_1 \rightarrow A_2 + ne$$

in which the chemical reaction is first order in both directions. In the absence of double layer effects, the limiting current is given by Eq. (287). If we assume that the chemical reaction takes place mostly at the outer Helmholtz plane and we consider the formal rate constant k of Eq. (287) as the product of a field independent rate constant k' by the exponential factor of Eq. (387), we have

$$\frac{\bar{i}_l}{\bar{i}_d - \bar{i}_l} = 0.886\sqrt{\frac{D_1}{D_3}\frac{k' t_d}{K}} \exp\left(-\frac{z_3 F}{2RT}\phi_2\right) \tag{407}$$

K has not been corrected for the diffuse layer structure because the forward and backward rate constants are equally influenced by ϕ_2. Comparison of Eq. (407) with Eq. (405), where ${}^{\delta}C_5$ is now ignored and $z_1 = z_3$, shows that Eq. (407) is erroneous because the proper correction factor is $\exp(-z_3 F\phi_2/RT)$. In conclusion, if $\mu \ll 1/\kappa$, it is always advisable to determine the factor G_I by following the simple procedure outlined in Eqs. (399) through (402), which can be easily extended to any given depolarization scheme.

3. Some Applications

The influence of the diffuse layer upon the chemical reaction (383b) was also considered by Hurwitz and Gierst (*118*, *120*), who limited themselves to the case of an uncharged electroinactive species $B_5(z_5 = 0)$. In the formalism adopted by these authors the correction for the diffuse layer structure is included in the expression for the "apparent rate constant" v^*, which is defined by the equation,

$$-\frac{i}{nFA} = D_1\left(\frac{\partial C_1}{\partial x} + \frac{z_1 F C_1}{RT}\frac{\partial \phi}{\partial x}\right)_{x=a_2} = v^{*}\,{}^{b,r}C_3 \tag{408}$$

On comparing Eqs. (393) and (408) it is readily seen that

$$v^* = \sqrt{\frac{D_1 k}{K^\delta C_5^\nu}}\, G_I^{-1}$$

If $\mu \ll 1/\kappa$, taking into account that now $z_1 = z_3$, from Eq. (398) we obtain

$$v^* = \sqrt{\frac{D_1 k}{K^\delta C_5^\nu}} \exp\left(-\frac{z_1 F}{RT}\phi_2\right) \tag{409}$$

Equation (409), where z_1 is the charge of the electroactive species A_1, has been widely used by Gierst and Hurwitz (*120*, *121*). It should be noted, however, that these authors improperly applied Eq. (409) to the reduction of cyanocadmium (II) and cyanomercury (II) ions on mercury, where the species B_5 is CN^-, and is therefore charged. At sufficiently large cyanide concentrations ($[CN^-] > 0.1$ *M*), $Cd(CN)_4^{2-}$ gives a polarographic wave exhibiting a maximum. According to Gierst and Hurwitz, the influence of the applied potential and of the ionic strength of the solution is fairly well described by Eq. (409), where $z_1 = -1$. From this experimental result these authors concluded that the rate-determining reaction preceding the charge-transfer proper is

$$Cd(CN)_4^{2-} \rightarrow Cd(CN)_3^- + CN^- \tag{410}$$

since this is the only antecedent reaction in which the supposed electroactive species, that is, $Cd(CN)_3^-$, has a negative unit charge. As a matter of fact the electroinactive species produced in the decomposition of cyanocadmium (II) is not uncharged and consequently v^* should be given by Eq. (402), in which the exponential factor accounting for the diffuse layer structure equals $\exp[3F\phi_2/(2RT)]$. Apart from quantitative considerations, the drop in current beyond the maximum of the cathodic wave of $Cd(CN)_4^{2-}$ is caused by a decrease in the rate of dissociation of $Cd(CN)_4^{2-}$. This is due to the increasingly negative value of ϕ_2, which produces a decrease in the concentration of $Cd(CN)_4^{2-}$ in the diffuse layer. An increase in the concentration of the supporting electrolyte at constant $E - E_z$ causes ϕ_2 to become less negative (cf. Eqs. (380) and (381)) and consequently reduces the electrostatic repulsion. Hence the maximum current grows and the experimental apparent rate constant v^* increases accordingly.

The influence of the diffuse layer structure on reaction (410) was also studied by Delahay and Kleinerman (*122*), who devised an ingenious method for changing ϕ_2. From Eq. (380) it is apparent that at constant applied potential E and electrolyte concentration C_s^*, ϕ_2 still depends on the potential of zero charge, E_z. This latter parameter can be varied independently by changing the nature of the electrode material. Thus all other conditions remaining unaltered, a gradual increase in the concentration of thallium in a thallium-amalgam electrode causes a progressive decrease in $|\phi_2|$ and consequently an increase in the limiting current of Cd^{II} in 0.1 M NaCN. According to Delahay and Kleinerman, the influence of ϕ_2 on v^* is expressed by the empirical relation,

$$v^* \propto \frac{1}{G_I} \propto \exp\left(\frac{2.6F\phi_2}{2RT}\right)$$

where the numerical factor within the exponential sign is less than the limiting one, 3/2. On the basis of the quantitative treatment of Matsuda (*117*), valid for any value of the ratio $1/\mu\kappa$, it is possible to see that the experimental slope $\log G_I/[-F\phi_2/(2 \times 2.3RT)] = 2.6$ corresponds to a value of $1/\mu\kappa$ equal to 10^2 (see Fig. (9)). Since $1/\kappa$ in 0.1 M NaCl is about 10^{-7} cm, μ should be $\cong 10^{-9}$ cm. This value of μ is physically unsound because it is lower than the diameter of an ordinary molecule by more than one order of magnitude. This seems to be the only case in which the quantitative treatment of diffuse layer effects on homogeneous chemical reactions developed by Matsuda (*117*), or also the analogous one developed by Hurwitz (*118*), has been applied. In fact only the limiting case of $\mu \ll 1/\kappa$ is usually considered. From the data available in the literature it seems that—provided depolarizer and supporting electrolyte are not specifically adsorbed to an appreciable extent—graphs of $\log[\bar{\imath}_l/(\bar{\imath}_d - \bar{\imath}_l)]$, or $\log G_I^{-1}$, or $\log v^*$, relative to an antecedent reaction, versus $F\phi_2/RT$ are often roughly linear although the slopes of these graphs only rarely satisfy the equations previously derived. A case in which Eq. (409) has been properly applied and suitably verified is the electroreduction of nickel ion on mercury. The limiting current of the polarographic wave of Ni^{2+} in not too concentrated noncomplexing media is diffusion controlled. With increasing ionic strength the limiting current diminishes with respect to the corresponding diffusion current, showing an increase when the applied potential is shifted toward negative values (*123*, *124*). The dependence of v^* on ϕ_2 is expressed by Eq. (409), with $z_1 = +2$. Since several noncomplexing electrolytes affect the wave in almost the same way,

Gierst (*125*) explains this behavior by the dehydration reaction,

$$Ni^{2+} \cdot 6\,H_2O \rightleftharpoons Ni^{2+}(6-m)\,H_2O + m\,H_2O \tag{411}$$

prior to charge transfer. In the present case an increase in ionic strength, by rendering ϕ_2 less negative, reduces the electrostatic attraction of the aquo complex and consequently the rate of reaction (411). A shift of potential towards more negative values partially compensates for this effect.

4. Limitations of the Theoretical Treatment of Homogeneous Chemical Reactions Influenced by the Diffuse Layer Structure

The numerous discrepancies between the theoretical relations previously described and experimental data may be due to several causes.

(1) The calculation of the potential difference across the diffuse layer is based on the application of the Gouy-Chapman theory, which is quite approximate especially at high salt concentrations. Thus the above theory does not take into account the variation of potential in planes parallel to the electrode surface since the Poisson equation (374) is solved for one dimension (*106*). As a matter of fact the local potential at a point of the plane of closest approach can be different from the average potential at this plane owing to the discrete nature of ions (*126*). This "local field effect" (*127*) is particularly evident when the ion being discharged at the plane of closest approach bears an ionic charge of the same sign as that of the ion of the indifferent electrolyte attracted in the diffuse layer but of different value. This heterogeneity of charge distribution in the plane of closest approach can explain why the correction for the diffuse layer structure, based on the identification of the local potential in this plane with the average potential in the outer Helmholtz plane, applies to the discharge of Ga^{3+} in presence of Al^{3+}, while it fails on replacement of Al^{3+} by Na^+ or Mg^{2+} (*127*).

(2) Specific chemical interactions between the discharging ion and the ions of the supporting electrolyte may occur in the double layer, leading to a sort of ion pairs of such transient stability as to be unimportant in the bulk of the solution (*128*). These interactions are not well distinguishable from local field effects. Thus Frumkin et al. (*129*) suggest that the discrete structure of the double layer in the electroreduction of $S_2O_8^{2-}$ can be taken into account by considering separately the interaction of the electroactive anion with the next neighboring cation, resulting in the formation of a sort of ion pair in the double layer, and by making use of a ϕ_2 value calculated on the

basis of the Gouy-Chapman theory in order to account for the action of other cations. The specific influence of the supporting electrolyte on electrode processes can also be partly attributed to ion pair formation in the bulk of the solution, which affects the transport of depolarizer particles to the electrode through the diffuse layer (*130*).

(3) Many anions of the supporting electrolyte can be specifically adsorbed at not too high cathodic potentials, causing a sensible deviation of ϕ_2 from the behavior expressed by Eq. (379) with $q = q_M$. Also, simple inorganic cations with large radius, like K^+ or Cs^+, are specifically adsorbed to an appreciable extent on the negative side of the electrocapillary maximum (*131*). Thus the absolute value of the negative ϕ_2 potential decreases when the cation of the supporting electrolyte is changed in the order Li^+, Na^+, K^+, Cs^+, causing the influence of the double layer structure on v^* to decrease in the same order. Although specific adsorption may be accounted for in the evaluation of ϕ_2 by setting $q = q_M + q_1$ in Eq. (379), a particular caution must be used in following this procedure when large organic cations are employed. Thus tetraalkylammonium cations are not only more effective than cesium cations in decreasing the absolute value of ϕ_2, but at the same time they reduce the free electrode surface on which the charge-transfer reaction takes place. Thus a progressive increase in the concentration of tetraalkylammonium ions causes an acceleration of the electroreduction of IO_3^- due to a decreased electrostatic repulsion of this anion, followed by a retardation attributable to the growing electrode coverage (*128*). The concentration of tetraalkylammonium ions for which inversion occurs is smaller, the larger the number of carbon atoms in the alkyl group. An analogous effect is exerted by tetraalkylammonium cations upon the cathodic limiting current of chromate ion, which is controlled by the rate of the antecedent reaction (*130*, *132*),

$$CrO_4^{2-} + H_2O \rightleftharpoons HCrO_4^- + OH^- \tag{412}$$

Small concentrations of tetraalkylammonium ions increase the height of the kinetic wave, whereas large concentrations decrease it. This latter effect can be attributed either to a slow penetration of the electroactive species $HCrO_4^-$ through the adsorbed layer of organic cations or to a displacement of CrO_4^{2-} from the electrode surface by the surface-active salt added to the solution. It is evident that the second explanation is acceptable only if we assume that the protonation reaction (412) is heterogeneous—that is, occurs in the adsorbed state. In fact the concentration of the negatively charged CrO_4^{2-}

ions in the outer Helmholtz plane, and more generally in the diffuse layer, is increased by tetraalkylammonium ions due to the decreased absolute value of the negative ϕ_2 potential. An analogous effect of tetraalkylammonium cations is observed in the reduction of the anions of phenyglycolic acid. Here the limiting current, which is controlled by an antecedent protonation, is increased at low concentrations and depressed until its disappearance at high concentrations of the organic cation (*116*).

(4) So far in considering the influence of the double layer on chemical reactions we have assumed that equilibrium constants as well as rate constants not embodying concentrations of species present in excess are not altered if the reaction takes place within the double layer. As a matter of fact a change in the reactivity of polarizable molecules in the double layer is to be expected (*9*, *133–135*). This is due to the high electric field existing in the neighborhood of the Helmholtz plane, which may seriously influence the electron density distribution in the reacting molecules. *p*-Amino, *p*-dimethylamino, and *p*-oxybenzaldehyde give cathodic limiting currents controlled by the rate of the antecedent protonation of the carbonyl group. The basicity of this group in comparison to the simple benzaldehyde is enhanced by the strong electron donors $-NH_2$, $-N(CH_3)_2$, and $-OH$ in para-position. Nevertheless, the amino groups in *p*-amino and *p*-dimethylamino-benzaldehyde are more basic than the carbonyl group, and the absorption spectra of these compounds clearly show that their protonation in the bulk of the solution takes place at the amino group. The change in reactivity occurring in the double layer is explained by considering that near the negatively charged electrode the molecules of substituted benzaldehyde are oriented with their positive end (the amino group) towards the electrode (*9*). Consequently an induced dipole moment is added to the permanent one and the basicity of the carbonyl group is increased. An expression for the equilibrium constant of the protonation reaction of the carbonyl group as a function of the electric field strength has been derived, under some simplifying assumptions, by Grabowski and Kemula (*9*). Another example of the influence of the potential gradient in the double layer on chemical reactions is provided by the cathodic reduction of benzil, leading to the formation of two isomeric stilbendiols, *cis* and *trans*. The ratio of the isomers formed depends on the electrode potential at which the reduction is accomplished. This unusual phenomenon has been explained by assuming that the initially formed *trans*-stilbendiol molecule isomerizes into the *cis* form at a rate that depends on the electric field existing in the inner double layer (*135*).

(5) For a uniunivalent supporting electrolyte, the diffuse layer thickness $1/\kappa$ is about 10^{-6} cm for $C_s^* = 10^{-3}\ M$ and $3{\cdot}10^{-8}$ cm for $C_s^* = 1\ M$. Thus for sufficiently concentrated solutions of supporting electrolyte ($C_s^* \geqslant 10^{-1}\ M$), the case $\mu \ll 1/\kappa$ corresponds to a thickness of the reaction layer of molecular dimensions. Then the concept of continuity of concentrations within this layer is hardly justified, and the same is true for the application of the diffusion equation (62). Furthermore the more the reaction layer thickness μ decreases, tending to molecular dimensions, the more its dependence on the concentrations of any of the reaction partners ceases to be felt. Ultimately when the "formal" value of μ is of the order of magnitude of the diameter of a solvated molecule, the reaction layer thickness is no longer concentration dependent and the rate and equilibrium constants of the chemical reaction retain their statistical significance only in a two-dimensional space, even if none of the reaction partners are specifically adsorbed to an appreciable extent. Then if we consider the depolarization scheme of Eq. (383) under limiting current conditions, the rate for the backward reaction is zero, since $C_1 = 0$ at $x = a_2$, and the expression for the current is given by

$$i = -nFAk_{\text{h}}\,{}^{a_2}C_3 = -nFAk_{\text{h}}\,{}^{b}C_3 \exp\left(-\frac{z_3 F\phi_2}{RT}\right)$$

where k_{h} is the formal heterogeneous rate constant for the forward reaction. Comparing this equation with Eqs. (408) and (409), we now see that

$$v^* = k_{\text{h}} \exp\left(-\frac{z_3 F\phi_2}{RT}\right) \tag{413}$$

If $z_5 = 0$, and consequently $z_1 = z_3$, the only difference between Eq. (409) and Eq. (413) is represented by the fact that in Eq. (413) the pre-exponential factor does not depend on the bulk concentration of B_5. According to the writer, this is an alternative way of explaining the results obtained by Dandoy and Gierst in the study of the reduction of nickel ion (*125*). Analogously if we consider the depolarization scheme of Eq. (403) and assume, as before, that the reaction layer thickness is about equal to the diameter of a solvated molecule, the limiting current is expressed by

$$-\frac{i}{nFA} = k_h\,{}^{a_2}C_3\,{}^{a_2}C_5^{\nu} = k_h\,{}^{b}C_3\,{}^{\delta}C_5^{\nu} \exp\left[-(z_3 + \nu z_5)\frac{F}{RT}\phi_2\right] \tag{414}$$

where k_h is the heterogeneous rate constant for the forward reaction. On comparing Eq. (414) with Eq. (404), based on the reaction layer concept, we see that in both equations the pre-exponential factor has the same dependence on ${}^bC_3$ and ${}^\delta C_5$, although the dependence of the current on ϕ_2 is different. However, the variation of the limiting current controlled by the antecedent reaction (406) with ϕ_2, as measured by Grabowski and Bartel (*119*), can equally well be interpreted by Eqs. (404) and (414). In fact here $z_3 + \nu z_5 = z_1 = 1$.

VI. HETEROGENEOUS CHEMICAL REACTIONS

So far we have assumed that none of the substances taking part in the general chemical reaction (185) are specifically adsorbed at the electrode surface. If the chemical reaction characterized by a nonzero electrochemical affinity occurs exclusively at the inner Helmholtz plane, $x = a_1$, with participation of adsorbed substances, it contributes to the overall charge-transfer overpotential as defined by Eq. (39) rather than to chemical overpotential.

The rate constants of many antecedent reactions of recombination of protons with anions of weak acids, as determined by the polarographic method, are greater than 10^{11} l mole^{-1} sec^{-1} (*54, 136–141*). Since a bimolecular reaction cannot proceed faster than the diffusion of reactants, in which case the rate constant is of about 10^{11} l mole^{-1} sec^{-1} (*142, 143*), the above values are physically unsound. Some plausible justifications for these experimental findings have already been considered. They are (a) the contribution of proton donors other than H_3O^+ to the formation of the undissociated acid, which leads to very high "apparent" rate constants, if neglected (*46, 47, 141*); (b) the alteration of the concentrations of charged species within the diffuse layer; and (c) the change in equilibrium constants due to the dissociation field effect (*144*). We must not forget, however, that the majority of organic substances are more or less specifically adsorbed at the mercury-water interphase and that the relatively high surface concentrations reached at $x = a_1$ may cause chemical reactions to occur predominantly in the adsorbed state. Consider a substance B taking part in a homogeneous chemical reaction characterized by a reaction layer thickness of 10^{-6} cm, and therefore scarcely influenced by the diffuse layer structure. If the concentration of B is practically uniform in the reaction layer and equals 10^{-7} moles cm^{-3}, the amount of B available for the reaction per 1 cm^2 of the electrode surface is 10^{-13} moles. Assume that B is weakly adsorbed, so that under the experimental conditions employed the ratio θ of the surface concentration Γ_b of B to the maximum surface concentration Γ_m corresponding to a complete monolayer coverage of the surface equals 0.01. If we attribute to Γ_m the reasonable value of 10^{-10} moles cm^{-2}, the amount of B adsorbed per 1 cm^2 is 10^{-12} moles—that is, by one order of magnitude larger than the amount of B available for the homogeneous reaction. Since, in reality, the fractional surface coverage with organic substances is usually much greater than 0.01, the chemical reaction tends to proceed mainly at the electrode surface proper. Thus Mairanovskii (*145*) claims that many antecedent protonation reactions that were originally considered to be of a "bulk" nature are instead of a purely "surface" nature. Obviously, owing to the specific interactions between adsorbed substances and electrode material, the mechanism of a surface reaction is generally different from that of the corresponding "bulk" reaction and the same can be true for the order of reaction with respect to each reactant. In this sense the previous comparison of the amount of B adsorbed with that contained in the reaction layer is meaningful only if the rate constant of the chemical reaction proceeding in the solution and that of the reaction in the adsorbed state are of the same order of magnitude. Usually we observe that the reactivities of adsorbed particles are increased by the strong electric field existing at $x = a_1$ with respect to the corresponding reactivities in the bulk of the solution (cf. Sec. V.C.4).

A. Some Diagnostic Criteria for the Distinction of Heterogeneous Chemical Reactions

Some diagonostic criteria for the distinction between "surface" reactions and "bulk" reactions, including those influenced by the diffuse layer structure, follow,

(1) In general "surface" reactions are less sensitive than "bulk" reactions to the alteration of the ionic atmosphere in the diffuse layer produced by a change in the concentration of a nonspecifically adsorbed supporting electrolyte (cf. Sec. V.B).

(2) The addition of a surface-active but eletroinactive species S has two effects. First, the added species S produces a change in the value of ϕ_2, especially if it is charged and, second, it displaces the adsorbed "surface" reaction partners partially or totally from the electrode surface. These two effects may influence the rate of the chemical reaction, and consequently

the height of the kinetic wave, in opposite directions. Thus the addition of tetraalkylammonium cations, which are strongly adsorbed on the negative side of the electrocapillary maximum, causes an appreciable reduction in the absolute value of the negative ϕ_2 potential. Hence the concentrations of negatively charged reactants in the diffuse layer and at the outer Helmholtz plane increase, while those of positively charged reactants decrease accordingly. An analogous situation is also encountered in the inner Hemholtz plane, $x = a_1$, provided the bulk concentration of the organic cations is low, so that their surface concentration is negligible with respect to the maximum value. The more the surface concentration of tetraalkylammonium cations approaches the maximum value corresponding to monolayer formation, the more the specifically adsorbed reactants are displaced from the plane $x = a_1$, independent of their ionic charge. Hence a progressive increase in the bulk concentration of tetraalkylammonium cations causes an initial increase in the surface concentration of a negatively charged adsorbed reactant, followed by an appreciable decrease of this latter. A different situation is encountered outside the outer Helmholtz plane, where an increase in the concentration of tetrasubstituted ammonium cations produces a gradual decrease in the electrostatic repulsion of negatively charged reactants as well as in the electrostatic attraction of positively charged reactants. Thus the initial increase and subsequent decrease in the kinetic current of CrO_4^{2-} ion, observed on the addition of increasing amounts of tetraalkylammonium ions (*130*), can be ascribed to the surface nature of the antecedent reaction. An analogous effect of tetrasubstituted ammonium cations is observed on the kinetic limiting current for the reduction of undissociated maleic acid, limited by the antecedent protonation of the monovalent anion. Here, too, the specifically adsorbed reactant is negatively charged and consequently an inversion effect is to be expected on addition of the organic cations (*146*). No such effect of tetrasubstituted ammonium ions is exerted on the kinetic current for the reduction of maleic acid monoanions, which is limited by the antecedent protonation of the divalent anion. Here the addition of the organic cations results only in an increase of the kinetic current. This fact has been explained by Mairanovskii (*146*) by assuming that the protonation reaction has a "bulk" nature and occurs in such a thin reaction layer it is influenced by the diffuse layer structure. Since the inner potentials within the diffuse layer become less negative on addition of tetraalkylammonium salts, the net effect is a decreased electrostatic repulsion of maleic acid dianions, which results in an increased kinetic limiting current. The different behavior of maleic acid monanion with respect to the corresponding dianion can be attributed to the lower tendency of the latter to be specifically adsorbed on a negatively charged mercury electrode.

(3) An increase in temperature has two effects on kinetic currents; (a) it causes an increase in the rate constant both of "bulk" and "surface" chemical reactions and (b) it reduces the surface concentrations of specifically adsorbed reactants (*145*). If the chemical reaction is of a bulk nature, only the first effect is felt and a plot of the logarithm of the homogeneous rate constant (determined by the usual procedure) against $1/T$ yields a straight line. If the chemical reaction is of a strictly surface nature, the effect of the increase in the heterogeneous rate constant on the kinetic current is partially, totally, or more than compensated for by the concomitant effect of the decrease in the surface concentrations of the adsorbed reaction partners. This may explain the low rate of change of some kinetic limiting currents with temperature (*48*) and the corresponding unacceptably low values for the activation energy of the rate-controlling chemical reaction as calculated on attributing a volume nature to this latter. If the chemical reaction has a "mixed bulk-surface" nature—that is, occurs both in the adsorbed state and in the reaction layer volume—the contribution of the surface reaction to the kinetic current usually prevails over that of the corresponding bulk reaction at low temperatures owing to the higher adsorptivity of reactants. An increase in temperature causes the gradual desorption of reactants and the consequent decrease in the surface reaction contribution, so that ultimately the kinetic current becomes of a purely bulk nature. Such behavior is shown by the kinetic current limited by the antecedent protonation of the maleic acid monoanion to undissociated acid (*114*, *115*). In this case the apparent protonation rate constant varies slightly at room temperature, showing an appreciable increase only at temperatures higher than 70°C. This sharp increase is attributed by Mairanovskii (*115*) to the transition from a kinetic current of a predominantly surface nature to the corresponding bulk kinetic current, which, as previously stated, is characterized by a higher rate of increase with temperature.

The simple fact that a kinetic limiting current changes with applied potential is not indicative of a surface reaction since bulk reactions occurring mainly in the diffuse layer also show this behavior. In this respect the effect of a change of potential on the kinetic current must be correlated with the effect of a change in the nature and concentration of the supporting electrolyte in order to draw some conclusions as to the bulk or surface nature of the chemical reaction. We shall note, however, that the majority of organic substances studied by the polarographic technique are electroreduced on mercury on the negative side of the electrocapillary maximum, E_z. The adsorptivity of neutral organic substances reaches its maximum value in the proximity of E_z. If one of these substances is the reactant in a heterogeneous chemical reaction preceding the charge-transfer proper, a shift of the applied potential toward increasingly negative values under limiting current conditions causes a decrease in the surface concentration of the adsorbed reactant. Hence the limiting current decreases and the polarographic wave exhibits a maximum that is not related to the tangential motion of the mercury surface. Specifically adsorbed organic reactants bearing a negative charge also show an analogous behavior, which, however, is not indicative of a surface reaction. In fact, in this particular case the fall in the kinetic limiting current at increasingly negative potentials could equally well be attributed to a purely electrostatic repulsion of nonspecifically adsorbed negatively charged particles from the diffuse layer. At sufficiently negative potentials, large organic cations (e.g., tetraalkylammonium ions) are also desorbed from the electrode surface. The desorption of organic cations, which occurs at more negative potentials than for neutral molecules, is caused by the energy gain due to the increase in the double layer capacity accompanying the replacement of organic species with water molecules at the interphase (*147*). At sufficiently negative potentials, such an energy gain prevails over the simultaneous energy loss resulting from the removal of the organic cation from the negatively charged electrode. In this respect there are no pronounced differences between the behaviors of surface kinetic currents involving organic cations and neutral molecules, respectively, whereas appreciable differences are to be expected if the kinetic currents are of a bulk nature.

In order to draw some conclusions as to the bulk or surface nature of kinetic currents on the basis of the changes of the limiting current with potential, it is convenient to consider the problem from a quantitative, or at least semiquantitative, point of view.

B. The Adsorption Isotherms

1. Adsorption Overpotential

Let us consider the charge-transfer reaction (41) together with the chemical reaction (185). If this latter reaction is assumed heterogeneous, there is no need for retaining the separation between the stoichiometric equations (41) and (185) since the corresponding reactions take place in approximately the same location—that is, at $x \leqslant a_2$. Furthermore, the charge-transfer reaction (41) may well be composed of both electrochemical and purely chemical elementary steps. Under the present conditions, therefore, the "overall" charge-transfer process is more conveniently expressed by the sum of the two stoichiometric equations (41) and (185),

$$\sum_{i=1}^{n} (\nu_i + \nu_i')A_i + \sum_{k=n+1}^{n'} \nu_k' \mathrm{B}_k + ne = 0 \tag{415}$$

It is readily seen that the electroactive species A_i characterized by $\nu_i + \nu_i' = 0$ represent intermediates and do not appear in the stoichiometric equation (415). In order to simplify notations we shall drop the distinction between electroactive species A_i and electroinactive ones, B_k, denoting both of them by A_j. Hence Eq. (415) is recasted as

$$\sum_{j=1}^{n'} \nu_j \mathrm{A}_j + ne = 0 \tag{416}$$

For a chemical reaction to be heterogeneous—that is, to be influenced by specific interactions of some of the reaction partners with the electrode material—it is sufficient that only one reactant be specifically adsorbed. Confining ohmic overpotential on the other side of the boundary $x = \delta$ of the diffusion layer with respect to the electrode, as done in Sec. I.B.2, the electrochemical affinity $\tilde{A}^{0\delta}$ of reaction (416) between $x = 0$ and $x = \delta$ can be formally separated as follows:

$$\begin{aligned}
\frac{\tilde{\mathrm{A}}^{0\delta}}{nF} = \eta^{0\delta} &= \frac{1}{nF}\left(-n\tilde{\mu}_e - \sum_j \nu_j{}^{\delta}\tilde{\mu}_j\right) \\
&= \frac{1}{nF}\left(-n\tilde{\mu}_e - \sum_{j'} \nu_{j'}{}^{a_1}\tilde{\mu}_{j'} - \sum_{j''} \nu_{j''}{}^{a_2}\tilde{\mu}_{j''}\right) \\
&\quad + \frac{1}{nF}\left(\sum_{j'} \nu_{j'}{}^{a_1}\tilde{\mu}_{j'} - \sum_{j'} \nu_{j'}{}^{b}\tilde{\mu}_{j'}\right) \\
&\quad + \frac{1}{nF}\left(\sum_{j''} \nu_{j''}{}^{a_2}\tilde{\mu}_{j''} - \sum_{j''} \nu_{j''}{}^{b}\tilde{\mu}_{j''}\right) \\
&\quad + \frac{1}{nF}\left(\sum_{j=j',\,j''} \nu_j{}^{b}\tilde{\mu}_j - \sum_j \nu_j{}^{\delta}\tilde{\mu}_j\right)
\end{aligned} \tag{417}$$

Here the subscripts j' and j'' designate, respectively, the species that are specificially and nonspecifically adsorbed, while the superscripts denote as usual the location at which the electrochemical potentials $\tilde{\mu}_j$ are considered. $\eta^{0\delta}$ is the overpotential between $x = 0$ and $x = \delta$. The separation carried out in Eq. (417) is analogous to that accomplished in Sec. I.B.2 in the absence of specific adsorption and aims at locating the various types of overpotential in different regions of the solution layer around the electrode. The first term within the round brackets on the right-hand side of Eq. (417) is the algebraic sum of the electrochemical potentials of the reacting species in the positions that they occupy immediately before or after the charge-transfer proper. Obviously, these positions are different for adsorbed ($x = a_1$) and non-adsorbed ($x = a_2$) reaction partners. The first term in Eq. (417) can be identified with the charge-transfer overpotential η_t. The second term within the round brackets in the same equation expresses the contribution to $\eta^{0\delta}$ due to the penetration of specifically adsorbed species into the diffuse layer and to the subsequent adsorption step. It can be termed "adsorption overpotential" (*5*) η_a. The third term is the "penetration overpotential," η_p, namely, the contribution to the total overpotential due to the penetration of nonspecifically adsorbed species into the diffuse layer. Finally, the last term in Eq. (417) expresses the diffusion overpotential η_d. Comparison of Eqs. (234), (415), (416), and (417) shows that η_d is practically identical with the diffusion overpotential expressed by Eq. (234) and relative to the charge-transfer process (41) coupled with the homogeneous chemical reaction (185). The only difference is represented by the fact that in the present case diffusion overpotential is imagined to be located in the region (b,δ)—that is, immediately outside the diffuse double layer. On the basis of the arguments put forth in Sec. I.B.2, we shall assume that η_p is practically zero. This amounts to saying that the concentrations of nonspecifically adsorbed species at $x = a_2$ are related to the corresponding concentrations at the outer boundary of the diffuse layer $x = b$ through the usual Boltzman factor, $\exp(-z_{j''}F\phi_2/RT)$.

As concerns the adsorption overpotential η_a, there is no sure evidence of a measurable contribution of this type of overpotential to the total overpotential $\eta^{0\delta}$, at least on mercury. In principle the contribution of η_a to $\eta^{0\delta}$ can be enhanced by choosing a "reversible" redox system—namely, a system characterized by a very low value of η_t at nonequilibrium potentials—and by reducing η_d by recourse to an ac technique; but even by doing this results are uncertain owing to the exiguity of the effect to be measured. Thus we can not always be sure whether it is the adsorption step or the charge-transfer step that determines the rate of the overall depolarization process

(*148*). From what was previously stated it can be concluded that adsorption at a mercury-solution interphase is a fast process ($\eta_a \cong 0$) and that in the time scale of polarographic measurements adsorption equilibrium can be assumed with full confidence. Upon equating η_a to zero, we have

$$^{a_1}\tilde{\mu}_{j'} = {}^{b}\tilde{\mu}_{j'}$$

or also

$$^{a_1}\tilde{\mu}^0_{j'} + RTf(\Gamma_{j'}) = {}^{b}\tilde{\mu}^0_{j'} + RT \ln {}^{b}a_{j'} \tag{418}$$

where $^{a_1}\,\tilde{\mu}^0_{j'}$, and $^{b}\tilde{\mu}^0_{j'}$, are the standard electrochemical potentials of the j'th species in the adsorbed state and at $x = b$, respectively. The quantity $f(\Gamma_{j'})$ is the activity of the j' species in the adsorbed state, expressed as a function of the corresponding surface concentration $\Gamma_{j'}$, while $^{b}a_{j'}$ is the activity of the j'th species at $x = b$.

There are two main differences encountered in setting η_p and η_a, respectively, equal to zero. (a) The standard chemical potentials $\mu^0_{j''}$ of the species of the j'' type—that is, those nonspecifically adsorbed—have the same values at $x = a_2$ and at $x = b$ or also in the bulk of the solution. Conversely, the values for the standard chemical potentials $\mu^0_{j'}$ of the species of the j' type at $x = a_1$ are generally different from the corresponding values at $x = b$. (b) The activity of a given species at $x \geqslant a_2$ can be taken as approximately equal to the corresponding concentration, while this cannot be done in general for the activity and the corresponding surface concentration at $x = a_1$. It follows that the relation between $\Gamma_{j'}$ and $^{b}a_{j'}$ requires specific assumptions as to the nature of particle-particle and particle-metal interactions in the layer $(0, a_1)$. It must be realized that the importance of this relation is outstanding. In fact, if it is true that the rate of diffusion depends as a first approximation on the difference between the volume concentrations of the various species at $x = b$ and the corresponding bulk concentrations (see the diffusion layer approximation in Sec. II.B.2.b), the rates of the concomitant charge-transfer steps depend on the surface concentrations $\Gamma_{j'}$ of specifically adsorbed species as well as on the volume concentrations at $x = a_2$, $^{a_2}C_{j''}$ of nonspecifically adsorbed species.

2. The Standard Free Energy of Adsorption $\widetilde{\Delta G^\circ}$

According to the writer, a convenient way of deriving the relation $\Gamma_{j'} = f(^{b}C_{j'})$ between the surface concentration of a given species $A_{j'}$ and the corresponding volume concentration $^{b}C_{j'}$ immediately outside the diffuse layer consists in expressing formally the adsorption of this species by the reaction,

$$A_{j'} + S \rightleftharpoons SA_{j'} \tag{419}$$

where S represents a " bare site " on the electrode surface and $SA_{j'}$ the species $A_{j'}$ in the adsorbed state. As a matter of fact reaction (419) is not merely formal if we consider that the adsorption of one molecule of $A_{j'}$ involves the displacement of an unspecified number of solvent molecules from the inner part of the double layer and the simultaneous establishment of specific interactions between $A_{j'}$ and the electrode material. In this respect, for aqueous media we could write the following displacement reaction,

$$(A_{j'})_{sol} + n(H_2O)_{ads} \rightleftharpoons (A_{j'})_{ads} + n(H_2O)_{sol}$$

in place of Eq. (419) (*149*). With the formalism of Eq. (419) the condition of adsorption equilibrium is expressed by

$$\tilde{\mu}_s + {}^b\tilde{\mu}_{j'} = {}^{a1}\tilde{\mu}_{j'}$$

or, more explicitly,

$$\tilde{\mu}_s^0 + RT \ln a_s + {}^b\tilde{\mu}_{j'}^0 + RT \ln {}^b a_{j'} = {}^{a1}\tilde{\mu}_{j'}^0 + RT \ln {}^{a1}a_{j'} \tag{420}$$

where a_s and $a_{j'}$ are the activities of the free sites and of $A_{j'}$, respectively. On rearrangement, Eq. (420) yields

$$\frac{{}^{a1}a_{j'}}{a_s} = {}^b a_{j'} \exp\left(-\frac{\widetilde{\Delta G}^\circ}{RT}\right) \tag{421}$$

where

$$\widetilde{\Delta G}^\circ \equiv {}^{a1}\tilde{\mu}_{j'}^0 - {}^b\tilde{\mu}_{j'}^0 - \tilde{\mu}_s^0$$

represents the standard free energy of adsorption. According to statistical mechanics, $\widetilde{\Delta G}^\circ$ can be written as the sum of the interaction energies between each adsorbed molecule and the surface plus the interaction energies between any two adsorbed molecules. In this regard, $\widetilde{\Delta G}^\circ$ can be considered as the sum of two contributions, $\widetilde{\Delta G}^\circ_{pp}$ and $\widetilde{\Delta G}^\circ_{pm}$, due to particle-particle and particle-metal interactions, respectively. A reasonable way of accounting for these two contributions consists in assuming that particle-metal interactions are those most directly affected by the "electrical state" of the system. Thus we can imagine maintaining the bulk activity of the adsorbing material at such low values as to ignore particle-particle interactions and at the same time considering the effect of a change in the electrical state of the system on particle-metal interactions. The next step consists in holding the electrical state constant but at the same time considering the effect of an increase in the bulk activity of the adsorbing material upon particle-particle interactions. This effect should be analogous to that observed with ordinary nonelectrochemical adsorption, for example, adsorption of gases on solids.

a. The Contribution of Particle-Metal Interactions to $\widetilde{\Delta G}°$. Let us consider a neutral molecule of $A_{j'}$ and let us calculate the work done against the electric forces in moving it from a position where the field strength is zero (e.g., the plane $x = b$) to the plane $x = a_1$, where the field strength is X. If this work is carried out reversibly and at constant temperature, we know from thermodynamics that it equals the opposite of the corresponding free energy change. The subsequent derivation follows Butler (*150*). A somewhat different approach leading to analogous results had been previously proposed by Frumkin (*147*). If we assume that the molecule of $A_{j'}$ behaves as a dielectric element of volume δv_A, we know from electrostatic theory that the work done when this neutral molecule is moved from a position where the field strength is zero to a position where it is X equals

$$w_A = \delta v_A \int_0^X P_A \, dX$$

where P_A is the polarization of $A_{j'}$ per unit volume. If the molecule has a permanent dipole moment with fixed orientation, both in the absence and in the presence of the electric field, P_A is given by $p_A + \alpha_A X$. Here p_A is the polarization per unit volume in the absence of the electric field, whereas $\alpha_A X$ is the induced polarization, which is proportional to X according to α_A (polarizability). It follows that $w_A = \delta v_A(p_A X + \alpha_A X^2/2)$. The adsorption of a molecule of $A_{j'}$ at $x = a_1$ causes an equal volume of molecules B of the solvent to move away from the electrode. Denoting by p_B and α_B, respectively, the permanent polarization and the polarizability of the solvent per unit volume, the overall electric work involved in the adsorption of one molecule of $A_{j'}$ is given by

$$w = w_A - w_B = [(p_A - p_B)X + \tfrac{1}{2}(\alpha_A - \alpha_B)X^2] \, \delta v_A \tag{422}$$

The effect of the polarization induced by the electric field is to attract molecules of both the solute $A_{j'}$ and the solvent B towards the electrode. Since, however, the polarizability α_B of the solvent is generally greater than that α_A of $A_{j'}$, the latter species is in effect "squeezed out" of the surface. This is readily seen from Eq. (422), where the electrical work $\frac{1}{2}(\alpha_A - \alpha_B)X^2\,\delta v_A$ due to the induced polarization is negative. It should be noted that in the derivation of Eq. (422) the interactions among adsorbed particles have not been considered. The contribution $\widetilde{\Delta G}^\circ_{pm}$ to the standard free energy of adsorption per mole of adsorbate is given by $\widetilde{\Delta G}^\circ_{pm,X=0} - Nw$, where N is Avogadro's number, $-Nw$ the electrostatic contribution to $\widetilde{\Delta G}^\circ_{pm}$, whereas $\widetilde{\Delta G}^\circ_{pm,X=0}$ is the value of $\widetilde{\Delta G}^\circ_{pm}$ at zero electric field strength. In practice $\widetilde{\Delta G}^\circ_{pm,X=0}$ measures the contribution of the specific noncoulombic forces, acting between adsorbed particles and metal, to the standard free energy of adsorption. From Eqs. (421) and (422) it follows that

$$\frac{{}^{a_1}a_{j'}}{a_s} = {}^{b}a_{j'} \exp[-(aX^2 + bX)] \exp\left[-\frac{\widetilde{\Delta G}^\circ_{pp} + \widetilde{\Delta G}^\circ_{pm,\,X=0}}{RT}\right] \tag{423}$$

where

$$a \equiv \frac{N(\alpha_B - \alpha_A)\,\delta v_A}{2RT}$$

and

$$b \equiv \frac{N(p_B - p_A)\,\delta v_A}{RT}$$

On noting that the electric field strength for which $aX^2 + bX$ reaches its minimum value is given by $X_m = -b/2a$, we can write

$$aX^2 + bX = a\left(X + \frac{b}{2a}\right) - \frac{b^2}{4a} = a(X - X_m)^2 - \frac{b^2}{4a} \tag{424}$$

If the adsorbed species $A_{j'}$ is a large organic $z_{j'}$ – valent ion, additional electrical work $z_{j'}e\phi_1$ is done by the ion in passing from the bulk of the solution into the adsorbed state. ϕ_1 is the inner potential at $x = 1$ and e is the electron charge.

b. The Contribution of Particle-Particle Interactions to $\widetilde{\Delta G}^\circ$. As concerns the expression for $\widetilde{\Delta G}^\circ_{pp}$, from nonelectrochemical adsorption we know that the heat of adsorption $-\Delta H_a$ decreases with increasing coverage of the surface by the adsorbate, so that an analogous behavior is to be expected for $-\widetilde{\Delta G}^\circ_{pp}$. Possible reasons for such a decrease are (a) "a priori heterogeneity" of the surface, (b) lateral interactions of the adsorbed species, and (c) work function effect.

A priori heterogeneity is encountered when the metal surface can be regarded as made up of a distribution of micropatches, each one with a characteristic local free energy of adsorption. The micropatches schematize the various kinds of defects and disorders existing on metal surfaces and also the different crystallographic planes exposed to the adsorbing substance by a polycrystalline metal. If the volume concentration of the adsorbed substance at $x = b$ is progressively increased, in the beginning adsorption will take place preferentially on the micropatches characterized by higher values of $-\widetilde{\Delta G}$, either because of the higher rate of adsorption on such sites or because the adsorbate will tend to diffuse from patches of lower to patches of higher $-\widetilde{\Delta G}$ values. The gradual increase in the volume concentration of the adsorbing species at $x=b$ will result in the progressive coverage of patches of increasingly lower $-\widetilde{\Delta G}$ values, and consequently the heat of adsorption will decrease. Only if the energy barrier height for surface diffusion is $\gg kT$ (immobile adsorption) and the activation energy of adsorption is low or zero (non activated adsorption), "a priori heterogeneity" does not result in a coverage dependent heat of adsorption since patches of different energy are then filled at random. In this case the experimental heat of adsorption is simply an average value. Assume that the standard free energy of adsorption $\widetilde{\Delta G}^\circ$ is constant over each patch, at constant electrical state, and that the activities in Eq. (421) can be replaced as a first approximation by the corresponding concentrations. If $A_{j'}$ is the only substance adsorbed at the metal surface and if a molecule of $A_{j'}$ occupies a single adsorption site, we then have

$$^{a_1}a_{j'} \cong \Gamma_{j'}, \qquad a_s \cong \Gamma_{m,j'} - \Gamma_{j'} \qquad \text{and} \qquad ^{b}a_{j'} \cong {}^{b}C_{j'}$$

where $\Gamma_{j'}$ is the surface concentration of $A_{j'}$ in moles cm^{-2} and $\Gamma_{m,j'}$ is the value of $\Gamma_{j'}$ when the surface is completely covered by a monomolecular layer. Hence on a given micropatch the following relation applies,

$$\frac{\theta_s}{(1-\theta_s)} = {}^{b}C_{j'} \exp\left(-\frac{\widetilde{\Delta G}^\circ}{RT}\right) \tag{425}$$

or

$$\theta_s = {}^{b}C_{j'} \exp\left(-\frac{\widetilde{\Delta G}^\circ}{RT}\right) \Big/ \left[1 + {}^{b}C_{j'} \exp\left(-\frac{\widetilde{\Delta G}^\circ}{RT}\right)\right]$$

where $\theta_s = \Gamma_{j'}/\Gamma_{m,j'}$ represents the fractional surface coverage on the particular micropatch under examination. If the metal surface of area $s = 1$ is composed of elementary patches of area ds, then the fraction θ of the whole surface covered with the adsorbate is obtained by adding up the various θ_s

$$\theta = \int_0^1 \theta_s \, ds = \int_0^1 {}^bC_{j'} \exp\left(-\frac{\widetilde{\Delta G}^\circ}{RT}\right) \Big/ \left[1 + {}^bC_{j'} \exp\left(-\frac{\widetilde{\Delta G}^\circ}{RT}\right)\right] ds \quad (426)$$

If we assume that $\widetilde{\Delta G}^\circ$ is the same for all the patches, Eq. (426) yields the relation,

$$\theta = {}^bC_{j'} \exp\left(-\frac{\widetilde{\Delta G}^\circ}{RT}\right) \Big/ \left[1 + {}^bC_{j'} \exp\left(-\frac{\widetilde{\Delta G}^\circ}{RT}\right)\right] \quad (427)$$

which expresses the Langmuir adsorption isotherm (*151*).

We have already stated that the experimental heat of adsorption $-\Delta H_a$ usually decreases with coverage. If we assume that the standard free energy of adsorption has a similar behavior, we can write approximately, in accordance with Temkin (*152*),

$$\widetilde{\Delta G}^\circ = \widetilde{\Delta G}_0^\circ + \alpha s \quad (428)$$

where $\widetilde{\Delta G}_0^\circ$ and $\widetilde{\Delta G}_0^\circ + \alpha$ are the standard free energies of adsorption on the free surface and on the last part of the surface covered, respectively. The plus sign in Eq. (428) is justified by the fact that both $\widetilde{\Delta G}^\circ$ and $\widetilde{\Delta G}_0^\circ$ are negative quantities. Equation (428) expresses a linear fall of the absolute value for the standard free energy of adsorption with coverage. On substituting $\widetilde{\Delta G}^\circ$ from Eq. (428) into Eq. (426) and integrating, we readily obtain the relation

$$\theta = \frac{RT}{\alpha} \ln \frac{1 + {}^bC_{j'} \exp(-\widetilde{\Delta G}_0^\circ/RT)}{1 + {}^bC_{j'} \exp[-(\widetilde{\Delta G}_0^\circ + \alpha)/RT]} \quad (429)$$

Equation (429) expresses the so-called Temkin isotherm (*152*, *153*). If the parameter α is sufficiently large, an intermediate range of values of ${}^bC_{j'}$ exists, where ${}^bC_{j'} \exp(-\widetilde{\Delta G}_0^\circ/RT) \gg 1 \gg {}^bC_{j'} \exp[-(\widetilde{\Delta G}_0^\circ + \alpha)/RT]$, so that Eq. (429) assumes the following simplified form;

$$\theta = \frac{RT}{\alpha} \ln \left[{}^bC_{j'} \exp\left(-\frac{\widetilde{\Delta G}_0^\circ}{RT}\right)\right] \quad (430)$$

Equation (430), which expresses the logarithmic Temkin isotherm, holds only at intermediate coverages.

Another possible reason for the decrease in the heat of adsorption with coverage is the lateral interaction of adsorbed molecules, which may be due to the mutual influence of the corresponding electric dipoles. If surface diffusion of adsorbate can take place (mobile adsorption) and the surface is assumed homogeneous, the particles will tend to distribute themselves at optimum distances in order to reduce dipole-dipole interactions to a minimum. An increase in the volume concentration of the adsorbing substance at $x = b$ will reduce the average distances among the adsorbed molecules, with the consequent increase in the dipole repulsion energy and the corresponding decrease in the absolute value of $\widetilde{\Delta G}^\circ$. A theoretical calculation of dipole interactions shows, however, that the entity of the effect of such interactions is quite small and is too insufficient to account for the relatively large fall in the heat of adsorption with coverage (*154*). It is evident that lateral interactions give rise to an "induced" heterogeneity since, prior to the adsorption, all free sites of the homogeneous surface have an equal a priory probability to be occupied.

Another possible cause of induced heterogeneity is represented by the "work function effect," which is encountered when the adsorbate is bound with the metal by a covalent chemisorption bond (*155*, *156*). In this case, at low values of the fractional coverage, the electrons shared by the metal come from the top of the conductivity band. With increasing coverage the energy expended on exciting an electron from the conductivity band to the surface orbitals grows and consequently the heat of adsorption decreases.

In the case of induced heterogeneity a convenient way of representing the relation between the surface concentration $\Gamma_{j'}$ and the corresponding volume concentration ${}^bC_{j'}$ at $x = b$ consists in applying the general equation (425) to the homogeneous surface of unit area $s = 1$ and in assuming that at constant electrical state $-\widetilde{\Delta G}^\circ$ decreases linearly with coverage (see also the analogous Eq. (428)),

$$\widetilde{\Delta G}^\circ = \widetilde{\Delta G}_0^\circ + \alpha\theta \tag{431}$$

Combination of Eqs. (425) and (431) yields

$$\frac{\theta}{1-\theta}\exp\left(\frac{\alpha\theta}{RT}\right) = {}^bC_{j'}\exp\left(-\frac{\widetilde{\Delta G}_0^\circ}{RT}\right) \tag{432}$$

Equation (432) has the form of the Frumkin isotherm. It must be noted, however, that Frumkin derived Eq. (432) in order to interpret the electrocapillary curves of fatty acids of high molecular weight (*157*). According to Frumkin the behavior of these organic compounds can be satisfactorily explained by postulating the existence of van der Waals' attraction forces among the adsorbed particles. Under these assumptions, $-\widetilde{\Delta G}^\circ$ increases with coverage and consequently the factor α is negative. The shape of the $\theta - {}^bC_{j'}$ curve differs considerably according to whether α is positive or negative. Thus for a positive value of α the rate of increase of θ with ${}^bC_{j'}$ is lower than that predicted by the Langmuir isotherm depending on whether the Temkin or the Frumkin isotherm is employed. Conversely, if α is negative, the $\theta - {}^bC_{j'}$ curve expressed by the Frumkin isotherm is steeper than the curve calculated from the corresponding Langmuir isotherm and exhibits an S shape (see Fig. (10)).

If two or more different substances are adsorbed, the activity of adsorption sites in the general Eq. (421) is approximately expressed by $1 - \sum_{j'} \theta_{j'} \equiv 1 - \sum_{j'} \Gamma_{j'}/\Gamma_{m,j'}$, where the summation extends to all the adsorbed

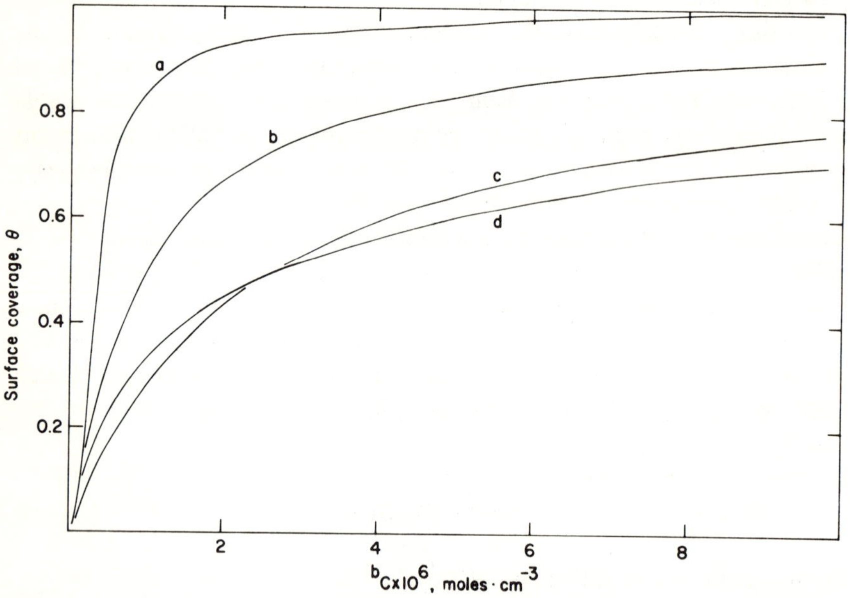

Fig. 10. Plot of θ against ${}^bC$ according to the Frumkin isotherm with negative α (curve *a*), the Langmuir isotherm (curve *b*), the Temkin isotherm (curve *c*), and the Frumkin isotherm with positive α, as obtained for $\exp(-\widetilde{\Delta G}_0^\circ/RT) = 10^{-6}$ cm^3 mole^{-1} and for $|\alpha| = 2RT$.

species. Obviously the surface activity ${}^{a_1}a_{j'}$ of a given species $A_{j'}$ is still expressed by $\theta_{j'} = \Gamma_{j'}/\Gamma_{m,j'}$. If the standard free energy of adsorption $\widetilde{\Delta G}^\circ_{j'}$ relative to a given species $A_{j'}$ is not independent of coverage at constant electrical state and an induced heterogeneity model can be assumed to account for the decrease of $|\widetilde{\Delta G}^\circ_{j'}|$ with coverage, we can approximately express $\widetilde{\Delta G}^\circ_{j'}$ in the form,

$$\widetilde{\Delta G}^\circ_{j'} = \widetilde{\Delta G}^\circ_{0,\,j'} + \sum_{j'} \alpha_{j'}\theta_{j'}$$

where different free energy parameters $\alpha_{j'}$ are assigned to the different adsorbates (*158*, *159*).

If it is desired to account for the influence of the electrical state on the adsorption isotherms of Eqs. (429), (430), or (432), in view of Eq. (423) and (424) it is sufficient to replace the parameter $\widetilde{\Delta G}^\circ_0$ in these equations by

$$\widetilde{\Delta G}^\circ_0 = \widetilde{\Delta G}^\circ_{\mathrm{pm},\,X=0} + RT\left[a(X - X_{\mathrm{m}})^2 - \frac{b^2}{4a}\right] \tag{433}$$

where $\widetilde{\Delta G}^\circ_{\mathrm{pm},\,X=0}$ is the standard free energy of adsorption at zero coverage (no particle-particle interactions) and at zero electric field. In the case of the Langmuir isotherm of Eq. (427), $\widetilde{\Delta G}^\circ$ is independent of coverage and can be directly identified with the expression of $\widetilde{\Delta G}^\circ_0$ given in Eq. (433).

c. The Parameter Defining the Electrical State. The major difficulty encountered in the practical application of the above isotherms to electrochemical adsorption consists in the fact that the electric field X in the inner double layer is not measurable directly, so that it is necessary to relate it to an accessible quantity. Thus if we assume a linear change of potential within the inner part ($0 \leqslant x \leqslant a_1$) of the double layer, the potential drop across this layer is proportional to X and, therefore, is a measure of the electric field strength. On the other hand by assuming the validity of Gauss' law (Eq. (378)) at $x = 0$ and a dielectric constant independent of x and of the amount of the adsorbate, it is the charge on the metal q_M that is proportional to X. According to Parsons (*160*) the best way of assigning an isotherm, especially for the adsorption of ions, consists in carrying out measurements of surface concentrations at constant q_M. One reason for this choice should be that the potential difference measured across the whole double layer includes the potential drop across the diffuse layer, which varies with the amount of ionic adsorption. A similar effect, although much less significant, should be expected with uncharged substances. According to Frumkin (*147*) and Damaskin (*161*), in the study of adsorption isotherms for uncharged substances the potential difference E across the whole double layer rather than the charge q_M on the metal should be chosen as the variable defining the electrical state. The question of which of the two electrical parameters q_M or E should be held constant in studying adsorption isotherms is still controversial.

C. Phenomenological Treatment of Polarographic Currents Controlled by Heterogeneous Reactions

In considering the total overpotential $\eta^{0\delta}$ expressed by Eq. (417), we can reasonably state that it consists of the two contributions η_t and η_d, since η_p and η_a are practically zero. Nevertheless the two equations $\eta_p = 0$ and $\eta_a = 0$ are important, since the former provides the relations between the concentrations of nonspecifically adsorbed species at $x = b$ and the corresponding concentrations at the outer Helmholtz plane, $x = a_2$ (cf. Eq. (372)), while the latter supplies the relations between the volume concentrations ${}^bC_{j'}$ of specifically adsorbed species at $x = b$ and the corresponding surface concentrations $\Gamma_{j'}$ at the inner Helmholtz plane under equilibrium conditions with respect to the adsorption steps (see the various adsorption isotherms). We have already stated in Sec. I.B.2 that, in general, an overall electrode reaction consists of a succession of heterogeneous steps. Some of these steps involve a charge transfer across the metal-solution interphase (electrochemical steps), while the others are purely chemical in nature. Under steady-state conditions the current density $I = i/A$ is related to the rate of any given elementary step by Eq. (48), provided the species involved are nonspecifically adsorbed. In the case of specific adsorption $J_{\rho'}$ is defined by Eq. (40) in terms of surface concentrations and is related to I through the obvious equation,

$$\frac{I}{nF} = -\frac{J_{\rho'}}{\nu_{\rho'}} \tag{434}$$

From a phenomenological point of view, the rate $J_{\rho'}$ of the ρ'th elementary step is expressed differently according to whether the step is electrochemical or chemical. Thus if we temporarily assume that Eq. (41) expresses an elementary electrochemical reaction, according to classical chemical kinetics the corresponding rate J is given by

$$J = k_f \prod_{\nu_i < 0} a_i^{-\nu_i} - k_b \prod_{\nu_i > 0} a_i^{\nu_i} \tag{435}$$

where a_i represents volume concentrations ${}^{a_2}C_i$ at $x = a_2$ in the case of nonspecifically adsorbed species and surface concentrations Γ_i at $x = a_1$ for specifically adsorbed species. The forward and backward rate constants k_f and k_b are assumed to depend exponentially on the difference between the electrical potential ϕ_0 at the electrode surface and that $\phi_{1,2}$ at the position occupied by the electrochemical reaction partners immediately before or after the electron transfer proper (*6*),

$$k_f = k_f^0 \exp\left[\frac{\alpha nF}{RT}(\phi_0 - \phi_{1,2})\right]; \; k_b = k_b^0 \exp\left[-(1-\alpha)\frac{nF}{RT}(\phi_0 - \phi_{1,2})\right] \tag{436}$$

$\phi_{1,2}$ is usually identified with the potential ϕ_2 at the outer Helmoltz plane, $x = a_2$, if all the species participating in the charge-transfer step are nonspecifically adsorbed, and with the potential ϕ_1 at the inner plane, $x = a_1$, if all the species are specifically adsorbed. If the electrochemical step involves both specifically and nonspecifically adsorbed substances, the choice of $\phi_{1,2}$ represents a stumbling block. According to Eq. (436) it is the electric field in the region between the electrode and the plane of closest approach for the reacting substances that affects the net rate of the electrochemical step by enhancing the rate in one direction and hindering the rate in the opposite. The potential difference $\phi_{1,2}$ between the plane of closest approach and the bulk of the solution affects kinetics only indirectly by altering the concentrations of reaction partners in this plane.

The expression for the rate of a purely chemical heterogeneous step is analogous to Eq. (435), but in the present case k_f and k_b do not depend directly on potential. However, an indirect effect of E on k_f and k_b is not to be excluded, owing to the influence of high electric fields on chemical equilibria (*162*) (cf. Sec. V.C.4). Equations (48) and (434) give the current as a function of the rates $J_{\rho'}$ of the various elementary steps. The expressions for these rates contain the concentrations at $x = a_2$, and/or $x = a_1$, or both, of various species, some of which do not appear in the stoichiometric equation of the overall electrode process of Eq. (416). According to the steady-state approach, the rate of change of the concentrations of these intermediates with time is taken as zero. If we express the rates of the various elementary steps of reaction (416) in terms of concentrations and we make use of Eqs. (48) or (434), it is straightforward to obtain by simple algebraic passages the current i as a function of the concentrations at $x = a_2$ and/or $x = a_1$ of all the reacting species, with the exclusion of intermediates. The surface concentrations of specifically adsorbed species are expressed successively in terms of the corresponding volume concentrations at $x = b$—that is, immediately outside the diffuse layer—by the use of the appropriate isotherm. Analogously the volume concentrations of nonspecifically adsorbed species at $x = a_2$ are expressed in terms of the corresponding concentrations at $x = b$ by using Eq. (372). In this connection tentative assumptions must be made about the nature of adsorption isotherms, the values of adsorption parameters, as well as about the value of the potential drop. $\phi_0 - \phi_{1,2}$ affecting the charge-transfer proper. The final step in the derivation of the theoretical current-potential characteristic consists in eliminating the various concentrations at $x = b$ from the expression of i by taking into due account diffusion overpotential (cf. Sec. II.B.2.b).

The fitting of the theoretical current-potential characteristic with the experimental one is cumbersome when using the steady-state approach owing to the high number of adjustable parameters contained in the theoretical expression for the current. Consequently it is often convenient to assume that all the elementary steps, with exclusion of one or at most two, have a practically zero electrochemical affinity. This amounts to postulating that in spite of the fact that all steps proceed at the same net rate $J_{\rho'}/v_{\rho'}$, the majority of them are characterized by forward and backward rates that are very high with respect to their difference. Consequently we can approximately equate the forward to the backward rate under nonequilibrium conditions, thus obtaining the law of mass action. According to this "quasi-equilibrium" approach, the law of mass action applies to almost all heterogeneous steps, with the exclusion of a very few. If the totality of steps has zero electrochemical affinity, we can apply Nernst's equation to the overall electrode process (416) so that the resultant current is diffusion controlled, ($\eta^{0\delta} = \eta_d$). In this case there is no need for making assumptions as to the double layer structure or the nature of adsorption isotherms, but obviously no conclusions at all can be drawn about the mechanism of the electrode process from experimental data.

An example of quasi-equilibrium approach has been given by Kůta and Koryta in the study of the reduction of oxygen to hydrogen peroxide at the mercury electrode (*163*). All possible mechanisms for the two-electron reduction of O_2 to H_2O_2 were examined by the authors. Thus the first step can be expressed by the following schemes:

$$\begin{cases} O_2 + e \rightleftharpoons O_2^- \\ O_2^- + H^+ \rightleftharpoons HO_2 \end{cases} \quad \begin{cases} O_2 + H^+ \rightleftharpoons HO_2^+ \\ HO_2^+ + e \rightleftharpoons HO_2 \end{cases}$$

whereas the second step can be

$$\begin{cases} O_2^- + e \rightleftharpoons O_2^{2-} \\ O_2^{2-} + 2H^+ \rightleftharpoons H_2O_2 \end{cases} \quad \begin{cases} HO_2 + e \rightleftharpoons HO_2^- \\ HO_2^- + H^+ \rightleftharpoons H_2O_2 \end{cases} \quad \begin{cases} HO_2 + H^+ \rightleftharpoons H_2O_2^+ \\ H_2O_2^+ + e \rightleftharpoons H_2O_2 \end{cases}$$

In determining the expression for the current as a function of the concentrations of O_2 and H_2O_2 at $x = a_2$ (specific adsorption was excluded for all reacting species), it was assumed that all chemical steps are at equilibrium and that only one of the two one-electron charge-transfer steps is rate determining. The concentrations at $x = a_2$ were expressed in terms of the corresponding concentrations at $x = b$ through the use of Eq. (372). The concentrations at $x = b$ of the unstable intermediates O_2^-, HO_2, $H_2O_2^+$ were eliminated from the expression of the current by applying Nernst's equation to that of the two charge-transfer steps assumed to be at equilibrium. From the experimental data it was concluded that the rate-determining step is the electrode reaction $O_2 + e \rightleftharpoons O_2^-$ proceeding at the outer Helmholtz plane.

1. General Formulation of the Diffusional Problem

The diffusion overpotential can be taken into account either by a rigorous numerical procedure or by an approximate procedure based on the diffusion layer approximation. In the preceding section we have seen that the rate v_t of the overall charge-transfer process can be expressed by a more or less complicated relation of the type,

$$v_t = -\frac{i}{nFA} = v_t(\{^bC_j\}, \{k, \alpha\}, \{\beta\}, \phi_{1,2}, E) \tag{437}$$

where $\{^bC_j\}$ designates the set of concentrations of the various species at $x = b$, $\{k, \alpha\}$ the set of kinetic parameters, and $\{\beta\}$ symbolizes the set of adsorption parameters. $\phi_{1,2}$ is the inner potential at the plane of closest approach and E is the whole potential drop ϕ_0 across the double layer as measured against a given reference electrode. Let us note that during one occurrence of the overall charge-transfer reaction (416), involving n Faradays, the number of moles of a given nonspecifically adsorbed species $A_{j''}$ that diffuse through the plane $x = b$ under steady-state conditions is $|\nu_{j''}|$. Hence the current is proportional to the number of $|\nu_{j''}|$ moles of $A_{j''}$ that diffuse through $x = b$ per unit of time per unit area of the electrode according to the proportionality factor nFA,

$$i = \frac{nFA}{\nu_{j''}} D_{j''}\left(\frac{\partial C_{j''}}{\partial x}\right)_{x=b} \qquad \text{for all species } A_{j''} \tag{438}$$

Equation (438) is entirely analogous to Eq. (243).

The number of moles $dN_{j'}/dt$ of a given specifically adsorbed species $A_{j'}$ that react at the electrode surface per unit time equals the number of moles of $A_{j'}$ diffusing through the plane $x = b$ minus that accumulating at the inner Helmholtz plane, $x = a_1$, in the same unit time,

$$\frac{dN_{j'}}{dt} = AD_{j'}\left(\frac{\partial C_{j'}}{\partial x}\right)_{x=b} - \frac{d}{dt}(A\Gamma_{j'}) \tag{439}$$

The second term on the right-hand side of Eq. (439) is the time derivative of the number of moles $A\Gamma_{j'}$, of $A_{j'}$ adsorbed on the electrode surface. Since the number of moles of $A_{j'}$ involved in one occurrence of the electrode reaction (416) and consequently in the transfer of n Faradays across the interphase is $|\nu_{j'}|$, the current is given by

$$\bar{i} = \frac{nF}{\nu_{j'}}\frac{dN_{j'}}{dt} = \frac{nF}{\nu_{j'}}\left\{AD_{j'}\left(\frac{\partial C_{j'}}{\partial x}\right)_{x=b} - \frac{d}{dt}(A\Gamma_{j'})\right\} \qquad \text{for all the species } A_{j'} \tag{440}$$

where $A = A_0 t^{2/3}$.

On expressing the surface concentrations $\Gamma_{j'}$ in Eq. (439) in terms of the volume concentrations at $x = b$ through the use of the appropriate adsorption isotherms, Eqs. (437), (438), and (440) constitute a set of $n' + 1$ relations between the current i and the concentrations ${}^bC_j$ of the n' diffusing species. Once i is eliminated from the above equations, we obtain n' relations among the concentrations ${}^bC_j$ and the corresponding gradients $(\partial C_j/\partial x)_{x=b}$. These relations represent the n' boundary conditions for the set of differential equations,

$$\frac{\partial C_j}{\partial t} = D_j \frac{\partial^2 C_j}{\partial x^2} + \frac{2x}{3t}\frac{\partial C_j}{\partial x} \qquad x \geq b, \quad j = 1, \ldots, n' \tag{441}$$

describing the diffusion of the various species A_j toward the growing drop. The initial conditions for the present boundary value problem are trivial and express the uniformity of the concentrations C_j from the bulk of the solution up to the plane $x = b$ at the start of electrolysis,

$$C_j = {}^{\delta}C_j \qquad \text{for} \begin{cases} t = 0, & x \geq b \\ t > 0, & x \to \infty \end{cases}, \quad j = 1, \ldots, n' \tag{442}$$

In general, the rigorous solution of the boundary value problem in Eqs. (437), (438), and (440) through (442) is quite involved, although in principle it can be carried out by the mathematical procedure outlined in the appendix.

2. Approximate Solution of the Diffusional Problem on the Basis of the Diffusion Layer Concept

A satisfactory approximate solution of the diffusional problem outlined in the preceding section can be readily obtained by using the diffusion layer approximation (cf. Sec. II.B.2.b). As in Sec. II.B.2 and IV.C.1, we shall assume that at constant potential the volume concentrations ${}^{b}C_j$ at $x = b$ remain constant during the whole drop life. It follows that owing to the instantaneous achievement of adsorption equilibrium, the various $\Gamma_{j'}$ are also time independent at constant potential. On setting $(\partial C_j/\partial x)_{x=b}$ equal to $({}^{\delta}C_j - {}^{b}C_j)/\sqrt{3\pi D_j t/7}$ in view of Eq. (110), and considering the various ${}^{b}C_j$ as independent of time, integration of Eq. (438) and (440) over the drop time yields

$$\bar{i} = \frac{1}{t_\mathrm{d}} \int_0^{t_\mathrm{d}} i\, dt = \sqrt{\frac{12 D_{j''}}{7\pi}}\, \frac{nFA_0}{\nu_{j''}}\, t_\mathrm{d}^{1/6}({}^{\delta}C_{j''} - {}^{b}C_{j''}) \qquad \text{for all } j'' \tag{443}$$

for nonspecifically adsorbed species, and

$$\bar{i} = \sqrt{\frac{12 D_{j'}}{7\pi}}\, \frac{nFA_0}{\nu_{j'}}\, t_\mathrm{d}^{1/6}({}^{\delta}C_{j'} - {}^{b}C_{j'}) - \frac{nFA_0}{\nu_{j'}}\, t_\mathrm{d}^{-1/3}\Gamma_{j'}(\{{}^{b}C_{j'}\}) \qquad \text{for all } j' \tag{444}$$

for specifically adsorbed species. In Eq. (444) we set $\Gamma_{j'} \equiv \Gamma_{j'}(\{{}^{b}C_{j'}\})$ in order to emphasize the fact that in the most general case a given surface concentration $\Gamma_{j'}$ depends upon the concentrations at $x = b$ of all adsorbed species $A_{j'}$.

As an example let us consider the form taken by Eq. (444) in the case that only two substances, A_1 and A_2, are specifically adsorbed. If we assume that

A_1 and A_2 are characterized by the same value of Γ_m, as a first approximation the activity a_s of adsorption sites in the general equation (421) can be expressed by $\Gamma_m - \Gamma_1 - \Gamma_2$. Upon identifying the surface activities with the corresponding surface concentrations and assuming that the standard free energies of adsorption for A_1 and A_2 are independent of coverage (Langmuir isotherm assumption), the application of Eq. (421) to the species A_1 and A_2 yields

$$\frac{\Gamma_1}{\Gamma_m - \Gamma_1 - \Gamma_2} = K_1 {}^bC_1; \qquad \frac{\Gamma_2}{\Gamma_m - \Gamma_1 - \Gamma_2} = K_2 {}^bC_2 \qquad (445)$$

where $K_1 \equiv \exp(-\widetilde{\Delta G_1^\circ}/RT)$ and $K_2 \equiv \exp(-\widetilde{\Delta G_2^\circ}/RT)$. Derivation of Γ_1 and Γ_2 from Eq. (445) and their replacement into Eq. (444) results in

$$\frac{\bar{i}}{nFA_0} = \sqrt{\frac{12D_{j'}}{7\pi}}\,\frac{t_d^{1/6}}{v_{j'}}({}^\delta C_{j'} - {}^bC_{j'}) - \frac{t_d^{-1/3}}{v_{j'}}\,\frac{K_{j'}\,{}^bC_{j'}}{1+\gamma} \qquad \text{with } j' = 1, 2 \qquad (446)$$

where $\gamma = K_1 {}^bC_1 + K_2 {}^bC_2$. Equation (446) has been used by the writer (*164*) in the study of the influence of Langmuirian adsorption of A_1 and A_2 upon the charge-transfer process $A_1 \rightleftharpoons A_2 + ne$. If one of the two species, for example, A_2, is not appreciably adsorbed, the parameter γ in Eq. (446) becomes equal to $K_1 {}^bC_1$. The resulting equation was used by Brdička in the theoretical study of adsorption waves (*165*).

Returning to the general problem, a further relation between the mean current $\bar{i}$ and $\{{}^bC_j\}$ can be obtained by integrating Eq. (437) over t_d,

$$-\frac{\bar{i}}{nFA_0} = -\frac{1}{nFA_0 t_d}\int_0^{t_d} i\,dt = v_t(\{{}^bC_j\}, E)\frac{1}{t_d}\int_0^{t_d} t^{2/3}\,dt$$

$$= \tfrac{3}{5} v_t(\{{}^bC_j\}, E)\, t_d^{2/3} \qquad (447)$$

Equations (443), (444), and (447) constitute a set of $n' + 1$ algebraic relations among the n' concentrations ${}^bC_j$, the current $\bar{i}$, and the applied potential E. It is therefore possible to eliminate the various ${}^bC_j$ by simple passages, thus obtaining the current-potential characteristic $\bar{i} = \bar{i}(E)$.

3. Limitations of Phenomenological Treatments of Surface Kinetic Currents

In principle the general approximate treatment previously outlined can be used for deriving the equation of any surface kinetic wave. It must be noted, however, that theoretical analysis of kinetic currents controlled by heterogeneous reactions is still quite rudimentary because of the great number of difficulties to be overcome. Thus definite assumptions must be made as to the mechanism of the overall electrode process. Detailed mathematical treatment of complex reaction mechanisms may not prove useful since the subsequent step consisting in the comparison of the theoretical current-potential curves with the experimental ones would seldom yield unambiguous answers. Furthermore the writing of the current in terms of volume concentrations of the reacting species at $x = b$ requires knowledge of the potential $\phi_{1,2}$ in the plane of closest approach. No serious difficulties are encountered if both the reaction partners and the supporting electrolyte are nonspecifically adsorbed. In fact, under these circumstances it is the supporting electrolyte that determines almost exclusively the value ϕ_2 of the inner potential at the outer Helmholtz plane. Consequently ϕ_2 can be obtained separately from electrocapillary measurements on solutions of the supporting electrolyte alone. An analogous procedure can be followed if it is only one of the constituent ions of the supporting electrolyte to be specifically adsorbed (*108*). In this case, however, some not wholly orthodox assumptions must be made in order to derive the value of ϕ_2 from electrocapillary data. If one or more reacting species are specifically adsorbed, the writing of the current in terms of the various ${}^{b}C_j$ requires selection of an adsorption isotherm for each adsorbed species and knowledge of the potential ϕ_1 in the adsorbed state. In this case no separate electrocapillary measurements at a polarized electrode can be made, since the substances adsorbed are also electroactive and the proper choice of adsorption isotherms is a major stumbling block. Furthermore, during electrolysis at constant potential the surface concentrations of the reacting substances specifically adsorbed vary somewhat, with the con-consequence that the inner potentials at $x = a_2$ and, more significantly, at $x = a_1$, change during the drop life.

This explains why in the few theoretical treatments of surface kinetic currents available in the literature, the potential drop ϕ_1, ineffective in either promoting or hindering the charge transfer, has never been subtracted from the applied potential E. Analogously the effect of a change in ϕ_2 on the concentrations at $x = a_2$ of nonspecifically adsorbed substances (e.g., hydrogen ions in antecedent protonation reactions) has always been considered only from a qualitative point of view.

As an example, the kinetic limiting current due to the reduction of 2-acetylthiophene semicarbazone (*145*), which seems to be controlled by a heterogeneous preprotonation reaction, exhibits a drop at more negative potentials due to desorption of the electroinactive form of the depolarizer. A decrease of the ionic strength μ at constant buffer capacity reduces the entity of this drop. This phenomenon can be qualitatively explained by considering that, independent of the presence of specifically adsorbed substances at $x = a_1$, the inner potential ϕ_2 at $x = a_2$ on the negative branch of the electrocapillary curve becomes more negative with decreasing μ. It follows that the volume concentration of hydrogen ions at $x = a_2$ increases, and since as a first approximation we may take the plane $x = a_2$ as the plane of closest approach for H_3O^+ ions, the kinetic current increases accordingly. Addition of tetraalkylammonium ions, which are specifically adsorbed, causes ϕ_2 to become less negative and consequently the hydrogen ion concentration at the outer Helmholtz plane to decrease. This results in a lowering of surface kinetic currents controlled by a preceding protonation reaction. However, partial replacement of the organic reactant by tetraalkylammonium ions at $x = a_1$ cannot be excluded as a possible cause of the above lowering. It must be emphasized that all the above considerations on double layer effects are only qualitative, owing to the practical impossibility of establishing the values of ϕ_2 and ϕ_1 with an acceptable degree of accuracy.

4. Application of the Approximate Procedure to Surface Kinetic Currents Controlled by a Preprotonation Reaction

In the present section we shall deal with the application of the general treatment of Sec. VI.C.2 to surface kinetic waves controlled by the rate of an antecedent protonation reaction. Incidentally, the above general treatment is based on the assumption that for an electroactive substance which is specifically adsorbed, adsorption must necessarily precede charge transfer. In other words parallel discharge of the unadsorbed substance is excluded. Although this position is not fully justified (*166*), several experimental facts seem to confirm its substantial correctness (*164*). The important class of surface catalytic hydrogen waves is outside the scope of this contribution and will not be considered here. For a detailed review of this topic the reader is referred to Mairanovkii's recent book, *Catalytic and Kinetic Waves in Polarography* (*167*).

Let us consider the depolarization scheme

$$H^+ + A^- \underset{k_s K_s}{\overset{k_s}{\rightleftharpoons}} HA; \qquad HA + ne \xrightarrow{k_f} \text{products} \tag{448}$$

where the antecedent protonation reaction occurs in the adsorbed state and the subsequent charge-transfer step is unidirectional. In the present case, the overall heterogeneous electrode process (416) is represented by the equation,

$$H^+ + A^- + ne \rightarrow \text{products}$$

and consists of the two consecutive steps of Eq. (448), both characterized by a nonzero electrochemical affinity. We shall limit ourselves to the case in which the protonation equilibrium is shifted to the left in the bulk of the solution, so that the depolarizer is present almost exclusively in its electro-inactive unprotonated form. We shall further assume that the solution is well buffered, so that the hydrogen ion concentration at $x = b$ can be considered to be equal to the corresponding bulk concentration during electrolysis. In view of Eqs. (434), (435) and (436) the instantaneous current i under steady-state conditions is expressed by the two equations,

$$\frac{i}{nFA} = k_s{}^{a_2}[H^+]\Gamma_A - k_s K_s \Gamma_{HA} \tag{449}$$

$$\frac{i}{nFA} = k_f^0 \Gamma_{HA} \exp\left[-\frac{\alpha nF}{RT}(E - E_m)\right] \tag{450}$$

In Eq. (449), expressing the rate of the chemical step, it has been assumed that both the undissociated and the dissociated forms of the acid react in the adsorbed state, whereas hydrogen ions take part in the reaction while occupying the outer Helmholtz plane. In Eq. (450), which represents the rate of the charge-transfer step, k_f° denotes the rate constant at the potential $E = E_m$ and embodies the correction term for the potential drop ϕ_1 across the layer (a_1, b). In fact, the charge-transfer involves the specifically adsorbed undissociated molecule HA and consequently is affected by the potential difference between $x = 0$ and $x = a_1$. We shall assume that the electric field strength X is proportional to the potential drop across the rigid part of the double layer, and we shall denote by E_m the applied potential at which $|\Delta\tilde{G}^\circ_{pm}|$ for the species A^- is a maximum. On combining Eqs. (449) and (450), we obtain

$$\frac{i}{nFA} = k_s' \Gamma_A \Big/ \left\{1 + \frac{k_s K_s}{k_f^0} \exp\left[\frac{\alpha nF}{RT}(E - E_m)\right]\right. \tag{451}$$

where $k_s' = k_s{}^{a_2}[H^+]$. It must be noted that k_s' and k_f^0 embody the double layer parameters ϕ_2 and ϕ_1, respectively, and consequently are potential dependent. We shall assume, however, that the changes of k'_s and k^0_f with potential are negligible in the potential range in which the kinetic wave develops. Integration of Eq. (451) over the drop time t_d results in

$$\frac{\bar{i}}{nFA_0} = \frac{\frac{3}{5}k'_s\Gamma_A({}^b[A^-], {}^b[HA])\, t_d^{2/3}}{1 + \dfrac{k_s K_s}{k^0_f}\exp\left[\dfrac{\alpha nF}{RT}(E - E_m)\right]} \tag{452}$$

Equation (452) is the application of the general equation (447) to the problem at hand. In it we have emphasized that, in the most general case, Γ_A is a function both of ${}^b[A^-]$ and ${}^b[HA]$. Diffusion of the species A^- towards the electrode is taken into account by using Eq. (444), which now reads

$$\frac{\bar{i}}{nFA_0} = \sqrt{\frac{12D_A}{7\pi}}\, t_d^{1/6}({}^\delta[A^-] - {}^b[A^-]) - t_d^{-1/3}\Gamma_A({}^b[A^-], {}^b[HA]) \tag{453}$$

Once the appropriate adsorption isotherm $\Gamma_A = \Gamma_A({}^b[A^-], {}^b[HA])$ is replaced in Eqs. (452) and (453), the combination of these latter equations leads immediately to the current-potential characteristic. If we assume that the adsorptivity of A^- is low, so that Γ_A is much smaller than its maximum value Γ_m, corresponding to monolayer formation, the average distance among adsorbed particles is large and particle-particle interactions can be neglected. Under these circumstances the application of Eqs. (423) and (424) to the species A^- yields

$$\frac{\Gamma_A}{\Gamma_m} = {}^b[A^-]\exp[-a(X - X_m)^2]\exp\left(\frac{b^2}{4a} - \frac{\widetilde{\Delta G}^\circ_{pm,\, X=0}}{RT}\right) \tag{454}$$

where the activity a_s of adsorption sites, which is approximately equal to $\Gamma_m - \Gamma_A - \Gamma_{HA}$, has been set equal to Γ_m. If the electrode potential is a measure of the electric field X, Eq. (454) can be written in the form,

$$\Gamma_A = \Gamma_m\, {}^b[A^-]K_0 \exp[-a'(E - E_m)^2] \tag{455}$$

Equation (455) expresses the linear (or Henry) adsorption isotherm. Under the simplifying assumption that the second term on the right-hand side of Eq. (453), denoting the amount of A^- adsorbed per unit of time, is negligible compared with the first term, representing the amount of A^- flowing through

$x = b$ per unit of time, the combination of Eqs. (452), (453), and (455) yields, after rearrangement,

$$\frac{\bar{i}}{\bar{i}_d - \bar{i}} = \frac{\frac{3}{5}k'_s t_d^{1/2} \Gamma_m K_0 \exp[-a'(E - E_m)^2]}{\sqrt{\frac{12D_A}{7\pi}} \left[1 + \frac{k_s K_s}{k_f^\circ} \exp\left[\frac{\alpha n F}{RT}(E - E_m)\right]\right]} \tag{456}$$

where

$$\bar{i}_d = \sqrt{\frac{12D_A}{7\pi}}\, nFA_0 t_d^{1/6}\, {}^{\delta}[A^-] \tag{457}$$

is the diffusion limiting current which would be observed if A^- were discharged directly at the electrode surface.

The theoretical characteristic of Eq. (456) is identical with that derived by Mairanovskii (*168*) through a less general procedure. Mairanovskii starts from an equation of Levich, Khainin, and Belokolos (*169*) expressing the surface concentration of a surface-active substance $A_{j'}$ at a dropping mercury electrode as a function of time, under conditions of diffusion-controlled adsorption. The surface-active substance is assumed not to participate in any chemical or electrochemical reaction. The equation of Levich *et al* can be derived by a procedure much simpler than that employed by these authors. Thus if Eq. (444) is applied to a nonreacting species, $\bar{i}$ equals zero and consequently one has

$$\sqrt{\frac{12D_{j'}}{7\pi}}\, t_d^{1/6}({}^{\delta}C_{j'} - {}^{b}C_{j'}) = t_d^{-1/3}\Gamma_{j'} \tag{458}$$

Equation (458) expresses the fact that the amount of $A_{j'}$ diffusing through $x = b$ equals the amount of $A_{j'}$ adsorbed at $x = a_1$. If we consider the adsorption isotherm for the species $A_{j'}$ in the form ${}^{b}C_{j'} = {}^{b}C_{j'}(\Gamma_{j'})$, expressing ${}^{b}C_{j'}$ as a function of $\Gamma_{j'}$, rather than in the reciprocal form $\Gamma_{j'} = \Gamma_{j'}({}^{b}C_{j'})$, Eq. (458) can be written as

$$\frac{\Gamma_{j'}}{{}^{b}C_{j'}(\Gamma_{j'}^{eq}) - {}^{b}C_{j'}(\Gamma_{j'})} = \sqrt{\frac{12D_{j'}}{7\pi}\, t_d} \tag{459}$$

where $\Gamma_{j'}^{eq}$ is the surface concentration, under equilibrium conditions with respect to the diffusion step ($\eta_d = 0$), that is, when the concentration of $A_{j'}$ is uniform throughout the solution up to $x = b({}^{b}C_{j'} = {}^{\delta}C_{j'})$. From Eq. (459) it is apparent that when t_d tends to ∞, Γ_j approaches $\Gamma_{j'}^{eq}$. Equation (459) applies approximately to any isotherm. Let us consider the particular case of the linear isotherm of Eq. (455), which we shall write in the form.

$$^{b}C_{j'} = \frac{\Gamma_{j'}}{\Gamma_m K} \tag{460}$$

where $K = K_0 \exp[-a'(E - E_m)^2]$. Replacement of $^{b}C_{j'}$ from Eq. (460) into Eq. (459) yields

$$\Gamma_{j'}(t_d) = \Gamma_m K\, {}^{\delta}C_{j'} \frac{\sqrt{12D_{j'}t_d/7\pi}}{\Gamma_m K + \sqrt{12D_{j'}t_d/7\pi}} \tag{461}$$

which expresses $\Gamma_{j'}$ as a function of electrolysis time, provided the adsorbing species is not consumed at the electrode either by a chemical or by an electrochemical reaction.

Mairanovskii (*168*) writes improperly that "adsorption equilibrium" is achieved only when $\Gamma_{j'} = \Gamma_m K^{\delta}C_{j'}$—that is, when $t_d \to \infty$. As a matter of fact the adsorption step is at equilibrium for any value of t_d, once the expression (460) for the adsorption isotherm is used ($\eta_a = 0$), and it is only the diffusion step that attains equilibrium when $t_d \to \infty$. Mairanovskii derives Eq. (456) on assuming that "adsorption equilibrium" is almost achieved even at low values of t_d and setting

$$\Gamma_A(t_d) = \Gamma_m K\, {}^{\delta}[A^-] \tag{462}$$

for any value of t_d in view of Eq. (461). In a further step of the derivation he equates the surface concentration Γ_A, appearing in Eq. (452), to $\Gamma_A(t_d) \cdot (1 - \bar{i}/\bar{i}_d)$, where $\Gamma_A(t_d)$ is given by Eq. (462), $\bar{i}_d$ by Eq. (457), and $\bar{i}$ by Eq. (453) in which the second term on the right-hand side is neglected. It is readily seen by simple passages that the expression of Γ_A inserted by Mairanovskii in Eq. (452) is none other than the value, $\Gamma_m K\, {}^{b}[A^-]$, of Γ_A in terms of the corresponding volume concentration at $x = b$ as given by the linear isotherm (455). It must be noted that Mairanovskii (*170*) proposes a further equation for the current-potential characteristic, which should be valid when "adsorption equilibrium" is far from being achieved. In this case, corresponding to low values of electrolysis time, the meaning is as follows: $\Gamma_m K$ is $\gg \sqrt{12 D_A t_d/(7\pi)}$ and Eq. (461) yields

$$\Gamma_A(t_d) = \sqrt{\frac{12 D_A t_d}{7\pi}}\, {}^{\delta}[A^-] \tag{463}$$

Here, too, Mairanovskii substitutes the surface concentration Γ_A in Eq. (452) by the expression,

$$\Gamma_A = \Gamma_A(t_d)\left(1 - \frac{\bar{i}}{\bar{i}_d}\right) = \sqrt{\frac{12D_A t_d}{7\pi}}\,{}^{b}[A^-] \tag{464}$$

Under the present circumstances the procedure followed by the above author leads to an erroneous result since, clearly, the relation $\Gamma_A = \Gamma_A({}^{b}[A^-])$ expressed by Eq. (464) is not an adsorption isotherm. The mistake is due to improper use of Eq. (461), which holds only in the case that the surface-active species $A_{j'}$ is not consumed at the electrode. From the procedure followed in this contribution for the derivation of Eq. (456) it is manifest that this equation holds for any value of t_d—that is, for both the cases examined by Mairanovskii.

Owing to the relatively high number of unknown parameters contained in Eq. (456), it has been found convenient to verify this equation by making different simplifying assumptions according to whether we examine the foot of the surface kinetic wave or its limiting current (*168*). Under limiting current conditions ($\bar{i} = \bar{i}_l$), the applied potential E is usually much more negative than E_m and the term $k_s K_s/k_f^0 \exp[\alpha n F(E - E_m)/RT]$ can be neglected with respect to unity in the denominator of Eq. (456). Hence we have

$$\frac{\bar{i}_l}{\bar{i}_d - \bar{i}_l} \cong \frac{3}{5}\sqrt{\frac{7\pi}{12D_A}}\,k_s' t_d^{1/2}\Gamma_m K_0 \exp[-a'(E - E_m)^2] \tag{465}$$

Equation (465) shows that the limiting current $\bar{i}_l$ decreases with a shift of potential toward more negative values, giving rise to a characteristic current maximum (see Fig. (11)). If reasonable assumptions are made as to the values of $\bar{i}_d$ and E_m, a plot of $\ln[\bar{i}_l/(\bar{i}_d - \bar{i}_l)]$ against $(E - E_m)^2$ must yield a straight line with a slope equal to $-a'$. The intersection of the straight line with the ordinate axis permits the determination of $k_s' \Gamma_m K_0$. At the foot of the wave, $(k_s K_s/k_f^0)\exp[\alpha n F(E - E_m)/RT]$ is $\gg 1$ and we can write

$$\frac{\bar{i}}{\bar{i}_d - \bar{i}} \cong \frac{3}{5}\sqrt{\frac{7\pi}{12D_A}}\,k_f^0\,\frac{{}^{a_2}[H^+]}{K_s}\,\Gamma_m K_0\, t_d^{1/2} \exp\left(-\frac{\alpha n F}{RT}\varphi - a'\varphi^2\right) \tag{466}$$

where $\varphi \equiv E - E_m$. Under these conditions the rate of the charge-transfer step becomes so low that the antecedent protonation reaction is practically at equilibrium. This explains why Eq. (466) does not contain the rate constant k_s' for the protonation reaction. If the potential $\varphi_{1/2}$ at which $\bar{i} = \bar{i}_d/2$

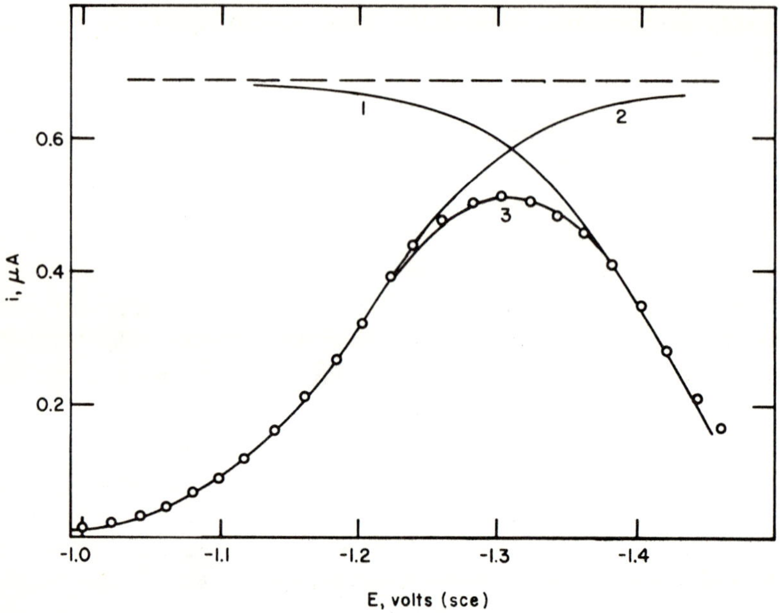

Fig. 11. The first reduction wave of 0.215 *mM* 5-bromo-2 acetyl thiophene in 0.1 *M* KCl + 0.1 *M* KOH at 25°C. Flowrate of mercury : 1.07 mg $\sec^{-1}$; drop time $t_d = 0.26$ sec. Curves 1, 2, and 3 were derived from Eqs. (465), (467), and (456), respectively. The dashed horizontal line expresses the diffusion limiting current. [From *172.*] (By permission of Pergamon Press.)

is strongly negative, then the potential φ along the rising portion of the polarographic wave can be expressed by the equation,

$$\varphi = \varphi_{1/2} + \Delta\varphi = \varphi_{1/2} + (\varphi - \varphi_{1/2})$$

where the absolute value of $\Delta\varphi$ is much less than that of $\varphi_{1/2}$. Then, as a first approximation,

$$\varphi^2 \cong \varphi_{1/2}^2 + 2\varphi_{1/2}\,\Delta\varphi = 2\varphi_{1/2}\,\varphi - \varphi_{1/2}^2$$

and Eq. (466) can be written in the simpler form,

$$\frac{\bar{\imath}}{\bar{\imath}_d - \bar{\imath}} = \frac{3}{5}\sqrt{\frac{7\pi}{12D_A}}\,k_f^0\,\frac{{}^{a_2}[\mathrm{H}^+]}{K_s}$$

$$\times\,\Gamma_m K_0\, t_d^{1/2}\exp(a'\varphi_{1/2}^2)\exp\left[-\left(\frac{\alpha nF}{RT} + 2a'\varphi_{1/2}\right)\varphi\right] \qquad (467)$$

It follows that a plot of $\ln[\bar{\imath}/(\bar{\imath}_d - \bar{\imath})]$ against φ at the foot of the wave yields a straight line of slope $-(\alpha nF/RT + 2a'\varphi_{1/2})$, from which the parameter $k_f^0\ {}^{a_2}[H^+]\Gamma_m K_0/K_s$ can be derived.

According to Mairanovskii (*168*, *171*, *172*) the first reduction wave of 5-bromo-2 acetyl thiophene in alkaline media, which is due to the electrochemical cleavage of the C—Br bond, is preceded by surface protonation of the carbonyl group by water. This wave satisfies Eqs. (465) and (467) in their fields of application as well as the more general Eq. (456). Thus curve 1 in Fig. (11) was derived from Eq. (465), curve 2 from Eq. (467), whereas curve 3, in good agreement with the experimental data indicated by circles, was calculated from Eq. (456). The dashed straight line expresses the diffusion limiting current $\bar{i}_d$. Other cases of kinetic currents controlled by a preceding surface protonation are encountered in the reduction of phenylacetaldehyde oxime in citrate-phosphate buffer, pH 6.4, in the presence of 10% ethyl alcohol (*168*) and in the reduction of 4-bromopyridine at pH 8.3 in 50% ethyl alcohol (*168*, *173*).

D. Mixed Volume-Surface Kinetic Currents

We have already mentioned in Sec. V.B that some purely chemical reactions can take place both in the adsorbed state and in a very thin solution layer around the electrode (*114*, *115*). In general, even if the homogeneous reaction and the corresponding heterogeneous one have identical stoichiometric equations, their mechanisms are different, and the same is true for their orders with respect to the various reaction partners. However, in some simple cases we can assume with some confidence that the surface and bulk reactions differ merely by the value of their rate constants. A general approximate procedure that can be profitably followed in order to derive the expression for the polarographic current is based on the diffusion and reaction layer approximations. We shall assume that the number of moles of each reaction partner, exchanging charges with the electrode material per unit of time, equals the number of moles of the same substance produced (or consumed) both by the bulk and the surface reactions in the same unit of time. Since this number is proportional to the current i, we can write

$$i = nFA[v_s(\{\Gamma_{j'}\}) + r\rho(\{{}^rC_{j'}\})] \tag{468}$$

Here ρ is the rate of the volume reaction, defined as in Eq. (239) and expressed in terms of the volume concentrations ${}^rC_{j'}$ within the reaction layer; r is the Brdička-Wiesner reaction layer; v_s is the rate of the surface reaction in terms of the surface concentrations $\Gamma_{j'}$. These latter can be expressed as functions of the corresponding volume concentrations ${}^{b}C_{j'}$ by making use of the appropriate adsorption isotherms: $v_S(\{\Gamma_{j'}\}) = v_s'(\{{}^{b}C_{j'}\})$. According to the Brdička-Wiesner model of reaction layer and to the diffusion layer approximation, the volume concentrations of the various species within the reaction layer (b, r) are uniform and do not depend on electrolysis time. Consequently, integration of Eq. (468) over the drop time t_d yields:

$$\frac{\bar{\imath}}{nFA_0} = \frac{3}{5} t_d^{2/3} [v_s'(\{{}^{b}C_{j'}\}) + r\rho(\{{}^{b}C_{j'}\})] \tag{469}$$

Diffusion overpotential is accounted for by using Eq. (444), which allows the concentrations ${}^{b}C_{j'}$ to be expressed in terms of the corresponding bulk concentrations ${}^{\delta}C_{j'}$.

As an example of the above approximate procedure let us consider a kinetic limiting current controlled by the antedecent protonation reaction in the body of the solution,

$$A^- + H^+ \underset{k_V K_V}{\overset{k_V}{\rightleftharpoons}} HA \tag{470}$$

and by the corresponding unidirectional surface reaction,

$$A^- + H^+ \overset{k_S}{\rightarrow} HA \tag{471}$$

We shall assume that the solution is well buffered so as to consider the forward reactions (470) and (471) as pseudo-first order. Let us set for simplicity $k_v' = k_v\,{}^{r}[H^+]$, $k_s' = k_s\,{}^{a_2}[H^+]$, and $K_v' = K_v/{}^{r}[H^+]$, recalling, however, that these constants are influenced by the double layer structure. If A^- is weakly adsorbed, the linear isotherm of Eq. (460) can be employed and application of Eq. (469) to the present problem leads to the relation,

$$\frac{\bar{\imath}_l}{nFA_0} = \frac{3}{5} t_d^{2/3} \left[k_s' \Gamma_m K + \left(\frac{k_v' D_{HA}}{K_v'} \right)^{1/2} \right] {}^{b}[A^-] \tag{472}$$

where use has been made of the reaction layer thickness expressed by Eq. (278); n is the number of Faradays involved in the reduction of one mole of undissociated acid. From Eq. (444) it follows that

$$\frac{\bar{\imath}_l}{nFA_0} = \sqrt{\frac{12D_A}{7\pi}}\, t_d^{1/6} ({}^{\delta}[A^-] - {}^{b}[A^-]) - \frac{K\Gamma_m\,{}^{b}[A^-]}{t_d^{1/3}} \tag{473}$$

Noting that the diffusion limiting current that would be observed if A^- were directly discharged at the electrode is given, as usual, by $\bar{\imath}_d = nFA_0$

$\times \sqrt{12D_A/(7\pi)}t_d^{1/6\,\delta}[A^-]$, combination of Eqs. (472) and (473) yields, after simple passages,

$$\frac{\bar{i}_l}{\bar{i}_d - \bar{i}_l}\frac{1}{t_d^{1/2}} = \frac{\frac{3}{5}(k_s' K\Gamma_m + \sqrt{k_v' D_{HA}/K_v'})t_d^{1/2}}{\sqrt{12D_A t_d/(7\pi)} + K\Gamma_m} \tag{474}$$

Equation (474) is identical with Eq. (12) of reference (*115*), derived by Mairanovskii and coworkers through a less general procedure, apart from the numerical factors 3/5 and $\sqrt{12/(7\pi)}$, which in Mairanovskii's paper are replaced by 0.86 and 0.60, respectively. In the absence of adsorption, and consequently of surface protonation, the term $K\Gamma_m$ vanishes and Eq. (474) reduces to Eq. (280), corresponding to a purely homogeneous reaction. If the protonation reaction is purely heterogeneous, the term $(k_v' D_{HA}/K_v')^{1/2}$ in Eq. (472) is dropped, and combination of this equation with Eq. (473) results in

$$\frac{\bar{i}_l}{\bar{i}_d - \bar{i}_l}\frac{1}{t_d^{1/2}} = \frac{\frac{3}{5}k_s' K\Gamma_m t_d^{1/2}}{\sqrt{12D_A t_d/(7\pi)} + K\Gamma_m} \tag{475}$$

If the chemical reaction is heterogeneous, but the species A^- is very weakly adsorbed, the adsorption coefficient K is very small and the term $K\Gamma_m$ can be neglected with respect to $\sqrt{12D_A t_d/(7\pi)}$ in the time scale of polarographic measurements. This is equivalent to saying that the amount of A^- that remains adsorbed on the growing drop is a negligible fraction of the total amount of A^- reaching the electrode by diffusion. Under these conditions Eq. (475) assumes the form,

$$\frac{\bar{i}_l}{\bar{i}_d - \bar{i}_l}\frac{1}{t_d^{1/2}} = \frac{3}{5}\sqrt{\frac{7\pi}{12D_A}}\,k_s' K\Gamma_m \tag{476}$$

It can be concluded that if the quantity $\bar{i}_l/(\bar{i}_d - \bar{i}_l)1/t_d^{1/2}$ is independent of the drop time t_d, the protonation reaction can be either homogeneous (Eq. (280)) or heterogeneous with a weakly adsorbed unprotonated form (Eq. (476)). The dependence of the kinetic current $\bar{i}_l$ upon temperature (cf. Sec. VI.A), or also the influence of a change in the double layer structure, can allow a distinction between the above two cases. According to Eqs. (474) and (475), the plot of $\bar{i}_l/(\bar{i}_d - \bar{i}_l)1/t_d^{1/2}$ against $t_d^{1/2}$ should yield a straight line passing through the origin of the coordinates at small drop times ($\sqrt{12D_A t_d/(7\pi)} \ll K\Gamma_m$). At relatively high drop times ($\sqrt{12D_A t_d/(7\pi)} \gg K\Gamma_m$), the rate of growth of $\bar{i}_l/(\bar{i}_d - \bar{i}_l)1/t_d^{1/2}$ with $t_d^{1/2}$ should decrease, and

$\bar{i}_l/(\bar{i}_d - \bar{i}_l)1/t_d^{1/2}$ should tend to a limiting value. Such a behavior is shown by the kinetic current controlled by the protonation of monoanions of maleic acid in acetate buffer solutions (*115*) and also by the kinetic waves due to the reduction of aromatic aldehydes and ketones in weakly acid solutions (*167*). It can be seen from Eqs. (474) and (475) that this behavior is typical both of purely heterogeneous protonation reactions and of mixed bulk-surface reactions, which makes a distinction between these two cases quite difficult.

In the case of the kinetic limiting current of maleic acid monoanions, the plot of $\log(k_s' K\Gamma_m + \sqrt{k_v' D_{HA}/K_v'})$, as determined through the use of a relation analogous to Eq. (474), against the reciprocal of the absolute temperature yields a curve consisting of two rectilinear segments (see Fig. (12)). According to Mairanovskii (*115*) this behavior seems to confirm the bulk-surface nature of the kinetic current in question. In fact, the lower the temperature, the larger the adsorptivity of A^-, and consequently the greater the contribution of the surface protonation to the current. An increase in temperature causes an increase in k_s' but a decrease in K, so that the "apparent" activation energy of the protonation reaction is low. As the temperature rises, the contribution of the homogeneous protonation reaction to the current becomes predominant, and it is the activation energy of this reaction which determines the upper slope of the curve in Fig. (12).

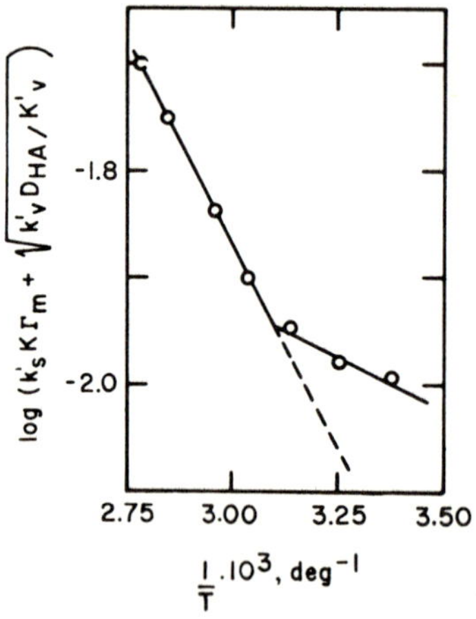

Fig. 12. Plot of $\log(k_s' K \Gamma_m + \sqrt{k_v' D_{HA}/K_v'})$ against $1/T$ for the preprotonation of maleic acid monoanions in acetate buffer solution. [From *167*.] (By permission of Plenum Press.)

VII. MATHEMATICAL APPENDIX

We have seen in Sec. IV.D that by the semirigorous procedure the solution of a set of diffusion equations complicated by kinetic terms is reduced to that of a single diffusion equation of the type,

$$\frac{\partial C}{\partial t} = D\frac{\partial^2 C}{\partial x^2} + \frac{2x}{3t}\frac{\partial C}{\partial x} \tag{A1}$$

This result is obtained at the expense of a major complexity of the boundary condition for Eq. (A1). The rigorous treatment of some simple kinetic problems can also be carried out by solving diffusion equations uncomplicated by kinetic terms as done in Sec. IV.B.

Equation (A1) can be solved through two different procedures, which make use of different substitutions of variables. One of these procedures, introduced by Koutecký (*22*, *23*, *174*), is usually referred to as the "method of dimensionless parameters." The other procedure, employed by Smutek and other researchers (*175–180*), is based on the use of Laplace transforms. The above two methods are perfectly equivalent since they lead to identical results. It should be noted that these results can be obtained leaving the particular type of substitution of variables and the particular mathematical procedure out of consideration. Since, however, in the literature, the substitution of variables performed is always clearly stated and the final expression for the polarographic current is given in a form that depends on the particular mathematical procedure followed, we shall examine the method of dimensionless parameters and that of Laplace transforms separately. Subsequently the equivalence of the two methods will be demonstrated and some general formulas suitable for direct application to the boundary conditions of the particular diffusional problem under study will be given.

A. The Substitution of Variables

Let us perform the following substitutions of variables in Eq. (A1),

$$z = K_1 \times t^n; \qquad y = K_2 t^m \tag{A2}$$

where K_1, K_2, n, and m are parameters to be determined. We recall that the total differential of a function f of two variables y and z is given by

$$df = \left(\frac{\partial f}{\partial y}\right) dy + \left(\frac{\partial f}{\partial z}\right) dz \tag{A3}$$

while the total differential of the second order of the same function is

$$d^2f = d\left(\frac{\partial f}{\partial y}\right) dy + \left(\frac{\partial f}{\partial y}\right) d^2y + d\left(\frac{\partial f}{\partial z}\right) dz + \left(\frac{\partial f}{\partial z}\right) d^2z \tag{A4}$$

with

$$d\left(\frac{\partial f}{\partial y}\right) = \frac{\partial^2 f}{\partial y^2} dy + \frac{\partial^2 f}{\partial y\,\partial z} dz$$

and

$$d\left(\frac{\partial f}{\partial z}\right) = \frac{\partial^2 f}{\partial y\,\partial z} dy + \frac{\partial^2 f}{\partial z^2} dz$$

In view of Eqs. (A3) and (A4) one has

$$\frac{\partial C}{\partial t} = \frac{\partial C}{\partial y}\frac{\partial y}{\partial t} + \frac{\partial C}{\partial z}\frac{\partial z}{\partial t} = \frac{\partial C}{\partial y} K_2 m t^{m-1} + \frac{\partial C}{\partial z} K_1 \times n t^{n-1}$$

$$\frac{\partial C}{\partial x} = \frac{\partial C}{\partial y}\frac{\partial y}{\partial x} + \frac{\partial C}{\partial z}\frac{\partial z}{\partial x} = \frac{\partial C}{\partial z} K_1 t^n$$

$$\frac{\partial^2 C}{\partial x^2} = \frac{\partial^2 C}{\partial y^2}\left(\frac{\partial y}{\partial x}\right)^2 + 2\frac{\partial^2 C}{\partial y\,\partial z}\frac{\partial y}{\partial x}\frac{\partial z}{\partial x} + \frac{\partial^2 C}{\partial z^2}\left(\frac{\partial z}{\partial x}\right)^2 + \frac{\partial C}{\partial y}\frac{\partial^2 y}{\partial x^2}$$

$$\frac{\partial C}{\partial z}\frac{\partial^2 z}{\partial x^2} = \frac{\partial^2 C}{\partial z^2} K_1^2 t^{2n}$$

On substituting $\partial C/\partial t$, $\partial C/\partial x$ and $\partial^2 C/\partial x^2$ into Eq. (A1) and multiplying both members by t, in view of Eq. (A2), we obtain:

$$DK_1^2 t^{2n+1}\frac{\partial^2 C}{\partial z^2} + \left(\frac{2}{3} - n\right) z\frac{\partial C}{\partial z} - my\frac{\partial C}{\partial y} = 0 \tag{A5}$$

Equation (A5) can be simplified in two different ways, namely (a) by eliminating the power t^{2n+1} in the first term through the substitution $n = -\frac{1}{2}$ or (b) by eliminating the whole second term through the substitution $n = \frac{2}{3}$. In the former case Eq. (A5) becomes

$$DK_1^2\frac{\partial^2 C}{\partial z^2} + \frac{7}{6} z\frac{\partial C}{\partial z} - my\frac{\partial C}{\partial y} = 0 \tag{A6}$$

which is the starting point for the method of dimensionless parameters; whereas, in the latter case, we obtain

$$DK_1^2 t^{7/3} \frac{\partial^2 C}{\partial z^2} - my \frac{\partial C}{\partial y} = 0 \tag{A7}$$

Recalling that $y = K_2 t^m$ and setting $m = 7/3$, Eq. (A7) simplifies as (*181*)

$$K_3 \frac{\partial^2 C}{\partial z^2} = \frac{\partial C}{\partial y} \tag{A8}$$

where $K_3 = 3DK_1^2/(7K_2)$. It is evident that if K_1 and K_2 are chosen so as to render dimensionless the variables y and z, K_3 is also dimensionless. Equation (A8), which is formally identical with the equation for diffusion towards a plane electrode, is the starting point for the method based on the use of Laplace transforms.

B. The Method of Integration in Series

The method of dimensionless parameters starts from Eq. (A6), which is written in the form,

$$\frac{\partial^2 C}{\partial z^2} + 2z \frac{\partial C}{\partial z} - \frac{12}{7} my \frac{\partial C}{\partial y} = 0 \tag{A9}$$

with

$$z = \frac{K_1 x}{t^{1/2}} = \sqrt{\frac{7}{12Dt}}\, x \tag{A10}$$

Because the dimensions of the diffusion coefficient D are [length^2/time], z is clearly dimensionless. The parameter K_2, which does not appear in Eq. (A9), is chosen so as to render y dimensionless and the boundary condition as simple as possible. The appellation "method of dimensionless parameters" is somewhat improper since the recourse to dimensionless variables is not determinant for the use of the method and, on the other hand, the method employing Laplace transforms can also make profitable use of dimensionless parameters. The name "method of integration in series" used in the titling of this section seems more appropriate. In fact the differential equation (A9) is solved by assuming that the concentration C can be expressed in the form of a power series of y,

$$C(z, y) = \sum_{i=0}^{\infty} f_i(z) y^{li} \tag{A11}$$

Here the coefficients f_i (z) are unknown functions of z, and l is a parameter which, according to the circumstances, can assume either positive or negative values. Providing that the series is convergent, solving the differential equation (A9) amounts to determining the coefficients f_i (z). Substituting the series (A11) into Eq. (A9) yields

$$\sum_{i=0}^{\infty} f_i''(z) y^{li} + 2z \sum_{i=0}^{\infty} f_i'(z) y^{li} - 12/7ml \sum_{i=0}^{\infty} i f_i(z) y^{li} = 0 \tag{A12}$$

where f_i' (z) and f_i'' (z) are first- and second-order derivatives of f_i (z) with respect to z. Equation (A12) must hold for any value of y and this is true only if the coefficient of every power of y is identically zero. It follows that the solution of Eq. (A12) reduces to that of the equation

$$f_i''(z) + 2z f_i'(z) - 2\alpha i f_i(z) = 0 \qquad i = 0, 1, \ldots \tag{A13}$$

where

$$\alpha \equiv \frac{6}{7} ml \tag{A14}$$

Equation (A13) can be solved by a further integration in series. In this connection let us express f_i (z) through a series in positive powers of z,

$$f_i(z) = \sum_{\lambda=0}^{\infty} a_{i,\lambda} z^{\beta+\lambda} \tag{A15}$$

where the lowest power occurring is z^{β}. When the positive series (A-15) is substituted in Eq. (A13), the result is

$$\sum_{\lambda=0}^{\infty} a_{i,\lambda}(\beta + \lambda)(\beta + \lambda - 1) z^{\beta+\lambda-2} + 2 \sum_{\lambda=0}^{\infty} a_{i,\lambda}(\beta + \lambda - \alpha i) z^{\beta+\lambda} = 0 \tag{A16}$$

For Eq. (A16) to be identically zero, the coefficients of all powers of z must be equated to zero. Thus on setting the coefficient of the lowest power occurring in Eq. (A16), that is, $z^{\beta-2}$, equal to zero we obtain

$$a_{i,o} \beta(\beta - 1) = 0 \tag{A17}$$

Equation (A17), called the "indicial equation," determines β. Clearly two values of β are allowed—namely, $\beta = 0$ and $\beta = 1$. Let us now consider the coefficient of the $(\beta + j)$th power of z, which is obtained by setting $\lambda = j + 2$ in the first summation of Eq. (A16) and $\lambda = j$ in the second,

$$a_{i,\,j+2}(\beta + j + 2)(\beta + j + 1) + 2a_{i,\,j}(\beta + j - \alpha i)$$

On equating the above coefficient to zero, we have

$$a_{i,\,j+2} = \frac{2(\alpha i - \beta - j)}{(\beta + j + 1)(\beta + j + 2)}\, a_{i,\,j} \qquad \text{(A18)}$$

If we choose to represent $f_i(z)$ through a series, $K_i = \sum_{\lambda=0}^{\infty} a_{i,\lambda} z^{\lambda}$, in even powers of z by setting $\beta = 0$ in Eq. (A15), the "recurrence equation" (A18) permits us to obtain $a_{i,\,2}$, $a_{i,\,4}$, etc., once $a_{i,\,0}$ has been established,

$$K_i(z) = \sum_{\lambda=0}^{\infty} a_{i,\,\lambda}\, z^{\lambda} \qquad \text{with} \qquad \frac{a_{i,\,\lambda+2}}{a_{i,\,\lambda}} = \frac{2(\alpha i - \lambda)}{(\lambda + 1)(\lambda + 2)} \qquad \text{(A19)}$$

Conversely if we set $\beta = 1$ in Eq. (A15), thus expressing $f_i(z)$ by a series, $L_i = \sum_{\lambda=0}^{\infty} c_{i,\lambda} z^{\lambda+1}$, in odd powers of z, Eq. (A18) permits us to obtain, successively, $c_{i,\,2}$, $c_{i,\,4}$, etc.; $c_{i,\,0}$, however, is arbitrary,

$$L_i(z) = \sum_{\lambda=0}^{\infty} c_{i,\,\lambda}\, z^{\lambda+1} \qquad \text{with} \qquad \frac{c_{i,\,\lambda+2}}{c_{i,\,\lambda}} = \frac{2(\alpha i - \lambda - 1)}{(\lambda + 2)(\lambda + 3)} \qquad \text{(A20)}$$

In both the summations $K_i(z)$ and $L_i(z)$, λ assumes only even values. The general solution of the second-order differential equation (A13) is a linear combination of the two power series (A19) and (A20),

$$f_i(z) = k_i K_i(z) + l_i L_i(z) = k_i a_{i,\,0} \sum_{\lambda=0}^{\infty} \frac{a_{i,\,\lambda}}{a_{i,\,0}} z^{\lambda} + l_i c_{i,\,0} \sum_{\lambda=0}^{\infty} \frac{c_{i,\,\lambda}}{c_{i,\,0}} z^{\lambda+1}$$

and contains, as expected, two arbitrary constants—namely, $k_i a_{i,0}$ and $l_i c_{i,0}$. Hence the general solution (A11) of the differential equation (A9) takes the form

$$c(z, y) = \sum_{i=0}^{\infty} [k_i K_i(z) + l_i L_i(z)] y^{li} \qquad \text{(A21)}$$

If we denote the value of C in the bulk of the solution by C^*, the initial condition for the diffusion equation (A1) reads

$$C(x, t) = C^* \qquad \text{for} \quad x \to \infty, t > 0 \tag{A22}$$

or also, in terms of the new variables z and y,

$$C(z, y) = C^* \qquad \text{for} \quad z \to \infty, y > 0 \tag{A23}$$

The initial condition of Eq. (A23) can be taken into account by adding to the general solution (A21) a particular solution independent of y, chosen so as to assume the value C^* for $z \to \infty$, at the same time choosing the values of $k_i a_{i,o}$ and $l_i c_{i,o}$ so as to cause the general solution to vanish for $z \to \infty$. A particular solution of Eq. (A9) independent of y is obtained on noting that the general solution of the differential equation,

$$\frac{\partial^2 C}{\partial z^2} + 2z \frac{\partial C}{\partial z} = 0 \tag{A24}$$

has the form,

$$c(z) = K_1 \int_0^z e^{-z^2} \, dz + K_2 \tag{A25}$$

as can be readily verified by substituting Eq. (A25) into Eq. (A24). Recalling that

$$\int_0^\infty e^{-z^2} \, dz = \frac{\pi^{1/2}}{2} \tag{A26}$$

Eq. (A25) will tend to C^* for $z \to \infty$ if we set $K_1 = 2C^*/\pi^{1/2}$ and $K_2 = 0$. The function $2/\pi^{1/2} \int_0^z e^{-z^2} \, dz$ is usually called the "error function" and is denoted by the symbol $erf(z)$. A suitable way for letting the general solution of Eq. (A21) approach zero when $z \to \infty$ consists in setting

$$-\frac{l_i c_{i,0}}{k_i a_{i,0}} = \lim_{z \to \infty} \frac{\overline{K_i(z)}}{\overline{L_i(z)}} \equiv p_i \tag{A27}$$

where $\overline{K_i(z)}$ and $\overline{L_i(z)}$ are, respectively, the normalized forms of the two series $K_i(z)$ and $L_i(z)$,

$$\overline{K_i(z)} = \frac{K_i(z)}{a_{i,0}}; \qquad \overline{L_i(z)} = \frac{L_i(z)}{c_{i,0}}$$

Temporarily we give without proof the formula for the evaluation of the coefficients p_i,

$$p_i = 2\,\frac{\Gamma(1 + \alpha i/2)}{\Gamma(1/2 + \alpha i/2)} \tag{A28}$$

Here the symbol Γ denotes the gamma function. In conclusion, the solution of Eq. (A9) accounting for the initial condition (A23) is

$$C(z, y) = \sum_{i=0}^{\infty} m_i[\overline{K_i(z)} - p_i\overline{L_i(z)}]y^{li} + C^* erf(z) \tag{A29}$$

where $m_i \equiv k_i a_{i,o}$.

In the above derivation we have intentionally omitted taking the boundary condition for the diffusion equation (A1) into account. In the most general case, a boundary condition is a relation among the volume concentrations $C_i(0, t)$ of the various diffusing species at $x = 0$ (or more exactly, at $x = b$ (cf. Sec. II.A.2)) and the corresponding gradients $(\partial C_i/\partial x)_{x=0}$. If some of the diffusing species react at the electrode in the adsorbed state, boundary conditions may contain surface concentrations Γ_i as well as their time derivatives $d\Gamma_i/dt$. Nevertheless, even in these more intricate cases, the various Γ_i and $d\Gamma_i/dt$ can be readily expressed in terms of the volume concentrations $C_i(0, t)$ and of the corresponding time derivatives $dC_i(0, t)/dt$, by having recourse to the appropriate adsorption isotherms (cf., for example, Eq. (445)).

As concerns the volume concentration C of Eq. (A1), we readily see from Eq. (A29) that its value at $x = 0$ is given by

$$C(x = 0, t) = C(z = 0, y) = \sum_{i=0}^{\infty} m_i y^{li} \tag{A30}$$

In fact all powers of z vanish when $z \to 0$ with the exclusion of the zeroth power of z contained in $\overline{K_i(z)}$. In view of Eq. (A10) the concentration gradient at $x = 0$ is given by

$$\left(\frac{\partial C}{\partial x}\right)_{x=0} = \left(\frac{\partial C}{\partial z}\right)_{z=0} \frac{\partial z}{\partial x} = \sqrt{\frac{7}{12Dt}} \left(\frac{\partial C}{\partial z}\right)_{z=0} \tag{A31}$$

Noting that $d\,erf(z)/dz = 2/\pi^{1/2} \exp(-z^2)$, differentiation of Eq. (A29) with respect to z results in

$$\frac{\partial C(z, y)}{\partial z} = \sum_{i=0}^{\infty} m_i \left[\sum_{\lambda=0}^{\infty} \frac{a_{i,\lambda}}{a_{i,0}} \lambda z^{\lambda-1} - p_i \sum_{\lambda=0}^{\infty} \frac{c_{i,\lambda}}{c_{i,0}} (\lambda + 1) z^{\lambda} \right] y^{li} + \frac{2}{\pi^{1/2}} C^* e^{-z^2} \tag{A32}$$

Since by definition λ assumes even values, the only zeroth power of z appearing in Eq. (A32) is that corresponding to $\lambda = 0$ in the second summation within the square brackets. Hence for $z = 0$,

$$\left(\frac{\partial C}{\partial z}\right)_{z=0} = \frac{2}{\pi^{1/2}} C^* - \sum_{i=0}^{\infty} m_i p_i y^{li} \tag{A33}$$

Once C(0, t) and $(\partial C/\partial x)_{x=0}$ are substituted from Eqs. (A30) and (A31), (A33) into the boundary condition (or into more boundary conditions if the diffusional problem requires the solution of a set of diffusion equations of the type (A1)), we readily obtain a recurrence relation that allows the evaluation of every coefficient m_i provided the so far undetermined constants m and l are given definite values. It is usually found convenient to set $m = \frac{1}{2}$, thus writing Eq. (A9) in the form,

$$\frac{\partial^2 C}{\partial z^2} + 2z \frac{\partial C}{\partial z} - \frac{6}{7} y \frac{\partial C}{\partial y} = 0$$

with $z = \sqrt{7/(12Dt)}x$, $y = K_2 t^{1/2}$. The reason for the above choice will be made clear later. For sufficiently low values of y the parameter l is set equal to 1, so that $C(z, y)$ and $(\partial C/\partial z)_{z=0}$ are expressed in the form of series in positive powers of y. For high values of y these series may converge very slowly or even diverge. In this case it is necessary to use series in negative powers of y by giving l a negative value, usually -1.

In general the polarographic current is expressed in terms of the fluxes $D_i(\partial C_i/\partial x)_{x=0}$ of some of the diffusing species at $x = 0$ and also, if adsorption phenomena take place, in terms of the time derivatives $d\Gamma_i/dt$ (cf. Eq. (440)). Therefore, once the coefficients p_i are calculated from Eq. (A28) and the coefficients m_i are derived from the boundary conditions, the values for the various $C_i(0, t)$ and $(\partial C_i/\partial x)_{x=0}$ as functions of time are known, and the same is true for the expression of the polarographic current.

C. The Method of Laplace Transforms

1. The Laplace Transformation

When a function $f(t)$ is multiplied by the exponential e^{-st} and integrated with respect to t from zero to ∞, the resulting function $g(s)$ of the variable s,

$$g(s) = \int_0^{\infty} e^{-st} f(t)\, dt$$

is called the "Laplace transform" of $f(t)$, whereas $f(t)$ is termed the inverse Laplace transform of $g(s)$. Symbolically we write

$$g(s) = \mathscr{L}\{f(t)\}$$

where the operator $\mathscr{L}$ denotes the Laplace transformation of $f(t)$, and also

$$f(t) = \mathscr{L}^{-1}\{g(s)\}$$

where $\mathscr{L}^{-1}\{g(s)\}$ denotes a function whose Laplace transform is $g(s)$. The Laplace transformation is of great importance in applied mathematics because it enables one to solve several partial differential equations with the corresponding boundary conditions by separation of variables. For the solution of the diffusion equation (A1), or better of the equivalent Eq. (A8), use will be made of three important theorems on Laplace transforms, which we shall demonstrate in a formal, manipulative manner. For the rigorous derivation of these theorems the reader is referred to the specialized treatises existing on the subject (*182*, *183*) .

Theorem I. If $f(t)$ is the derivative $dF(t)/dt$ of another function $F(t)$, then

$$\mathscr{L}\{f(t)\} = \int_0^\infty e^{-st} \frac{dF(t)}{dt}\, dt$$

On integrating by parts we have

$$\mathscr{L}\left\{\frac{dF(t)}{dt}\right\} = \Big[e^{-st}F(t)\Big]_0^\infty + s\int_0^\infty e^{-st}F(t)\, dt = s\mathscr{L}\{F(t)\} - F(0) \quad \text{(A34)}$$

where $F(0)$ is the value of $F(t)$ for $t = 0$. Equation (A34) is of fundamental importance because it enables us to replace the operation of differentiation of $F(t)$ by a simple algebraic operation on the corresponding transform.

Theorem II. Let $f(t)$ be equal to $\int_0^t F(\tau)\, d\tau$. Then

$$\mathscr{L}\{f(t)\} = \int_0^\infty e^{-st} \int_0^t F(\tau)\, d\tau\, dt = -\frac{1}{s}\int_0^\infty \frac{d}{dt}(e^{-st}) \int_0^t F(\tau)\, d\tau\, dt$$

By a formal integration by parts we have

$$\mathscr{L}\left\{\int_0^t F(\tau)\, d\tau\right\} = \left[-\frac{1}{s}e^{-st}\int_0^t F(\tau)\, d\tau\right]_0^\infty + \frac{1}{s}\int_0^\infty e^{-st}F(t)\, dt = \frac{1}{s}\,\mathscr{L}\{F(t)\}$$

(A35)

in view of the fact that the integrated part vanishes. Equation (A35) permits us to replace the operation of integration on the function $F(t)$ by a division by s on the corresponding transform. Equation (A35) can also be written in the inverse form

$$\int_0^t F(\tau)\,d\tau = \mathscr{L}^{-1}\left\{\frac{1}{s}\,\mathscr{L}\{F(t)\}\right\} \tag{A36}$$

Theorem III. If $f_1(t)$ and $f_2(t)$ are two functions defined to be zero for negative arguments, the convolution of $f_1(t)$ and $f_2(t)$ is expressed by the integral,

$$\int_0^t f_1(\tau)f_2(t-\tau)\,d\tau \tag{A37}$$

and is symbolized by $f_1(t) * f_2(t)$. By definition the Laplace transform of the convolution reads

$$\mathscr{L}\{f_1(t) * f_2(t)\} = \int_0^\infty e^{-st}\int_0^t f_1(\tau)f_2(t-\tau)\,d\tau\,dt$$

Replacing the integration variable t by $\tau' = t - \tau$ and interchanging the order of integration, we find that

$$\begin{aligned}\mathscr{L}\{f_1(t) * f_2(t)\} &= \int_{-\tau}^\infty \int_0^t e^{-s(\tau+\tau')}f_1(\tau)f_2(\tau')\,d\tau\,d\tau' \\ &= \int_0^\infty e^{-s\tau'}f_2(\tau')\,d\tau' \int_0^\infty e^{-s\tau}f_1(\tau)\,d\tau \\ &= \mathscr{L}\{f_1(t)\}\mathscr{L}\{f_2(t)\} \\ &\equiv g_1(s)g_2(s)\end{aligned} \tag{A38}$$

The replacement of the lower limit of integration, $-\tau$, by zero can be justified by noting that

$$\int_{-\tau}^\infty e^{-s\tau}f_1(\tau)\,d\tau = \int_{-\tau}^0 e^{-s\tau}f_1(\tau)\,d\tau + \int_0^\infty e^{-s\tau}f_1(\tau)\,d\tau$$

and by recalling that $f_1(t)$ is zero for negative arguments. By following an analogous line of reasoning, it is readily seen that the upper limit t in Eqs. (A37) and (A38) can be replaced by ∞, provided $f_2(t)$ is defined to be zero for negative values of the argument. The convolution theorem (A38) is more frequently employed in the inverse form,

$$\mathscr{L}^{-1}\{g_1(s) \cdot g_2(s)\} = \int_0^t f_1(\tau) f_2(t-\tau)\, d\tau \tag{A39}$$

2. Solution of the Equation for Diffusion Towards a Growing Drop by Laplace Transforms

Let us make use of Laplace transforms for the solution of the diffusion equation (A8),

$$K_3 \frac{\partial^2 C}{\partial z^2} = \frac{\partial C}{\partial y} \tag{A40}$$

in which

$$z = K_1 \times t^{2/3}; \quad y = K_2 t^{7/3}; \quad K_3 = \frac{3}{7} D \frac{K_1^2}{K_2} \tag{A41}$$

taking into account the initial condition (A22), which now reads,

$$C(z, y) = C^* \qquad \text{for} \begin{cases} z \geqslant 0, & y = 0 \\ z \to \infty, & y > 0 \end{cases} \tag{A42}$$

In view of Theorem I, the Laplace transformation of Eq. (A40) with regard to the variable y yields

$$\begin{aligned} s\mathscr{L}\{C(z, y)\} - C(z, 0) &= \int_0^\infty e^{-sy} K_3 \frac{\partial^2 C(z, y)}{\partial z^2}\, dy \\ &= K_3 \frac{\partial^2}{\partial z^2} \int_0^\infty e^{-sy} C(z, y)\, dy \\ &= K_3 \frac{\partial^2}{\partial z^2} \mathscr{L}\{C(z, y)\} \end{aligned} \tag{A43}$$

For simplicity of notations, let us set $\mathscr{L}\{C(z, y)\} \equiv \bar{C}$. Noting from Eq. (A42) that $C(z, 0) = C^*$, Eq. (A43) can be written in the form,

$$K_3 \frac{\partial^2 \bar{C}}{\partial z^2} - s\bar{C} + C^* = 0 \tag{A44}$$

The use of Laplace transforms has allowed the partial differential equation (A40) to be converted into the ordinary differential equation (A44), which can be solved by standard methods. Substitution of the variable $\bar{C}$ by $\bar{u} = \bar{C} - C^*/s$ in Eq. (A44) results in

$$K_3\,\partial^2\bar{u}/\partial z^2 - s\bar{u} = 0 \tag{A45}$$

Equation (A45) is a linear equation with constant coefficients and zero right-hand member. It can be solved by having recourse to the corresponding "auxiliary equation" (*184*),

$$K_3 r^2 - s = 0 \tag{A46}$$

The solutions of Eq. (A46) are $\pm s^{1/2}/K_3^{1/2}$, so that the general solution of Eq. (A45) is

$$\bar{u} = \bar{C} - \frac{C^*}{s} = M(s)\exp\left(-\frac{s^{1/2}z}{K_3^{1/2}}\right) + N(s)\exp\left(\frac{s^{1/2}z}{K_3^{1/2}}\right)$$

where the integration constants $M(s)$ and $N(s)$ are, in general, functions of s Since $C(z, y)$ and consequently its Laplace transform have finite values for $z \to \infty$ in view of the initial condition (A42), $N(s)$ must necessarily be zero. Hence

$$\bar{C}(z, s) = \frac{C^*}{s} + M(s)\exp\left(-\frac{s^{1/2}z}{K_3^{1/2}}\right) \tag{A47}$$

For $z = 0$, Eq. (A47) becomes

$$\bar{C}(0, s) = \frac{C^*}{s} + M(s) \tag{A48}$$

The inverse transformation of Eq. (A48) yields

$$\mathscr{L}^{-1}\{\bar{C}(0, s)\} = C(0, y) = \mathscr{L}^{-1}\left\{\frac{C^*}{s}\right\} + \mathscr{L}^{-1}\{M(s)\} \tag{A49}$$

Noting that by definition the Laplace transform of a constant C^* is

$$\mathscr{L}\{C^*\} = \int_0^\infty e^{-sy}C^*\,dy = \left[-\frac{C^*}{se^{-sy}}\right]_0^\infty = \frac{C^*}{s}$$

we also have

$$\mathscr{L}^{-1}\{C^*/s\} = C^*$$

so that Eq. (A49) becomes

$$C(0, y) = C^* + \mathscr{L}^{-1}\{M(s)\} \tag{A50}$$

On differentiating Eq. (A47) with respect to z, we obtain

$$\frac{\partial}{\partial z}\mathscr{L}\{C(z, y)\} = \frac{\partial}{\partial z}\int_0^\infty e^{-sy}C(z, y)\,dy = \int_0^\infty e^{-sy}\frac{\partial C(z, y)}{\partial z}\,dy$$

$$= \mathscr{L}\left\{\frac{\partial C(z, y)}{\partial z}\right\} = -\frac{s^{1/2}}{K_3^{1/2}}M(s)\exp\left(-\frac{s^{1/2}z}{K_3^{1/2}}\right)$$

or for $z = 0$,

$$\mathscr{L}\left\{\frac{\partial C(z, y)}{(\partial z)_{z=0}}\right\} = -\frac{s^{1/2}}{K_3^{1/2}}M(s) \tag{A51}$$

At this point two slightly different procedures can be followed, according to whether we prefer to expand either $C(0, y)$ *(180)* or $\partial C(z, y)/(\partial z)_{z=0}$ *(175)* in a power series of y.

a. Expansion of $C(0, y)$ ***in a Power Series of*** y. Application of Theorem II in the form of Eq. (A36) to the function $(\partial C(z, y)/\partial z)_{z=0}$ yields, in view of Eq. (A51):

$$K_3^{1/2}\int_0^y \frac{\partial C(z, \tau)}{(\partial z)_{z=0}}\,d\tau = -\mathscr{L}^{-1}\left\{\frac{M(s)}{s^{1/2}}\right\} \tag{A52}$$

The inverse transformation on the right-hand side of Eq. (A52) can be carried out by making use of the convolution theorem (A39), where now $g_1(s)$ and $g_2(s)$ are represented by $M(s)$ and $1/s^{1/2}$, respectively,

$$K_3^{1/2}\int_0^y \frac{\partial C(z, \tau)}{(\partial z)_{z=0}}\,d\tau = -\int_0^y f_1(\tau)f_2(y-\tau)\,d\tau \tag{A53}$$

Here $f_1(y) = \mathscr{L}^{-1}\{M(s)\}$ and $f_2(y) = \mathscr{L}^{-1}\{1/s^{1/2}\}$. Incidentally we note that

$$\mathscr{L}^{-1}\left\{\frac{1}{s^{1/2}}\right\} = \frac{1}{\pi^{1/2}y^{1/2}} \tag{A54}$$

In fact, by definition, the Laplace transform of $y^{-1/2}$ is given by

$$\mathscr{L}\left\{\frac{1}{y^{1/2}}\right\} = \int_0^\infty e^{-sy} y^{-1/2}\, dy \tag{A55}$$

Substituting $sy = x^2$ we have $dy = 2x/s\, dx$, and the integral in Eq. (A55) can be written in the form,

$$\mathscr{L}\left\{\frac{1}{y^{1/2}}\right\} = \frac{2}{s^{1/2}} \int_0^\infty e^{-x^2}\, dx$$

Taking Eq. (A26) into account we have

$$\mathscr{L}\left\{\frac{1}{y^{1/2}}\right\} = \frac{\pi^{1/2}}{s^{1/2}}$$

from which Eq. (A54) is immediately derived. Combination of Eqs. (A50) (A53), and (A54) results in

$$K_3^{1/2} \int_0^y \frac{\partial C(z, \tau)}{(\partial z)_{z=0}}\, d\tau = \frac{1}{\pi^{1/2}} \int_0^y \frac{C^* - C(0, \tau)}{(y - \tau)^{1/2}}\, d\tau \tag{A56}$$

Equation (A56) is a form of Duhamel's theorem.

An important result can be immediately derived from Eq. (A56) if we take into account the substitutions of variables in Eq. (A41) by setting

$$\tau = K_2 t^{7/3}, \quad y = K_2 t_{\mathrm{d}}^{7/3} \quad \text{and} \quad z = K_1 x t^{2/3}$$

Then

$$\frac{\partial C(z, \tau)}{(\partial z)_{z=0}} = \frac{\partial C(x, t)}{(\partial x)_{x=0}} \frac{1}{K_1 t^{2/3}}; \qquad d\tau = \frac{7}{3} K_2 t^{4/3}\, dt$$

and Eq. (A56) becomes

$$\frac{K_3^{1/2} K_2^{1/2}}{K_1} \int_0^{t_1} \frac{\partial C(x, t)}{(\partial x)_{x=0}} t^{2/3}\, dt = \frac{1}{\pi^{1/2}} \int_0^{t_1} \frac{C^* - C(0, t)}{(t_d^{7/3} - t^{7/3})^{1/2}} t^{4/3}\, dt \tag{A57}$$

Noting from Eq. (A41) that $\sqrt{K_3 K_2/K_1} = \sqrt{3D/7}$ and substituting the variable t by $\lambda = (t_{\mathrm{d}}^{7/3} - t^{7/3})$ in the integral on the right-hand side of Eq. (A57), this latter equation takes the form,

$$\left(\frac{7\pi D}{3}\right)^{1/2} \int_0^{t_d} \frac{\partial C(x, t)}{(\partial x)_{x=0}} t^{2/3}\, dt = \int_0^{t_d^{7/3}} \frac{C^* - C(0, \lambda)}{\lambda^{1/2}}\, d\lambda \tag{A58}$$

It is interesting to observe that the two arbitrary constants K_1 and K_2 introduced in Eq. (A41) cancel out from Eq. (A58). If the volume concentration $C(0, t)$ at $x = 0$ is independent of electrolysis time (a condition which we have seen in Sec. II.B to be satisfied during a reversible charge-transfer process not complicated either with chemical reactions or with adsorption phenomena) the term $[C^* - C(0, \lambda)]$ can be brought out of the integral sign

$$\int_0^{t_d} \frac{\partial C(x, t)}{(\partial x)_{x=0}} t^{2/3}\, dt = \left(\frac{12}{7\pi D}\right)^{1/2} t_d^{7/6}[C^* - C(0, t_d)] \tag{A59}$$

In the case of the reversible charge-transfer process $0 + ne \rightleftharpoons \mathrm{R}$, the mean current is expressed by the equation

$$\bar{i} = \frac{nFA_0 D}{t_d} \int_0^{t_d} \left(\frac{\partial C}{\partial x}\right)_{x=0} t^{2/3}\, dt \tag{A60}$$

where C is the concentration of the oxidized species, O. On combining Eq. (A59) and (A60) we obtain

$$\bar{i} = \left(\frac{12D}{7\pi}\right)^{1/2} nFA_0 t_d^{1/6}[C^* - C(0, t_d)]$$

expressing the well-known Ilkovic equation. In Sec. IV.C.1 we have seen that on the basis of the diffusion layer approximation, Eq. (A59) is applied to cases in which $C(0, t)$ is not rigorously time independent. Now we can better visualize the simplifying assumptions on which the above approximation relies, as well as its limitations. Thus if the rate of change of $C(0, t)$ with time is much less than that of the ratio $t^{4/3}/(t_d^{7/3} - t^{7/3})^{1/2}$ (a circumstance that is frequently encountered in practice), $[C^* - C(0, \lambda)]$ can be brought out of the integral sign in Eq. (A57) with some confidence. This explains why the diffusion layer approximation gives results that are often in fairly good agreement with those obtained by the rigorous or semirigorous procedure.

Returning to Eq. (A56), we can attempt to solve it by expanding $C(0, y)$ in a power series of y,

$$C(0, y) = \sum_{i=0}^{\infty} r_i y^{bi} \tag{A61}$$

In this connection we let $u = \tau/y$ replace τ as integration variable on the right-hand side of Eq. (A56),

$$K_3^{1/2} \int_0^y \frac{\partial C(z, \tau)}{(\partial z)_{z=0}}\, d\tau = \frac{y^{1/2}}{\pi^{1/2}} \int_0^1 \frac{C^* - C(0, yu)}{(1-u)^{1/2}}\, du \tag{A62}$$

On substituting $C(0, y)$ from Eq. (A61) into the right-hand member of Eq. (A62), this latter becomes

$$\frac{C^* y^{1/2}}{\pi^{1/2}} \int_0^1 \frac{du}{(1-u)^{1/2}} - \frac{y^{1/2}}{\pi^{1/2}} \int_0^1 \frac{\sum_{i=0}^{\infty} r_i y^{bi} u^{bi}}{(1-u)^{1/2}} du$$

$$= \frac{2C^* y^{1/2}}{\pi^{1/2}} - \frac{1}{\pi^{1/2}} \sum_{i=0}^{\infty} r_i y^{bi+1/2} \int_0^1 \frac{u}{(1-u)^{1/2}} du \quad \text{(A63)}$$

Incidentally, we note that the integral,

$$\int_0^1 \frac{u^{r-1}}{(1-u)^{1-s}} du \quad \text{(A64)}$$

is a function of r and s which is known as B function or β function and symbolyzed by $B(r, s)$. A useful formula for the calculation of the B function, for the derivation of which the reader is referred to mathematical treatises (*185*), expresses $B(r, s)$ in terms of gamma functions,

$$B(r, s) = \frac{\Gamma(r)\Gamma(s)}{\Gamma(r+s)} \quad \text{(A65)}$$

Eq. (A62) then becomes

$$K_3^{1/2} \int_0^y \frac{\partial C(z, \tau)}{(\partial z)_{z=0}} d\tau = \frac{2C^* y^{1/2}}{\pi^{1/2}} - \frac{1}{\pi^{1/2}} \sum_{i=0}^{\infty} r_i \frac{\Gamma(bi+1)\Gamma(\frac{1}{2})}{\Gamma(bi+3/2)} y^{bi+1/2} \quad \text{(A66)}$$

Differentiation of Eq. (A66) with respect to y yields

$$K_3^{1/2} \frac{\partial C(z, y)}{(\partial z)_{z=0}} = \frac{C^*}{\pi^{1/2} y^{1/2}} - \sum_{i=0}^{\infty} r_i \frac{\Gamma(bi+1)}{\Gamma(bi+3/2)} (bi+1/2) y^{bi-1/2} \quad \text{(A67)}$$

where use has been made of the equation $\Gamma(\frac{1}{2}) = \pi^{1/2}$. Recalling the important property of Γ functions,

$$\Gamma(z+1) = z\Gamma(z)$$

Eq. (A67) can be written in the more compact form

$$\frac{\partial C(z, y)}{(\partial z)_{z=0}} = \frac{1}{K_3^{1/2} y^{1/2}} \left[\frac{C^*}{\pi^{1/2}} - \sum_{i=0}^{\infty} r_i \frac{\Gamma(bi+1)}{\Gamma(bi+\frac{1}{2})} y^{bi}\right] \quad \text{(A68)}$$

The substitution of $C(0, y)$ and $\partial C(z, y)/(\partial z)_{z=0}$ from Eqs. (A61) and (A68) into the boundary conditions leads to recurrent relations that allow the computation of the coefficients r_i, in a way entirely analogous to that described in connection with the method of integration in series.

b. Expansion of $\partial C(z, y)/(\partial z)_{z=0}$ ***in a power Series of y.*** An alternative path leading to Eqs. (A61) and (A68) consists in carrying out the inverse transformation $\mathscr{L}^{-1}\{M(s)\}$ by the use of the convolution theorem. In this connection we shall write

$$\mathscr{L}^{-1}\{M(s)\} = -K_3^{1/2}\mathscr{L}^{-1}\left\{-\frac{s^{1/2}M(s)}{K_3^{1/2}}\cdot\left(\frac{1}{s^{1/2}}\right)\right\}$$

where in view of Eq. (A51) $-s^{1/2}\ M(s)/K_3^{1/2}$ is the Laplace transform of $\partial C(z, y)/(\partial z)_{z=0}$. Taking Eqs. (A50) and (A54) into account, the application of the convolution theorem (cf. Eq. (A39)) yields the relation,

$$\mathscr{L}^{-1}\{M(s)\} = C(0, y) - C^* = -\frac{K_3^{1/2}}{\pi^{1/2}}\int_0^y \frac{\partial C(z, \tau)}{(\partial z)_{z=0}}\frac{1}{(y-\tau)^{1/2}}\,d\tau \tag{A69}$$

which is another form of Duhamel's theorem. With the substitution $u = \tau/y$, Eq. (A69) reads

$$C(0, y) = C^* - \frac{K_3^{1/2}y^{1/2}}{\pi^{1/2}}\int_0^1 \frac{\partial C(z, yu)}{(\partial z)_{z=0}}\frac{1}{(1-u)^{1/2}}\,du \tag{A70}$$

Let us expand $\partial C(z, y)/(\partial z)_{z=0}$ in the following power series of y,

$$\frac{\partial C(z, y)}{(\partial z)_{z=0}} = \sum_{i=0}^{\infty} q_i\, y^{bi+a} \tag{A71}$$

where the parameters b and a are to be determined. Combination of Eqs. (A70) and (A71) leads to the relation,

$$\begin{aligned} C(0, y) &= C^* - \frac{K_3^{1/2}y^{1/2}}{\pi^{1/2}}\sum_{i=0}^{\infty} q_i\, y^{bi+a}\int_0^1 \frac{u^{bi+a}}{(1-u)^{1/2}}\,du \\ &= C^* - K_3^{1/2}y^{1/2}\sum_{i=0}^{\infty} q_i\,\frac{\Gamma(bi+a+1)}{\Gamma(bi+a+3/2)}\,y^{bi+a} \end{aligned} \tag{A72}$$

in which use has been made of Eq. (A65) for the B function, as well as of the equation $\Gamma(\frac{1}{2}) = \pi^{1/2}$. Setting for convenience $a = -\frac{1}{2}$, Eqs. (A71) and (A72) become

$$\frac{\partial C(z, y)}{(\partial z)_{z=0}} = \frac{1}{y^{1/2}} \sum_{i=0}^{\infty} q_i y^{bi} \tag{A73}$$

and

$$C(0, y) = C^* - K_3^{1/2} \sum_{i=0}^{\infty} q_i \frac{\Gamma(bi + \frac{1}{2})}{\Gamma(bi + 1)} y^{bi} \tag{A74}$$

respectively. The equivalence of Eqs. (A73) and (A74) with Eqs. (A68) and (A61), respectively, is readily demonstrated if we note that for $y = 0$ Eq. (A74) becomes

$$C(0, 0) = C^* - K_3^{1/2} q_0 \frac{\Gamma(\frac{1}{2})}{\Gamma(1)}$$

Since from the initial condition (A42) $C(0, 0) = C^*$, it follows that $q_0 = 0$. Hence setting $r_i \equiv -K_3^{1/2} q_i \Gamma(bi + \frac{1}{2})/\Gamma(bi + 1)$, Eqs. (A73) and (A74) may be written

$$\frac{\partial C(z, y)}{(\partial z)_{z=0}} = -\frac{1}{K_3^{1/2} y^{1/2}} \sum_{i=1}^{\infty} r_i \frac{\Gamma(bi + 1)}{\Gamma(bi + \frac{1}{2})} y^{bi} \tag{A75}$$

and

$$C(0, y) = C^* + \sum_{i=1}^{\infty} r_i y^{bi} \tag{A76}$$

Noting that $\Gamma(1) \equiv 1$ and $\Gamma(\frac{1}{2}) \equiv \pi^{1/2}$, addition and subtraction of $C^*/(\pi^{1/2} K_3^{1/2} y^{1/2}) = \Gamma(1) C^*/[\Gamma(\frac{1}{2}) K_3^{1/2} y^{1/2}]$ to the right-hand side of Eq. (A75) results in Eq. (A68), provided we set $r_0 \equiv C^*$. With the above identity Eq. (A67) can be expressed in the more compact form of Eq. (A61). The perfect equivalence of the two procedures based, respectively, on the series expansion of $C(0, y)$ and $\partial C(z, y)/(\partial z)_{z=0}$ is therefore proved.

D. Equivalence of the Method of Integration in Series with the Method of Laplace Transforms

The various researchers who have employed either the procedure based on Laplace transforms (*175–180*) or the method of integration in series (*22*, *23*, *174*) have started from the diffusion equations written in the form of Eqs.

(A40) or (A9), respectively, considering initial and boundary conditions at the same time. As a matter of fact if all the conclusions that can be drawn from the diffusional problem without accounting for the boundary conditions are examined, as has been done in the previous sections, the above two mathematical procedures appear identical from the point of view of their practical application. The unification of these procedures may prove useful to the uninitiated researcher willing to obtain recurrence relations without getting entangled in the intricacies of applied mathematics.

Considering first the method of integration in series, let us expand $C(x = 0, t)$ and $\partial C(x, t)/(\partial x)_{x=0}$ in power series of t by making use of Eqs.(A30) and (A33). If we recall that $z = \sqrt{7/(12Dt)}\ x$ and $y = K_2 t^m$, from Eq. (A30) we immediately have

$$C(x = 0, t) = C(z = 0, y) = \sum_{i=0}^{\infty} s_i t^{7/6\alpha i} \tag{A77}$$

where $s_i \equiv m_i K_2^{li}$ and $7/6\alpha \equiv ml$ (see Eq. (A14)). Analogously, from Eq. (A33) it follows that

$$\frac{\partial C(x, t)}{(\partial x)_{x=0}} = \frac{\partial C(z, y)}{(\partial z)_{z=0}} \frac{\partial z}{\partial x} = \sqrt{\frac{7}{12Dt}} \left[\frac{2C^*}{\pi^{1/2}} - \sum_{i=0}^{\infty} s_i p_i t^{7/6\alpha i}\right] \tag{A78}$$

where p_i is given by Eq. (A28).

Let us now consider the method of Laplace transforms. Recalling that in this case $z = K_1 t^{2/3}$ and $y = K_2 t^{7/3}$, Eq. (A61) yields

$$C(x = 0, t) = C(z = 0, y) = \sum_{i=0}^{\infty} s_i t^{7/3bi} \tag{A79}$$

where $s_i \equiv r_i K_2^{bi}$. The expression of $\partial C(x, t)/(\partial x)_{x=0}$ as a function of time t can be readily obtained from Eq. (A68)

$$\frac{\partial C(x, t)}{(\partial x)_{x=0}} = \frac{\partial C(z, y)}{(\partial z)_{z=0}} \frac{\partial z}{\partial x} = \sqrt{\frac{7}{3Dt}} \left[\frac{C^*}{\pi^{1/2}} - \sum_{i=0}^{\infty} s_i \frac{\Gamma(bi + 1)}{\Gamma(bi + \frac{1}{2})} t^{7/3bi}\right] \tag{A80}$$

In the derivation of Eq. (A80) use has been made of the equation $K_1/\sqrt{K_2 K_3} = \sqrt{7/(3D)}$ (cf. Eq. (A41)). Equations (A77) and (A78) are clearly identical with Eqs. (A79) and (A80), provided we set $\alpha \equiv 2b$. The equivalence of the method of integration in series with the method of Laplace transforms is therefore proved. Furthermore formula (A28), which was given without proof, is indirectly demonstrated.

The most direct procedure leading to solutions in power series of time consists in substituting $C(0, t)$ and $\partial C(x, t)/(\partial x)_{x=0}$ from Eqs. (A79) and (A80) into the boundary condition. By so doing a recurrence relation is obtained, which allows the computation of the coefficients s_i. In some cases it is more convenient to use Eqs. (A80) and (A79) in the modified forms,

$$\frac{\partial C(x, t)}{(\partial x)_{x=0}} = -\sqrt{\frac{7}{3Dt}} \sum_{i=1}^{\infty} v_i t^{7/3bi} \tag{A81}$$

$$C(0, t) = C^* + \sum_{i=1}^{\infty} v_i \frac{\Gamma(bi + \frac{1}{2})}{\Gamma(bi + 1)} t^{7/3bi} \tag{A82}$$

which are readily obtained from Eq. (A75) and (A76) upon substitution of variables. Let us now examine the application of the unified procedure outlined in this section to some boundary value problems.

E. Examples

Example I

Consider the reversible electrode process $O + ne \rightleftharpoons R$, without kinetic complications. Under these conditions, the volume concentrations $C_o\,(0, t)$ and $C_R(0, t)$ of O and R at the electrode surface remain constant during electrolysis at constant potential (cf. Sec. II.B). The diffusional problem is therefore expressed by the differential equation,

$$\frac{\partial C_O}{\partial t} = D\frac{\partial^2 C_O}{\partial x^2} + \frac{2x}{3t}\frac{\partial C_O}{\partial x} \tag{A83}$$

with the initial condition

$$C_O = C_O^* \qquad \text{for} \begin{cases} x \geqslant 0, & t = 0 \\ x \to \infty, & t > 0 \end{cases} \tag{A84}$$

and the boundary condition

$$C_O = \bar{C}_O \qquad \text{for } x = 0, t > 0 \tag{A85}$$

The polarographic current is obtained from the relation,

$$i = nFAD_O\left(\frac{\partial C_O}{\partial x}\right)_{x=0} \tag{A86}$$

Substitution of C_O (0, t) from Eq. (A79) into the boundary condition (A85) yields

$$\bar{C}_O = \sum_{i=0}^{\infty} s_i t^{7/3bi} \tag{A87}$$

The time-independence of C_O (0, t) is realized only if we set $s_O = \bar{C}_O$ and $s_i = 0$ for $i \geq 1$, on the right-hand side of Eq. (A87). From Eq. (A80) it follows that

$$\frac{\partial C_O(x, t)}{(\partial x)_{x=0}} = \sqrt{\frac{7}{3D_O t}}\left[\frac{C_O^*}{\pi^{1/2}} - \bar{C}_O \frac{\Gamma(1)}{\Gamma(\frac{1}{2})}\right] = \sqrt{\frac{7}{3\pi D_O t}}(C_O^* - \bar{C}_O)$$

Hence the instantaneous current is given by

$$i = \sqrt{\frac{7D_O}{3\pi t}}\, nFA(C_O^* - \bar{C}_O) \tag{A88}$$

which expresses the well-known Ilkovic equation (*7*).

Example II

Assume that the electrode process $0 + ne \rightarrow \mathrm{R}$ is totally irreversible. Under these conditions the diffusional problem is expressed by the diffusion equation (A83), with the initial condition of Eq. (A84) and the boundary condition,

$$D_O\left(\frac{\partial C_O}{\partial x}\right)_{x=0} = k_f C_O(0, t) \tag{A89}$$

Here k_f is the potential dependent rate constant for the forward charge-transfer process. Combination of Eqs. (A81), (A82), and (A89) results in

$$-\sqrt{\frac{7D_O}{3t}} \sum_{i=1}^{\infty} v_i t^{7/3bi} = k_f C_O^* + k_f \sum_{i=1}^{\infty} v_i \frac{\Gamma(bi + \frac{1}{2})}{\Gamma(bi + 1)} t^{7/3bi} \tag{A90}$$

For the derivation of a recurrence relation it is convenient to set $b = 3/14$, so as to represent both members of Eq. (A90) as power series of t of the form $\sum_{i=1}^{\infty} a_i t^{i/2}$. Equation (A90) must hold for every value of t and this is true only if the coefficient of any given power of t in the left member equals the coefficient of the same power of t in the right member. Thus for the zeroth power $t^0 = 1$, we have

$$-\sqrt{\frac{7D_O}{3}}\, v_1 = k_f C^* \tag{A91}$$

Analogously, for the general power $t^{i/2}$, we obtain

$$-\sqrt{\frac{7D_O}{3}}\, v_{i+1} = k_f v_i \frac{\Gamma(3/14i + 1/2)}{\Gamma(3/14i + 1)} \tag{A92}$$

Equation (A91) and the recurrence relation of Eq. (A92) permit the evaluation of every coefficient v_i and, consequently, in view of Eqs. (A81) and (A86), the determination of the polarographic current i.

A more compact expression of i is obtained if, instead of t we make use of a dimensionless variable. This result can be achieved by noting that the recurrence relation of Eq. (A92) can be written in the form,

$$-\beta_{i+1} = \beta_i \frac{\Gamma(3/14i + 1/2)}{\Gamma(3/14i + 1)} \tag{A93}$$

where $\beta_i = v_i(1/k_f\sqrt{7D_O/3})^i$. Taking Eq. (A81) into account, the concentration gradient of O at $x = 0$ then writes

$$\left(\frac{\partial C_O}{\partial x}\right)_{x=0} = -\sqrt{\frac{7}{3D_O t}}\, \beta_1 \sum_{i=1}^{\infty} \gamma_i \xi^i$$

where

$$\xi = k_f \sqrt{\frac{3t}{7D_O}} \qquad \text{and} \qquad \gamma_i = \frac{\beta_i}{\beta_1}$$

Obviously $\gamma_1 = 1$ and furthermore from Eqs. (A91) and (A93) it follows that

$$\beta_1 = -C_O^*; \qquad \frac{\gamma_{i+1}}{\gamma_i} = -\frac{\Gamma(3/14i + 1/2)}{\Gamma(3/14i + 1)} \tag{A94}$$

Hence the polarographic current is given by

$$i = \sqrt{\frac{7D_O}{3t}}\, nFA\, C_O^* \sum_{i=1}^{\infty} \gamma_i \xi^i$$

Noting that the diffusion limiting current i_d is expressed by the equation $i_d = \sqrt{7D_O/(3\pi t)}\; nFA\; C_O^*$ (see for instance Eq. (A88) in which $\bar{C}_O$ is equated to zero), we have

$$\frac{i}{i_d} = \pi^{1/2} \sum_{i=1}^{\infty} \gamma_i \xi^i \tag{A95}$$

An equation analogous to (A95) was derived by Koutecký by using the method of integration in series (*22*). The ratio $\bar{i}/\bar{i}_d$ of the corresponding mean currents is obtained by noting that i_d is proportional to $t^{1/6}$ according to the time independent factor $\sqrt{7D_O/(3\pi)}\; nFA_0\; C_O^* \equiv$ const. Hence

$$\frac{\bar{i}}{\bar{i}_d} = \frac{\text{const} \int_0^{t_d} i(t^{1/6}/i_d)\, dt}{\text{const} \int_0^{t_d} t^{1/6}\, dt}$$

Replacing t by ξ as integration variable, we obtain

$$\frac{\bar{i}}{\bar{i}_d} = \frac{7}{3} \xi_1^{-7/3} \int_0^{\xi_1} \left(\frac{i}{i_d}\right) \xi^{4/3}\, d\xi \tag{A96}$$

where $\xi_1 = k_f\sqrt{3t_d/(7D_O)}$. Clearly, on substituting (i/i_d) from Eq. (A95) into Eq. (A96), the ratio $\bar{i}/\bar{i}_d$ is obtained in the form of a power series of ξ_1.

A boundary value problem formally identical with the one previously considered is that relative to a very fast first-order antecedent chemical reaction (cf. Sec. IV.F.1.a). In fact Eq. (A89) is identical with the boundary condition of Eq. (276) satisfied by the electroinactive species B_3, provided we replace the heterogeneous rate constant k_f by the term $\sqrt{D_1 k/K}$. Hence the ratio of the kinetic limiting current to the corresponding diffusion limiting current is expressed by Eq. (A95), where now $\xi = \sqrt{3D_1\, kt/(7D_3 K)}$. The recurrence relation for the coefficients γ_i is still given by Eq. (A94).

Example III

Consider an electrode process $O + ne \rightleftharpoons R$ and assume that it is "polarographically reversible." In this case both the rate of the forward and that of the backward electrode process must be taken into account and the boundary condition (A89) becomes

$$D_O\left(\frac{\partial C_O}{\partial x}\right)_{x=0} = k_f\, C_O(0, t) - k_b\, C_R(0, t) \tag{A97}$$

where k_b is the rate constant for the backward process. The boundary condition (A97) involves the concentration of the reduced species, so that a further boundary condition is needed in order that the boundary value problem be determined. In the absence of adsorption phenomena, we can write the relation,

$$D_O\left(\frac{\partial C_O}{\partial x}\right)_{x=0} = -D_R\left(\frac{\partial C_R}{\partial x}\right)_{x=0} \tag{A98}$$

according to which the amount of O reaching the electrode surface per unit time equals the amount of R leaving the electrode in the same unit time. If C_O^*, C_R^* and D_O, D_R designate the bulk concentrations and the diffusion coefficients of O and R, respectively, the application of Eqs. (A81) and (A82) to these two species yields

$$\frac{\partial C_O(x, t)}{(\partial x)_{x=0}} = -\sqrt{\frac{7}{3D_O t}} \sum_{i=1}^{\infty} v_i^O t^{7/3bi}; \qquad \frac{\partial C_R(x, t)}{(\partial x)_{x=0}} = -\sqrt{\frac{7}{3D_R t}} \sum_{i=1}^{\infty} v_i^R t^{7/3bi} \tag{A99}$$

$$\left.\begin{aligned} C_O(0, t) &= C_O^* + \sum_{i=1}^{\infty} v_i^O \frac{\Gamma(bi + \frac{1}{2})}{\Gamma(bi + 1)} t^{7/3bi} \\ C_R(0, t) &= C_R^* + \sum_{i=1}^{\infty} v_i^R \frac{\Gamma(bi + \frac{1}{2})}{\Gamma(bi + 1)} t^{7/3bi} \end{aligned}\right\} \tag{A100}$$

Substitution of $\partial C_O\,(x, t)/(\partial x)_{x=0}$ and $\partial C_R(x, t)/(\partial x)_{x=0}$ from Eq. (A99) into Eq. (A98) results in

$$-\sqrt{D_O} \sum_{i=1}^{\infty} v_i^O t^{7/3bi} = \sqrt{D_R} \sum_{i=1}^{\infty} v_i^R t^{7/3bi}$$

from which it follows that

$$v_i^R = -\sqrt{\frac{D_O}{D_R}}\, v_i^O \tag{A101}$$

Combination of Eqs. (A100) and (A101) yields the relation,

$$\sqrt{D_O}[C_O^* - C_O(0, t)] = -\sqrt{D_R}[C_R^* - C_R(0, t)] \tag{A102}$$

Equation (A102) holds whenever the flux of O at $x = 0$ is the opposite of that of R (boundary condition of Eq. (A98)) and it is often reported in the polarographic literature without proof (*186*). If we substitute $(\partial C_O/\partial x)_{x=0}$, $C_O(0, t)$, and $C_R(0, t)$ from Eqs. (A99) and (A100) into Eq. (A97) taking Eq. (A101) into account and setting $b = 3/14$, we obtain

$$\sqrt{\frac{7D_O}{3}} \sum_{i=1}^{\infty} v_i^O t^{i/2-1/2} + k_f C_O^* - k_b C_R^*$$

$$+ \left(k_f + k_b \sqrt{\frac{D_O}{D_R}}\right) \sum_{i=1}^{\infty} v_i^O \frac{\Gamma(3/14i + \frac{1}{2})}{\Gamma(3/14i + 1)} t^{i/2} = 0$$

On equating the coefficient of the zeroth power of t to zero, we have

$$v_i^O = \frac{k_b C_R^* - k_f C_O^*}{\sqrt{7D_O/3}} \tag{A103}$$

whereas, for the coefficient of the general power $t^{i/2}$,

$$v_{i+1}^O = -\sqrt{\frac{3}{7}}\left(\frac{k_f}{\sqrt{D_O}} + \frac{k_b}{\sqrt{D_R}}\right)\frac{\Gamma(3/14i + 1/2)}{\Gamma(3/14i + 1)} v_i^O \tag{A104}$$

Equation (A103) and the recurrence relation of Eq. (A104) permit the evaluation of $(\partial C_O/\partial x)_{x=0}$ and, consequently, that of the polarographic current.

A compact expression for i in terms of dimensionless parameters is readily obtained on setting

$$\beta_i = \frac{v_i^O}{\left[\sqrt{\frac{3}{7}}\left(\frac{k_f}{\sqrt{D_O}} + \frac{k_b}{\sqrt{D_R}}\right)\right]^i} \tag{A105}$$

Equation (A104) is then written as

$$\beta_{i+1} = -\frac{\Gamma(3/14i + \frac{1}{2})}{\Gamma(3/14i + 1)} \beta_i \tag{A106}$$

In view of Eqs. (A86) and (A99) i is given by

$$i = -\sqrt{\frac{7D_O}{3t}}\, nFA\,\beta_1 \sum_{i=1}^{\infty} \gamma_i \xi^i \tag{A107}$$

where $\xi = (k_f/\sqrt{D_O} + k_b/\sqrt{D_R})\,\sqrt{3t/7}$ and $\gamma_i = \beta_i/\beta_1$. Obviously γ_i obeys the same recurrent relation (A106) as β_i. Furthermore, from Eqs. (A103) and (A105) we obtain

$$\beta_1 = \frac{(k_b/k_f)C_R^* - C_O^*}{1 + \sqrt{D_O/D_R}\,k_b/k_f}$$

Suppose we let k_f and k_b tend to ∞ while keeping their ratio k_b/k_f constant. This mathematical procedure amounts to passing from a "quasi-reversible" charge-transfer process of the type $O + ne \underset{k_b}{\overset{k_f}{\rightleftharpoons}} R$ to the corresponding reversible one. In the case of reversibility the rate for the forward charge-transfer process is practically equal to that for the backward process,

$$k_f\, C_O(0, t) = k_b\, C_R(0, t)$$

so that Nernst's equation

$$\frac{k_b}{k_f} = \theta = \exp\left[\frac{nF}{RT}(E - E^\circ)\right] = \frac{C_O(0, t)}{C_R(0, t)} \tag{A108}$$

can be applied at $x = 0$. Under these conditions the current is expressed by the Ilkovic equation (A88), in which the time-independent surface concentration $\bar{C}_O$ is immediately obtained as a function of the applied potential E by combining Eqs. (A102) and (A108),

$$\bar{C}_O = \frac{C_O^* + \sqrt{D_R/D_O}\,C_R^*}{1 + 1/\theta\sqrt{D_R/D_O}}$$

From Eq. (A88) it follows that

$$i_\infty = \sqrt{\frac{7D_O}{3\pi t}}\, nFA\, \frac{C_O^* - \theta C_R^*}{1 + \theta\sqrt{D_O/D_R}} = -\sqrt{\frac{7D_O}{3\pi t}}\, nFA\,\beta_1 \tag{A109}$$

where i_∞ is the instantaneous current that would be observed at a given potential E if the charge-transfer process were reversible. On dividing Eq. (A107) by Eq. (A109) we obtain the relation,

$$\frac{i}{i_\infty} = \pi^{1/2} \sum_{i=1}^{\infty} \gamma_i \xi^i$$

which is formally identical with Eq. (A95), valid under conditions of total irreversibility (*22*).

Example IV

Consider the boundary value problem of Eqs. (203) through (207) of Sec. IV.B.1 and expand $\varphi(0, t)$ in the power series,

$$\varphi(0, t) = \sum_{i=0}^{\infty} s_i t^{7/3bi}$$

From Eq. (A80) and the boundary condition of Eq. (207a), it follows that

$$\left(\frac{\partial \varphi}{\partial x}\right)_{x=0} = \sqrt{\frac{7}{3Dt}} \left[\frac{\varphi^*}{\pi^{1/2}} - \sum_{i=0}^{\infty} s_i \frac{\Gamma(bi + 1)}{\Gamma(bi + \frac{1}{2})} t^{7/3bi}\right] = 0 \qquad \text{(A110)}$$

If we equate to zero the coefficients of every power of t in Eq. (A110), for the $t^{-1/2}$ power we obtain

$$\frac{\varphi^*}{\pi^{1/2}} - s_0 \frac{\Gamma(1)}{\Gamma(\frac{1}{2})} = \frac{\varphi^*}{\pi^{1/2}} - \frac{s_0}{\pi^{1/2}} = 0$$

that is, $s_0 = \varphi^*$. For all other powers of t, we have $s_i = 0$. Consequently:

$$\varphi(0, t) = \varphi^* \qquad \text{(A111)}$$

If we expand $\psi(0, t)$ in the power series,

$$\psi(0, t) = \sum_{i=0}^{\infty} s_i' t^{7/3bi} \qquad \text{(A112)}$$

substitution of $\psi(0, t)$ from Eq. (A112) and of $\varphi(0, t)$ from Eq. (A111) into the boundary condition (207b) results in

$$\sum_{i=0}^{\infty} s_i' t^{7/3bi} = \frac{\theta - K}{1 + \theta} \varphi^* \exp(-\lambda t)$$

Expanding $\exp(-\lambda t)$ in the well-known power series $\sum_{i=0}^{\infty}(-\lambda t)^i/i!$ and setting for convenience $b = \frac{3}{7}$, we obtain the relation,

$$\sum_{i=0}^{\infty} s_i' t^i = \frac{\theta - K}{1 + \theta} \varphi^* \sum_{i=0}^{\infty} \frac{(-\lambda t)^i}{i!} \tag{A113}$$

from which it follows that

$$s_i' = \frac{\theta - K}{1 + \theta} \frac{\varphi^*}{i!} (-\lambda)^i \tag{A114}$$

In view of Eqs. (206a), (A80), and (A114), the concentration gradient of ψ at $x = 0$ is given by

$$\left(\frac{\partial \psi}{\partial x}\right)_{x=0} = \sqrt{\frac{7}{3Dt}} \left[\frac{K - \theta}{1 + \theta} \varphi^* \sum_{i=0}^{\infty} \frac{(-\lambda)^i}{i!} \frac{\Gamma(\frac{3}{7}i + 1)}{\Gamma(\frac{3}{7}i + \frac{1}{2})} t^i\right] \tag{A115}$$

As shown in Sec. IV.B.1 the expression for the polarographic current is immediately obtained (*23*) from Eq. (A115).

Example V

Consider the boundary value problem of Eqs. (365) through (367), and expand $\varphi(0, t)$ in the power series,

$$\varphi(0, t) = \frac{C_3^*}{\sqrt{\frac{3}{7}\pi k t}} = \sum_{i=1}^{\infty} v_i \frac{\Gamma(bi + \frac{1}{2})}{\Gamma(bi + 1)} t^{7/3bi} \tag{A116}$$

in accordance with Eq. (A81). In the present case it is convenient to set $b = -3/14$. With this substitution, on equating the coefficients of every power of t on the right-hand side of Eq. (A116) to the corresponding coefficients on the left, we obtain

$$v_1 = \frac{C_3^*}{\sqrt{\frac{3}{7}\pi k}} \frac{\Gamma(11/14)}{\Gamma(4/14)}$$

and

$$v_i = 0 \qquad \text{for } i > 1$$

The flux of φ at $x = 0$ is immediately obtained upon application of Eq. (A82)

$$\left(\frac{\partial\varphi}{\partial x}\right)_{x=0} = -\sqrt{\frac{7}{3Dt}}\sum_{i=1}^{\infty} v_i t^{-i/2} = -\frac{C_3^*}{\sqrt{\frac{3}{7}\pi Dt}}\frac{1}{\sqrt{kt}}\sqrt{\frac{7}{3}}\frac{\Gamma(11/14)}{\Gamma(4/14)}$$

$$= -\frac{C_3^*}{\sqrt{\frac{3}{7}\pi Dt}}\frac{0.573}{\sqrt{kt}} \qquad \text{(A117)}$$

REFERENCES

1. H. S. Harned and B. B. Owen, *The Physical Chemistry of Electrolytic Solutions*, 3rd ed., Reinhold, New York, 1958, p. 2.

2. J. Koryta, *Collection Czech. Chem. Commun.*, **24**, 3057 (1959).

3. I. Prigogine, *Introduction to Thermodynamics of Irreversible Processes*, 2d ed., Wiley (Interscience) New York, 1962.

4. S. R. De Groot, *Thermodynamics of Irreversible Processes*, North Holland Publ., Amsterdam, 1966, p. 9.

5. R. Guidelli, *Trans. Faraday Soc.*, **66**, 1185, 1194 (1970).

6. A. N. Frumkin, *Z. Physik. Chem.*, **164A**, 121 (1933).

7. D. Ilkovič, *J. Chim. Phys.*, **35**, 129 (1938).

8. R. Guidelli and D. Cozzi, *J. Phys. Chem.*, **71**, 3020 (1967); **71**, 3027 (1967).

9. Z. Grabowski and W. Kemula, in *Polarographie in der Chemotherapie, Biochemie und Biologie* (I. Jenaer Simposium, 1962), Akademie Verlag, Berlin, 1964, p. 377.

10. J. Heyrovský and J. Kůta, *Principles of Polarography*, Academic Press, New York, 1966, pp. 121–198.

11. K. Vetter, *Electrochemical Kinetics*, Academic Press, New York, 1967, pp. 157–158.

12. H. Matsuda and Y. Ayabe, *Z. Elektrochem.*, **66**, 469 (1962).

13. J. Tirouflet, in *Advances in Polarography* (I. S. Longmuir ed.), Vol. 2, Pergamon Press, Oxford, 1960, p. 740.

14. J. Koutecký and J. Koryta, *Electrochim. Acta*, **3**, 318 (1961).

15. I. M. Kolthoff and J. Jordan, *J. Am. Chem. Soc.*, **75**, 1571 (1953).

16. A. L. Beilby and A. L. Crittenden, *J. Phys. Chem.*, **64**, 177 (1960).

17. G. Piccardi and R. Guidelli, *J. Phys. Chem.*, **72**, 2782 (1968).

18. I. Smoler, *Collection Czech. Chem. Commun.*, **19**, 238 (1954).

19. W. Hans, W. Henne, and E. Meurer, *Z. Elektrochem.*, **58**, 836 (1954).

20. K. Micka and I. Smoler, *Chem. Listy*, **50**, 988 (1956).

21. L. Němec and I. Smoler, *Chem. Listy*, **51**, 1958 (1957).

22. J. Koutecký, *Collection Czech. Chem. Commun.*, **18**, 597 (1953).

23. J. Koutecký, *Collection Czech. Chem. Commun.*, **18**, 311 (1953).

24. J. Koutecký, R. Brdička, and V. Hanuš, *Collection Czech. Chem. Commun.*, **18**, 611 (1953).

24(1). R. S. Nicholson, J. M. Wilson, and M. L. Olmstead, *Anal. Chem.*, **38**, 542 (1966).

24(2). B. Kastening and L. Holleck, *Z. Elektrockem.*, **63**, 166 (1959).

24(3). B. Kastening, *Anal. Chem.*, **41**, 1142 (1969).
24(4). G. S. Alberts and I. Shain, *Anal. Chem.*, **35**, 1859 (1963).
24(5). H. B. Herman and A. J. Bard, *J. Phys. Chem.*, **70**, 396 (1966).
25. J. Koutecký and J. Koryta, *Collection Czech. Chem. Commun.*, **19**, 845 (1954).
26. R. Brdička and K. Wiesner, *Collection Czech. Chem. Commun.*, **12**, 139 (1947).
27. V. Hanuš, *Chem. Zvesti*, **8**, 702 (1954).
28. K. Wiesner, *Chem. listy*, **41**, 6 (1947).
29. K. H. Henke and W. Hans, *Z. Elektrochem.*, **57**, 591 (1953).
30. R. Brdička, V. Hanuš, and J. Koutecký, in *Progress in Polarography* (P. Zuman, ed., with the collaboration of I. M. Kolthoff), Vol. 1, Addison-Wesley, Reading, Mass., 1962, p. 145.
31. J. Koutecký, *Collection Czech. Chem. Commun.*, **19**, 857 (1954).
32. J. Cizek, J. Koryta, and J. Koutecký *Collection Czech. Chem. Commun.*, **24**, 3844 (1959).
33. R. Bieber and G. Trümpler, *Helv. Chim. Acta*, **30**, 706, 971, 1109, 1286, 1534, 2000 (1947).
34. K. Veselý and R. Brdička, *Collection Czech. Chem. Commun.*, **12**, 313 (1947).
35. N. B. Neiman and M. I. Gerber, *Zh. Anal. Khim.*, **2**, 135 (1947).
36. P. Valenta, *Collection Czech. Chem. Commun.*, **25**, 855 (1960).
37. R. Brdička, *Collection Czech. Chem. Commun.*, **20**, 387 (1955).
38. R. Bieber and G. Trümpler, *Helv. Chim. Acta*, **31**, 5 (1948).
39. S. Ono, M. Takagi, and T. Wasa, *J. Am. Chem. Soc.*, **75**, 4369 (1953).
40. P. J. Elving and C. J. Bennett, *J. Am. Chem. Soc.*, **76**, 1412 (1954).
41. D. R. Norton and N. H. Furman, *Anal. Chem.*, **26**, 1116 (1954).
42. S. Ono, M. Takagi, and T. Wasa, *Bull. Chem. Soc. Japan*, **31**, 356 (1958).
43. J. Krupička and J. J. K. Novák, *Collection Czech. Chem. Commun.*, **25**, 1275 (1960).
44. R. Brdička, *Collection Czech. Chem. Commun.*, **12**, 212 (1947).
45. E. G. Clair and K. Wiesner, *Nature*, **165**, 202 (1950).
46. K. Wiesner, M. Wheatley, and M. Los, *J. Am. Chem. Soc.*, **76**, 4858 (1954).
47. J. H. Green and A. Walkley, *Australian J. Chem.*, **8**, 51 (1955).
48. W. Kemula, Z. Grabowski, and E. Bartel, *Roczniki Chem.*, **33**, 1125 (1959).
49. M. Becker and H. Strehlow, *Z. Elektrochem.*, **64**, 42, 818 (1960).
50. V. D. Bezuglyi, V. N. Dmitrieva, T. S. Tarasyuk, and N. A. Izmailov, *Zh. Obshch. Khim.*, **30**, 2415 (1960).
51. V. D. Bezuglyi and E. Yu. Novik, *Zavodsk. Lab.*, **27**, 544 (1961).
52. Yu. A. Vakhrushev and Ya. I. Tur'yan, *Zh. Fiz. Khim.*, **37**, 1650 (1963).
53. J. Volke and V. Volková, *Chem. Listy*, **49**, 490 (1955).
54. J. Volke and V. Volková, *Collection Czech. Chem. Commun.*, **22**, 1777 (1957).
55. E. T. Bartel, Z. R. Grabowski, and W. Kemula, *Roczniki Chem.*, **34**, 345 (1960).
56. I. M. Kolthoff and A. Liberti, *J. Am. Chem. Soc.*, **70**, 1885 (1948).
57. H. Lund, *Acta Chem. Scand.*, **13**, 249 (1959).
58. H. J. Gardner and W. P. Georgans, *J. Chem. Soc.*, 4180 (1956).
59. V. Hanuš and R. Brdička, *Chem. Listy*, **44**, 291 (1950).
60. J. Koutecký, *Collection Czech. Chem. Commun.*, **19**, 1093 (1954).
61. J. Kůta, *Collection Czech. Chem. Commun.*, **22**, 1411 (1957).

62. S. G. Mairanovskii, *Dokl. Akad. Nauk SSSR*, **149**, 1373 (1963).
63. V. Čermák, *Chem. Zvesti*, **8**, 714 (1954).
64. V. Čermák, *Collection Czech. Chem. Commun.*, **23**, 1471, 1871 (1958).
65. Wang Er-Kong and A. A. Vlček, *Collection Czech. Chem. Commun.*, **25**, 2082 (1960).
66. J. Koutecký and V. Hanuš, *Collection Czech. Chem. Commun.*, **20**, 124 (1955).
67. M. Březina, *Collection Czech. Chem. Commun.*, **22**, 339 (1957).
68. J. Koutecký, *Collection Czech. Chem. Commun.*, **22**, 160 (1957).
69. I. M. Kolthoff and E. P. Parry, *J. Am. Chem. Soc.*, **73**, 3718 (1951).
70. R. Brdička and K. Wiesner, *Naturwiss*, **31**, 247 (1943).
71. R. Brdička and K. Wiesner, *Collection Czech. Chem. Commun.*, **12**, 39 (1947).
72. R. Sellner and M. Bucher, *Acta Biol. Med. Ger.*, **7**, 427 (1961).
73. J. Koryta, *Collection Czech. Chem. Commun.*, **20**, 1125 (1955).
74. B. Matyska and D. Dušková, *Collection Czech. Chem. Commun.*, **22**, 1747 (1957).
75. I. M. Kolthoff and E. P. Parry, *J. Am. Chem. Soc.*, **73**, 5315 (1951).
76. K. Fülöp and L. J. Csányi, *Acta Chim. Acad. Sci. Hung.*, **38**, 193 (1963).
77. A. A. Vlček, *Collection Czech. Chem. Commun.*, **25**, 2685 (1960).
78. J. Koryta, *Collection Czech. Chem. Commun.*, **19**, 666 (1954).
79. A. Blažek and J. Koryta, *Collection Czech. Chem. Commun.*, **18**, 326 (1953).
80. J. Koryta, *Chem. Zvesti*, **8**, 723 (1954).
81. J. Koryta and J. Tenygl, *Collection Czech. Chem. Commun.*, **20**, 423 (1955).
82. F. Mánok and B. Tokés, *Studia Univ., Babes-Bolyai*, Ser. 1, **2**, 35 (1959).
83. I. M. Kolthoff and I. Hodara, *J. Electroanal. Chem.*, **5**, 2 (1963).
84. I. M. Kolthoff and I. Hodara, *J. Electroanal. Chem.*, **4**, 369 (1962).
85. I. M. Kolthoff and I. Hodara, *Bull. Res. Council Israel*, Sec. A, **11A**, 203 (1962).
86. H. A. Laitinen and W. A. Ziegler, *J. Am. Chem. Soc.*, **75**, 3045 (1953).
87. R. Höltje and R. Geyer, *Z. Anorg. Allgem. Chem.*, **246**, 258 (1941).
88. G. P. Haight, Jr., *Anal. Chem.*, **23**, 1505 (1951).
89. G. P. Haight, Jr., and W. F. Sager, *J. Am. Chem. Soc.*, **74**, 6056 (1952).
90. G. P. Haight, Jr., *J. Am. Chem. Soc.*, **76**, 4718 (1954).
91. G. A. Rechnitz and H. A. Laitinen, *Anal. Chem.*, **33**, 1473 (1961).
92. M. G. Johnson and R. J. Robinson, *Anal. Chem.*, **24**, 366 (1952).
93. J. Koryta, *Collection Czech. Chem. Commun.*, **20**, 667 (1955).
94. K. Wiesner, *Z. Elektrochem.*, **49**, 164 (1943).
95. R. C. Bower and I. M. Kolthoff, *J. Am. Chem. Soc.*, **81**, 1836 (1959).
96. W. M. Schwarz and I. Shain, *J. Phys. Chem.*, **70**, 845 (1966).
97. D. M. H. Kern, *J. Am. Chem. Soc.*, **75**, 2473 (1953).
98. J. Koutecký, *Collection Czech. Chem. Commun.*, **20**, 116 (1955).
99. D. M. H. Kern, *J. Am. Chem. Soc.*, **76**, 1011 (1954).
100. J. Koryta and Z. Zábranský, *Collection Czech. Chem. Commun.*, **25**, 3153 (1960).
101. J. Čižek, J. Koryta and J. Koutecký, *Collection Czech. Chem. Commun.*, **24**, 663 (1959).
102. S. G. Mairanovskii, *Dokl. Akad. Nauk. SSSR*, **110**, 593 (1956).
103. P. Delahay, *Double Layer and Electrode Kinetics*, Wiley (Interscience), New York, 1965, p. 33.
104. D. C. Grahame, *Chem. Rev.*, **41**, 441 (1947).
105. D. C. Grahame, *J. Am. Chem. Soc.*, **76**, 4819 (1954); **79**, 2093 (1957).

106. R. Parsons in *Advances in Electrochemistry and Electrochemical Engineering* (P. Delahay ed.), Vol. 1, Wiley (Interscience), New York, 1961, pp. 1-64.
107. A. N. Frumkin, V. S. Bagotskii, Z. A. Iofa, and B. N. Kabanov, *Kinetika elektrodnych protsesov*, Izv. Mosk. Univ., Moskva, 1952.
108. D. C. Grahame and B. A. Soderberg, *J. Chem. Phys.*, **22**, 449 (1954).
109. J. Koryta in *Advances in Polarography* (I. S. Longmuir ed.), Vol. 1, Pergamon Press, Oxford, 1960, p. 359.
110. H. Gerisher, *Z. Physik. Chem.*, N. F., **2**, 4, 1954.
111. L. Gierst, *Cinétique d'approche. Inaugural Dissertation*, Univ. Libre de Bruxelles, 1958.
112. M. Breiter, M. Kleinerman, and P. Delahay, *J. Am. Chem. Soc.*, **80**, 5111 (1958).
113. J. Koryta, *Z. Elecktrochem.*, **61**, 423 (1957).
114. S. G. Mairanovskii and L. I. Lishcheta, *Izv. Akad. Nauk SSSR*, Otd. Khim. Nauk, 1984 (1962).
115. S. G. Mairanovskii, E. D. Belokolos, V. P. Gul'tyai, and L. I. Lishcheta, *Elektrokhimiya*, **2**, 693 (1966).
116. J. Koryta, *Rev. Polarog.* (Kyoto), **13**, 1 (1965).
117. H. Matsuda, *J. Phys. Chem.*, **64**, 336 (1960).
118. H. Hurwitz, *Z. Elektrochem.*, **65**, 178 (1961).
119. Z. R. Grabowski and E. T. Bartel, *Roczniki Chem.*, **34**, 611 (1960).
120. L. Gierst and H. Hurwitz, *Z. Elektrochem.*, **64**, 36 (1960).
121. L. Gierst in *Transactions of the Symposium on Electrode Processes* (E. Yeager ed.), Wiley, New York, 1961, p. 109.
122. P. Delahay and M. Kleinerman, *J. Am. Chem. Soc.*, **82**, 4509 (1960).
123. A. A. Vlček, *Chem. Listy*, **50**, 828 (1956).
124. E. Orlemann and R. Sanborn, *J. Am. Chem. Soc.*, **78**, 4852 (1956).
125. J. Dandoy and L. Gierst, *J. Electroanal. Chem.*, **2**, 116 (1961).
126. A. N. Frumkin, N. Nikolajeva and R. Ivanova, *Can. J. Chem.*, **37**, 253 (1959).
127. K. Asada, P. Delahay, and A. K. Sundarem, *J. Am. Chem. Soc.*, **83**, 3396 (1961).
128. Reference 121, pp. 142–143.
129. A. N. Frumkin, O. A. Petry, and N. V. Nikolaeva-Fedorovich, *Electrochim. Acta*, **8**, 177 (1963).
130. L. Gierst, J. Tondeur, R. Cornelissen, and F. Lamy. Report of the 14th CITCE Meeting, Moscow, 1963.
131. A. N. Frumkin, *Ber. Akad. Wiss. UdSSR*, **115**, 751 (1957).
132. J. J. Tondeur, A. Dombret, and L. Gierst, *J. Electroanal. Chem.*, **3**, 225 (1962).
133. J. Koryta, *Z. Elektrochem.*, **64**, 26 (1960).
134. G. Hoijtink, *Rec. Trav. Chim.*, **76**, 869, 887 (19657).
135. A. Vincenz-Chodkowska and Z. R. Grabowski, *Electrochim. Acta*, **9**, 789 (1964).
136. J. Volke and V. Volková, *Collection Czech. Chem. Commun.*, **20**, 1332 (1955).
137. Z. Vodrážka, *Chem. Listy*, **45**, 293 (1951).
138. R. Zahradnik, E. Svatek, and M. Chvapil, *Collection Czech. Chem. Commun.*, **24**, 347 (1959).
139. O. Hrdý, *Chem. Listy*, **52**, 1058 (1958).
140. P. Rüetschi and G. Trümpler, *Helv. Chim. Acta*, **35**, 1947 (1953).

141. E. T. Bartel, Z. R. Grabowski, W. Kemula, and W. Turnowska-Rubaszewska, *Roczniki Chem.*, **31**, 13 (1957).
142. L. Onsager, *J. Chem. Phys.*, **2**, 599 (1934).
143. P. Debye, *Trans. Electrochem. Soc.*, **82**, 265 (1942).
144. P. Delahay and W. Vielstich, *J. Am. Chem. Soc.*, **77**, 4955 (1955).
145. S. G. Mairanovskii, *J. Electroanal. Chem.*, **4**, 166 (1962).
146. S. G. Mairanovskii and L. I. Lishcheta, *Collection Czech. Chem. Commun.*, **25**, 3025 (1960).
147. A. N. Frumkin, *Z. Physik*, **35**, 792 (1926).
148. B. Kastening, H. Gartmann, and L. Holleck, *Electrochim. Acta*, **9**, 741 (1964).
149. J. O'M Bockris and D. A. J. Swinkels, *J. Electrochem. Soc.*, **111**, 736 (1964).
150. J. A. V. Butler, *Proc. Roy. Soc.*, (London), **122A**, 399 (1929).
151. I. Langmuir, *J. Am. Chem. Soc.*, **40**, 1361 (1918).
152. M. I. Temkin, *Zh. Fiz. Khim.*, **15**, 296 (1941).
153. S. Brunauer, K. S. Love, and R. G. Keenan, *J. Am. Chem. Soc.*, **64**, 751 (1942).
154. B. M. W. Trapnell, *Chemisorption*, Butterworth, London, 1955.
155. J. H. de Boer in *Chemisorption* (W. E. Garner ed.), Butterworth, London, 1957, p. 27.
156. J. H. de Boer, *Adv. Catalysis*, **8**, 18 (1956).
157. A. N. Frumkin, *Z. Physik. Chem.*, **116**, 466 (1925).
158. P. Rehbinder and E. Wenström, *Zh. Fiz. Khim.*, **19**, 1 (1945).
159. M. Siddiqui and F. C. Tompkins, *Proc. Roy. Soc.* (London), **A268**, 452 (1962).
160. R. Parsons, *J. Electroanal. Chem.*, **5**, 397 (1963).
161. B. B. Damaskin, *J. Electroanal. Chem.*, **7**, 155 (1964).
162. B. L. Timan, *Zh. Fiz. Khim.*, **31**, 2143 (1957); **33**, 1189 (1959).
163. J. Kůta and J. Koryta, *Collection Czech. Chem. Commun.*, **30**, 4095 (1965).
164. R. Guidelli, *J. Phys. Chem.*, **74**, 95 (1970).
165. R. Brdička, *Z. Elektrochem.*, **48**, 278 (1942).
166. A. B. Ershler, G. A. Tedoradze, and S. G. Mairanovskii, *Dokl. Akad. Nauk. SSSR*, **145**, 1324 (1962).
167. S. G. Mairanovskii, *Catalytic and Kinetic Waves in Polarography*, Plenum Press, New York, 1968.
168. Reference 167, pp. 187–200.
169. V. G. Levich, B. I. Khaikin, and E. D. Belokolos, *Elektrokhimiya*, **1**, 1273 (1965).
170. Reference 167, p. 213.
171. S. G. Mairanovskii, N. V. Barashkova, and Yu. B. Vol'kenshtein, *Izv. Akad. Nauk SSSR*, Ser. Khim,, **9**, 1539 (1965).
172. A. N. Frumkin, *Electrochim. Acta*, **9**, 465 (1964).
173. J. Holubek and J. Volke, *Collection Czech. Chem. Commun.*, **27**, 680 (1962).
174. J. Koutecký, *Czech. J. Phys.*, **2**, 50 (1953).
175. M. Smutek, *Collection Czech. Chem. Commun.*, **20**, 247 (1955).
176. K. H. Henke and W. Hans, *Z. Elektrochem.*, **59**, 676 (1955).
177. A. Matsuda and Y. Ayabe, *Bull. Chem. Soc. Japan*, **28**, 422 (1955).
178. M. Senda, *Rev. Polarog.* (Kyoto), **4**, 89 (1956).
179. W. H. Reinmuth, *J. Phys. Chem.*, **65**, 473 (1961).
180. K. Holub, *Collection Czech. Chem. Commun.*, **31**, 1461 (1966).

181. N. Meiman, *Zh. Fiz. Khim.*, **22**, 1454 (1948).
182. R. V. Churchill, *Operational Mathematics*, 2d ed., McGraw-Hill, New York, 1958.
183. I. N. Sneddon, *Fourier Transforms*, McGraw-Hill, New York, 1951.
184. H. Margenau and G. M. Murphy, *The Mathematics of Physics and Chemistry*, 2d Ed., Van Nostrand, Princeton, New Jersey, 1956, p. 49.
185. E. T. Whittaker and G. N. Watson, *A Course of Modern Analysis*, 4d Ed., Cambridge Univ. Press, 1940.
186. P. Delahay and G. W. Stiehl, *J. Am. Chem. Soc.*, **74**, 3500 (1952).

Author Index

Numbers in parentheses are reference numbers and indicate that an author's work is referred to although his name is not cited in the text. Numbers in italics give the page on which the complete reference is listed.

C

D

E

F

G

H

L

M

N

Subject Index

L

M

N

O

P

R